CHAOTIC SIGNALS IN DIGITAL COMMUNICATIONS

THE ELECTRICAL ENGINEERING
AND APPLIED SIGNAL PROCESSING SERIES

Edited by Alexander D. Poularikas

CHAOTIC SIGNALS IN
DIGITAL
COMMUNICATIONS

Marcio Eisencraft • Romis Attux • Ricardo Suyama

CRC Press
Taylor & Francis Group
Boca Raton London New York

CRC Press is an imprint of the
Taylor & Francis Group, an **informa** business

CRC Press
Taylor & Francis Group
6000 Broken Sound Parkway NW, Suite 300
Boca Raton, FL 33487-2742

First issued in paperback 2017

© 2014 by Taylor & Francis Group, LLC
CRC Press is an imprint of Taylor & Francis Group, an Informa business

No claim to original U.S. Government works

ISBN-13: 978-1-4665-5722-2 (hbk)
ISBN-13: 978-1-138-07683-9 (pbk)

Library of Congress Cataloging-in-Publication Data

Chaotic signals in digital communications / editors, Marcio Eisencraft, Romis Attux,
    Ricardo Suyama.
        pages cm. -- (Electrical engineering & applied signal processing series ; 26)
    Includes bibliographical references and index.
    ISBN 978-1-4665-5722-2 (hardback : alk. paper)
    1. Signal processing. 2. Digital communications. 3. Chaotic behavior in systems. I.
Eisencraft, Marcio, editor of compilation. II. Attux, Romis R., editor of compilation. III.
Suyama, Ricardo, editor of compilation.

TK5102.9.C4715 2013
621.382'2--dc23   2013025263

Visit the Taylor & Francis Web site at
http://www.taylorandfrancis.com

and the CRC Press Web site at
http://www.crcpress.com

To our families.

To the lovely memory of Prof. Max Gerken.

Contents

List of Figures

List of Tables

Foreword

As an aperiodic bounded dynamics in a deterministic system, chaos has been found in various fields of science and technology for the past 50 years. In the last two decades, we have witnessed a rapid development in chaos control (synchronization) theory and its application, that since the seminal works referred to as the OGY control and PC synchronization method for chaotic systems. As a result, chaos is recognized as a fundamental and useful phenomenon because of its special properties. For instance, ergodicity, aperiodicity, and sensitive dependence on initial conditions are basic concepts from chaotic dynamics utilized in control and synchronization. Chaos has in fact been used in many engineering fields from spacecraft targeting, fusion, to high resolution radar imaging, chaotic road rollers, analog to digital converters, weak periodic signal detection, encryption, and communication. Among those applications, chaos communication is one of the most challenging fields, due to its promising security, broadband, orthogonality, and the fact that chaotic signals can be generated by low power, low cost, and small size electronic circuits. Moreover, the chaos-based communication scheme promises to integrate source encoding, channel encoding, security, and noise robustness to achieve higher efficiency compared to the conventional ones.

Chaos communication work can be split roughly into four areas: chaos masking, chaotic modulation, chaos shift key, and symbolic message bearing method. Chaos is also used as pseudorandom number generators in some communication schemes. Initially, chaos communication research was dedicated to the feasibility, security, and its weakness to the attacker in an ideal communication channel. The physical constraint of the channel considered at this stage was mainly due to the additional white noise. Recently, as more researchers from engineering got involved in chaos communication research, the more practical communication channels, such as laser optical channel, satellite communication channel, radio frequency (RF) channel, and underwater acoustic channel, are being considered in the chaos communication schemes. Typically, in addition to noise, the physical constraints of these channels are limited bandwidth (or phase shift), fading, and multipath propagation. To deal with such kinds of channel constraints, many new methods have been proposed recently from the engineering application viewpoint.

The book *Chaotic Signals in Digital Communications* is dedicated to present the recent progress in the chaos communication field. The book has three main features: the *first feature* is that the book gives a comprehensive review of conventional digital communication schemes and concepts from chaotic dynamics that are relevant to chaos-based communication that covers the first three chapters. These reviews, together with references therein, provide a very useful background knowledge to understand the subsequent chapters. The *second feature* is that the book covers typical chaos communication methods and their variances. For example, the chaotic masking is covered in Chapter 5, chaotic modulation is addressed in Chapters 5 and 8, chaotic shift key is used in Chapter 4, and symbolic message bearing method is given in Chapters 11 and 16. In addition to these typical methods, the bidirectional communication, based on the isochronal synchronization, and the secure communication method, based on the temporal coupled map, are given in Chapters 14 and 15, respectively. The *third feature* is that the book offers some novel methodologies to deal with the communication channel imperfections. For example, Chapters 8 and 9 deal with bandlimited channel chaos communication, Chapter 4 deals with the radio channel with fading, and Chapter 16 addresses the resistance of a special chaotic signal to multipath propagations. Most chapters (including 6, 10, 11, 12, and 13) take the noise of the channel into account. Besides these features, the book also addresses many topics that are closely related to engineering applications, such as the topics about the optical communication in Chapter 5, chaotic match filter and its circuits implementations in Chapter 7, and microwave FM-DCSK system in Chapter 4, which are an additional bonus in the book. In summary, the book is not only helpful for the uninformed readers in the field, but also helpful for paving the way to full-fledged concepts and techniques in chaos communication at a practical engineering level.

Celso Grebogi
University of Aberdeen

Hai-Peng Ren
Xi'an University of Technology

Preface

Chaotic signals are characterized by an aperiodic and limited behavior, which is generated by deterministic dynamical systems presenting sensitive dependence on initial conditions. Applications involving chaotic signals have been investigated in several areas, including signal processing and telecommunications.

Due to their properties, chaotic signals usually occupy a wide bandwidth and have an impulse-like autocorrelation function. Furthermore, the cross-correlation between signals generated from different initial conditions tends to be small. These features support the use of chaotic signals in spread spectrum techniques.

Since the beginning of the 1990s decade, many papers and books describing interesting and promising chaos-based applications in communications have been published. Although the proposed schemes work well in almost ideal environments, the presence of additive noise, distortion, or delay, usually found in practical channels, brings unsatisfactory results in terms of bit error rate, when compared to conventional communication systems.

Hence, in the last years, much research has been conducted with the objective of bridging the gap between the performance of chaos-based communication systems and that of conventional communication systems in realistic environments.

Bearing this in mind, the aim of this book is to review some of these new techniques that can allow chaotic signals to be used in practical applications in the near future, serving as a "toolbox" of methods to be employed by researchers working on theoretic and experimental aspects of this field.

Given the broad expected audience, the first four chapters have a tutorial approach to the main aspects further developed in the subsequent ones. The first chapter presents an introductory historical overview of the chaos-based communication field. It is followed by Chapter 2, which provides an overview of the conventional and practical techniques applied in digital communications, including a section on recent trends in the area. Chapter 3 gives an overview of dynamical systems, with the fundamental concepts and notations employed in the following chapters. Finally, Chapter 4 presents the principles of chaos-based communications. Afterwards, from Chapters 5 to 16, the reader is presented with contributions authored by researchers working on real-world applications of chaos theory that encompass subjects as diverse as optical communications, channel equalization, noise/spectral distortion mitigation, synchronization, statistical estimation, blind source separation, cryptography, and wireless systems.

Marcio Eisencraft
Romis Attux
Ricardo Suyama

Acknowledgments

First of all, the editors would like to thank very much all the involved authors, Prof. João Marcos T. Romano, CRC Press, and the National Council for Scientific and Technological Development (CNPq–Brazil), who have made this work possible.

Marcio would like to thank his wife Ana Carolina and his son David for the patience and everyday inspiration and incentive; his mother Helena and his grandmother Rê (Regina) (in memoriam) for the examples to be followed and all his family and friends for the continuous support in the good and bad times. He is also indebted to his professors, colleagues, students, former students, and friends at Universidade de São Paulo, Universidade Federal do ABC, and Universidade Presbiteriana Mackenzie, especially Profs. Luiz A. Baccalá, Max Gerken (in memoriam), Maria D. Miranda, Luiz H. A. Monteiro, Jose R. C. Piqueira, Magno T. M. Silva, Moyses Szajnbok, and Haydée F. Wertzner.

Romis would like to thank Dilmara, Clara, Marina, Dina, Cecília, João Gabriel, and Flora for their love and their patience; his father (in memoriam) for all he continues to do for him; Afrânio, Isabel, Beth, Toninho, Naby (in memoriam), Sônia, Ramsa, Dedê, and his whole family for their warm affection; his students and former students for the inspiration they bring to his life; Cristiano, Dr. Danilo, Rafael, and Ricardo for their attention in some quite difficult moments; Diogo and Fazanaro for having taught him so much about dynamic systems and chaos; Alice, Arthur (in memoriam), Angela, Flávia, Taciana, and Vasco for their support in his first academic years; his friends from DSPCOM for all the happy moments spent together; his friends from G6; Prof. João Marcos for his constant encouragement; all his friends and colleagues from UNICAMP and all his friends.

Ricardo would like to thank his lovely wife Gislaine for all her love and constant support; Jorge, Cecília, Bruna, and many others in the family for their warm affection and support; all colleagues from UFABC and DSPCom, for the enriching conversations and friendship.

Editors

Marcio Eisencraft is an Assistant Professor at the Polytechnic School of the University of São Paulo (USP). He received the B.S., M.S., and Ph.D. degrees in Electrical Engineering from the University of São Paulo in 1998, 2001, and 2006, respectively. His research interests include digital signal processing, communication systems, neuronal signal and chaos and nonlinear systems applied to communication systems.

Romis Attux is an Assistant Professor at the University of Campinas (UNICAMP). He received the B.S., M.S., and Ph.D. degrees in Electrical Engineering from the University of Campinas (UNICAMP) in 1999, 2001, and 2005, respectively. His main research interests are information processing, dynamical systems/chaos, and computational intelligence.

Ricardo Suyama is an Assistant Professor at the Federal University of ABC (UFABC). He received the B.S., M.S., and Ph.D. degrees in Electrical Engineering from the University of Campinas in 2001, 2003, and 2007, respectively. His main research interests include unsupervised signal processing, computational intelligence, and applications in communication systems.

Contributors

Greta A. Abib
Centro de Engenharia, Modelagem e
  Ciências Sociais Aplicadas
  (CECS)
Universidade Federal do ABC
  (UFABC)
Santo André, SP, Brazil

Rafael A. Ando
School of Electrical and Computer
  Engineering (FEEC)
University of Campinas (UNICAMP)
Campinas, SP, Brazil

Apostolos Argyris
Department of Informatics and
  Telecommunications
National and Kapodistrian
  University of Athens
Panepistimiopolis, Ilisia, Greece

Romis Attux
School of Electrical and Computer
  Engineering (FEEC)
University of Campinas (UNICAMP)
Campinas, SP, Brazil

Murilo S. Baptista
Institute for Complex Systems and
  Mathematical Biology
King's College, University of
  Aberdeen
Aberdeen, UK

Antonio M. Batista
Departamento de Matemática e
  Estatística
Universidade Estadual de Ponta
  Grossa
PR, Brazil

Jonathan N. Blakely
Charles M. Bowden Laboratory, US
  Army Aviation & Missile
  Research, Development, and
  Engineering Center
Redstone Arsenal, AL, USA

Renato Candido
Escola Politécnica
Universidade de São Paulo
São Paulo, SP, Brazil

Ivan R. S. Casella
Centro de Engenharia, Modelagem e
  Ciências Sociais Aplicadas
  (CECS)
Universidade Federal do ABC
  (UFABC)
Santo André, SP, Brazil

Ned J. Corron
Charles M. Bowden Laboratory, US
  Army Aviation & Missile
  Research, Development, and
  Engineering Center
Redstone Arsenal, AL, USA

Leonardo T. Duarte
School of Applied Sciences (FCA)
University of Campinas (UNICAMP)
Limeira, SP, Brazil

Marcio Eisencraft
Escola Politécnica
Universidade de São Paulo
São Paulo, SP, Brazil

Renato D. Fanganiello
Escola de Engenharia da
    Universidade Presbiteriana
    Mackenzie
São Paulo, SP, Brazil

Fabiano A. S. Ferrari
Department of Physics
Universidade Estadual de Ponta
    Grossa
Ponta Grossa, Brazil

Rodrigo T. Fontes
Escola Politécnica
Universidade de São Paulo
São Paulo, SP, Brazil

Celso Grebogi
Institute for Complex System and
    Mathematical Biology
King's College, University of
    Aberdeen
Aberdeen, UK

José M. V. Grzybowski
Technological Institute of
    Aeronautics (ITA)
São José dos Campos, SP, Brazil

Géza Kolumbán
Pázmány Péter Catholic University
The Faculty of Information
    Technology
Budapest, Hungary

Tamás Krébesz
Budapest University of Technology
    and Economics
Budapest, Hungary

Francis C. M. Lau
The Hong Kong Polytechnic
    University
Hong Kong SAR, China

Murilo B. Loiola
Centro de Engenharia, Modelagem e
    Ciências Sociais Aplicadas
    (CECS)
Universidade Federal do ABC
    (UFABC)
Santo André, SP, Brazil

David Luengo
Universidad Politécnica de Madrid
Madrid, Spain

Elbert E. N. Macau
Laboratory for Applied Mathematics
    and Computing (LAC)
National Institute for Space
    Research (INPE)
São José dos Campos, SP, Brazil

Luiz H. A. Monteiro
Escola de Engenharia da
    Universidade Presbiteriana
    Mackenzie
São Paulo, SP, Brazil
Escola Politécnica da Universidade
    de São Paulo
São Paulo, SP, Brazil

Everton Z. Nadalin
School of Electrical and Computer
    Engineering (FEEC)
University of Campinas (UNICAMP)
Campinas, SP, Brazil

Vanessa B. Olivatto
School of Electrical and Computer
    Engineering (FEEC)
University of Campinas (UNICAMP)
Campinas, SP, Brazil

Aline de O. N. Panazio
Centro de Engenharia, Modelagem e
Ciências Sociais Aplicadas
(CECS)
Universidade Federal do ABC
(UFABC)
Santo André, SP, Brazil

Sandro E. de S. Pinto
Department of Physics
Universidade Estadual de Ponta
Grossa
Ponta Grossa, Brazil

Hai-Peng Ren
Department of Information and
Control Engineering
Xi'an University of Technology
Xi'an, China

João Marcos T. Romano
School of Electrical and Computer
Engineering (FEEC)
University of Campinas
(UNICAMP)
Campinas, SP, Brazil

Ignacio Santamaría
Universidad de Cantabria
Santander, Spain

Magno T. M. Silva
Escola Politécnica
Universidade de São Paulo
São Paulo, SP, Brazil

Diogo C. Soriano
Centro de Engenharia, Modelagem e
Ciências Sociais Aplicadas
(CECS)
Universidade Federal do ABC
(UFABC)
Santo André, SP, Brazil

Ricardo Suyama
Centro de Engenharia, Modelagem e
Ciências Sociais Aplicadas
(CECS)
Universidade Federal do ABC
(UFABC)
Santo André, SP, Brazil

Dimitris Syvridis
Department of Informatics and
Telecommunications
National and Kapodistrian
University of Athens
Panepistimiopolis, Ilisia, Greece

Romeu M. Szmoski
Department of Physics
Universidade Estadual de Ponta
Grossa
Ponta Grossa, Brazil

Chi K. Tse
The Hong Kong Polytechnic
University
Hong Kong SAR, China

Ricardo L. Viana
Department of Physics
Universidade Federal do Paraná
Curitiba, PR, Brazil

Takashi Yoneyama
Technological Institute of
Aeronautics (ITA)
São José dos Campos, SP, Brazil

1

Introduction and main concepts

Elbert E. N. Macau

Laboratory for Applied Mathematics and Computing (LAC)
National Institute for Space Research (INPE)

CONTENTS

1.1 Introduction

The main propose of *communication systems* is to reliably transmit information between two players [12], i.e., the sender and the receiver. Throughout history, these systems behave as essential to foster and support the development and integration of people, besides being the support for their defense initiatives. Thus, efforts towards having systems more efficient, reliable, and safe are constant and strategic [18]. In this direction, the *chaotic dynamics* is proving to be an appropriate direction to be seriously considered.

The chaotic behavior in physical systems implies a signal that presents an erratic and irregular time evolution [8]. Over time, its presence was thought to be a kind of an unwanted noise that should be avoided or eliminated. This picture radically changed in the last decades of the past century, and today we have a better understanding of the chaotic behavior and its profound consequences for science and technology [21]. Actually, the chaotic dynamics happens in deterministic nonlinear systems and its main characteristic is the sensitive dependence on initial conditions [8], the so-called "butterfly effect" [14]. This characteristic implies that even a very small perturbation applied on a system trajectory is able to change dramatically its subsequent time history. It is this property that practically rules out long-term forecast for chaotic systems if the initial condition is not exactly known.

A chaotic signal is aperiodic and presents a "complex-like" behavior, while its spectral analysis unveils a broad-band structure [21]. These last characteristics may suggest that in theory it could be somehow used to bear a great amount of information and so, in principle, it may be used as a building block for communication systems [11,17]. However, the accomplishment of this idea would not be that easy. The main problem is how to deal with the "butterfly effect" and so exploit these favorable characteristics to encoding information in a chaotic waveform. Would it be feasible? If this is possible, how can a chaotic waveform be transmitted over a communication channel? Would it be possible for the receiver to properly recover the information that the sender is transmitting?

Actually, two prominent theoretical achievements have shed light over these issues. The first one is related to the control of "butterfly effect." This term was coined by Edward Lorenz to better title the talk he gave at the December 1972 meeting of the American Association for Advanced of Science, which was "*Predictability: Does the Flap of a Butterfly's Wings in Brazil Set Off a Tornado in Texas?*". Embedded on this term is the idea that even a very small perturbation applied to a chaotic trajectory is enough to make it deviate from its original track so that, after some time, this new trajectory is completely uncorrelated with the original one [14]. This behavior means that a chaotic system can be viewed as a system that is "locally unstable" in the sense that it never sets on a regular motion and is radically influenced by small perturbations acting on it. In this scenario of extreme "local instability," would it be possible to control a chaotic evolution? This issue has motived intensive discussions in which several acclaimed researchers affirmed that it would not be possible at all. However, in 1990, a revolutionary work entitled *Controlling Chaos* [22] shows that chaotic evolution can in fact be controlled buy using very small perturbations judiciously chosen and in a way that preserves the system's chaotic dynamics. This outstanding result led to an intensive research effort, which still goes on today. As a consequence, now we know that the "butterfly effect" can be tamed, resulting from it technological applications that take advantage from the intrinsic complexity of the chaotic evolution to get a unique level of efficiency [16]. This result is obtained by expending very little energy, because the control action is built on perturbations [15,16]. This approach can also be used to codify with efficiency desired information in the chaotic dynamics, as will be explain later.

The second achievement that opened the way for chaotic communication systems is related to *synchronization*. The cornerstone for standard communication systems is the synchronization among oscillators [12]. Sender and receiver must have their clock oscillators synchronized to allow the recovery by the receiver of the information codified in a communication waveform. Even today, considerable research efforts are devoted to the development of better and more efficient techniques to synchronize regular oscillators over a communication network [18].

This requirement also holds if one wants to use a chaotic signal for communication proposes: the receiver must somehow recover the chaotic evolution of the sender to decode the information codified in a chaotic waveform [23]. For a dissipative system, chaotic trajectories are located on a chaotic attractor in state space [8]. Recovering the chaotic evolution means that the state trajectory on the receiver's chaotic attractor must reproduce exactly the one that takes place in the sender's attractor, even if not all the state variables that characterize the system's dynamics are transmitted through the communication channel. It means that the chaotic dynamics of the receiver must somehow be synchronized with the one in the sender. However, if one thinks about the "butterfly effect" [14], at first this apparently seems to be not possible or even counterintuitive. This impression is heightened when one considers that not all variables should preferably be transmitted. However, this impression is wrong. In 1990 a remarkable work from Pecora and Carroll [25], in which they extended a previous result obtained by Yamada and Fujisaka [30,31], used geometric arguments to show how it is possible to couple chaotic systems to get them synchronized. In their approach, two chaotic systems may be synchronized by driving a "subsystem" of the receiver with a scalar signal from the full sender system. Later in this text, this approach will be explained. In this same work, the authors also have suggested that their finding could be used as a basing block for a chaotic communication system.

Following these remarkable achievements, a very intensive research effort was started to implement and develop communication systems based on chaos [26]. This intense effort of research continues today, as seen by the increasing number of research papers that are published each year in scientific journals [13,24]. All of them are trying to get systems with more efficiency and flexibility, and also to give a kind of increase in security and robustness to the message that is transmitted through it. In this direction, one can see impressive works ranging from digital modulation schemes, generation of spreading codes, encryption, performance optimization, etc. [24]. The more prominent result obtained so far was the one in which an optical carrier wave generated by a chaotic laser was used to encode a message and transmit it over 120 km of optical fibre in the metropolitan area network of Athens, Greece [2], in which transmission rates in gigabits per second were achieved with very small bit-error rates. This experiment indubitably has proven the feasibility of communication of chaos for daily life and commercial use.

The use of chaos in communication has evolved around two approaches: the coherent and the noncoherent one. Historically, the former was first introduced and is based on the outstanding results regarding chaotic dynamics previously displayed. Thus, one has information codified in the chaotic waveform that is transmitted through the communication channel so that the receiver uses a synchronization approach to recover the chaotic waveform and a kind of "chaotic correlation approach" to recover the message [3]. This approach is now seen as having great potential and appeal in areas where the standard communication approach is not consolidated especially because it does not meet the desired requirements of efficiency and even efficacy, as in cases of space communication, deep ocean communication, wireless communication, etc. Moreover, relative to the noncoherent approach, the main point is that the receiver does not need to exactly recover the chaotic waveform that is being generated by the sender [24]. This provides a robustness of the system to the inherent problems in the communication channel.

Another very active and successful area of research is spectrum spreading with chaotic sequence [5]. Here, the spreading code is a stored chaotic sequence, which means that there are potentially infinitely many difference sequences that can form a unique ensemble according to the ergodic invariant measures. Furthermore, a performance comparable to conventional systems can be achieved.

In the next sections we review with sufficient level of detail the main points which provide the basis for the development of chaos-based communication systems. The main purpose here is to convince the reader that these systems may incorporate a potential level of efficiency and flexibility which makes them the best solution, in particular for nontrivial applications in areas where the traditional approach is not widespread and so consolidated.

1.2   Controlling chaos and the codification by feedback control of symbolic dynamics

The concept of *chaos control* was coined and conceived in the last decade of the twentieth century [22]. Very recently, the original paper named "Controlling Chaos" [22] that introduced this idea was considered to be one of the most important and significant of the twentieth century. In this article, chaos control was presented as a method for *stabilizing chaotic behavior*. It wisely exploited the key properties presented on chaotic dynamics. Basically, a chaotic invariant set has embedded on it an infinite but enumerable set of *unstable periodic orbits (UPOs)* of any period [8]. In dissipative chaotic systems, these *UPOs* have associated with them directions along which trajectories converge or diverge to them, called *stable* and *unstable manifolds*, respectively [1], i.e., they are saddle points. Given a trajectory, one counts on the transitive property of the chaotic invariant set so that this trajectory came very close to a specific unstable periodic orbit. As the trajectory is sufficiently close to this orbit, small judiciously chosen perturbations are applied to keep the trajectory on the unstable periodic orbit. As so, the chaotic behavior is stabilized and a former "irregular" trajectory starts to behave "regularly" (periodically). As the chaotic invariant set presents infinite periodic orbits of any period, in theory infinite periodic motions can be achieved.

An enormous flow of theoretical and experimental research has followed this seminal work. The consequences of it for science and technology are profound and impressive. It shows that chaos can be tamed by properly taking advantage of its main characteristics, the "butterfly effect": the same perturbation that is able to dramatically change a chaotic trajectory if applied judiciously can be used to control the chaotic evolution. This procedure was subsequently extended so that the flexibility inherent of the chaotic evolution could then be properly exploited in an unprecedented level, especially in regard to applications. As so, the so-called *targeting* procedure came about in which properly applied perturbations are used to also take advantage of the "butterfly effect" to rapidly drive a chaotic trajectory between the neighborhood of two points of the chaotic invariant set [15,16]. This methodology, that actually implies an optimized solution, is today applied in several technological applications, including in guiding spacecrafts in space travels [16].

A similar procedure can be used to control a system trajectory so that a message can be conveniently codified on it. A chaotic signal is characterized by an aperiodic and apparently random waveform that produces a sequence of maxima and minima [2]. One can associate the digital symbol 1 each time the wave overpasses a previously specified value, while the symbol 0 is associated with signal excursions that underpass another prior established value [3,4,28]. By doing so, one can view a chaotic system as a binary sequence generator. Furthermore, this generator can be controlled by using small perturbations in a framework of an implemented control of chaos strategy [9,10,17]. Thus, the chaotic dynamics can be controlled to produce desired digital sequences. The control of chaos methodology used to accomplish the symbolic codification is named *feedback control of symbolic dynamics*. Let us discuss how it works.

As the control of chaos philosophy just allows one to take advantage of the inherent complexity of the chaotic dynamics to control it by using "small perturbations" [15] (and so with expenses of low energy), typically not all the sequences can be produced. Thus, there are sequences that the system dynamics is unable to produce. These sequences are named by forbidden sequences. For a specific system, we call its *grammar* the set of rules that specify the allowable sequences that its dynamics can produce. For a given system, there are various proposed methods for the determination of its grammar.

In a communication scenario, the source of information must be allowed to produce any desired sequence. To codify this sequence in a chaotic waveform, some conversion algorithm must be used on the original source stream. This algorithm is usually easily implemented by using a conversion table [9,10,17, 19].

Let us now apply these ideas to the *Chua's circuit* [20] that appears in Figure 1.1. The mathematical equations that describe its dynamics are the

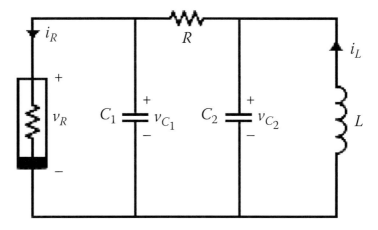

FIGURE 1.1
Chua circuit used to illustrate our communication strategy.

following:

$$\begin{cases} C_1 \dot{v_{C_1}} = G\left(v_{C_2} - v_{C_1}\right) - g\left(v_{C_1}\right) \\[2mm] C_2 \dot{v_{C_2}} = G\left(v_{C_1} - v_{C_2}\right) + i_L \\[2mm] L\dot{i_L} = -v_{C_2} \end{cases} \qquad (1.1)$$

in which C_1, C_2, $G = 1/R$ and L are given constants.

The nonlinear resistor (Chua's diode) presents a 3-segment odd-symmetric voltage-current characteristic that can be described by the following equation:

$$g\left(v_{C_1}\right) = G_b v_{C_1} + \frac{1}{2}\left(G_a - G_b\right)\left(|v_{C_1} + B_p| - |v_{C_1} - B_p|\right). \qquad (1.2)$$

Here, we use the normalize parameter values: $C_1 = 1/9$, $C_2 = 1$, $L = 1/7$, $G = 0.7$, $G_b = -0.5$, $G_a = -0.8$ and $B_p = 1$.

Let us now define a Poincaré surface of section at $i_L = \pm GF$, $|v_{C_1}| \le F$, where $F = B_p\left(G_b - G_a\right)/\left(G + G_b\right)$, so that half planes intersect the attractor with edges at the unstable fixed points at the center of the attractor lobes. As appears in Figure 1.2, these planes allow us to associate the digital symbols 0 or 1 to the system dynamic, which depends on each of the two planes the trajectory crosses.

The system's grammar is determined by allowing the system to evolve freely, following its intrinsic dynamics [9, 10]. Each time the system crosses the Poincaré section, we register the x value and the corresponding sequence of symbols. Thus, suppose that the system generates the sequence of symbols $0.b_1b_2b_3\cdots$, which corresponds to the real number $r = \sum_{\infty}^{n=1} b_n 2^{-n}$. In this schema, r represents the symbolic state of the system. By following the system dynamics and its intersection with the defined Poincaré section, we obtain

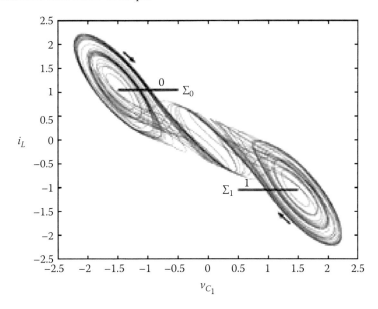

FIGURE 1.2
State-space trajectory of the Chua's circuit in double scroll mode projected on the $v_{C_1} x i_L$ and the defined Poincaré section.

the relation between x and its symbolic codification. This relation is called *codification function* and for our case, it can be seen in Figure 1.3.

From the analysis of the symbolic sequence, we can realize that the grammar for this system is very simple [1,14]: any sequence of bits can be generated with the exception of two consecutive maximums. Thus, we can codify any sequence of bits generated by the source if a symbol 1 is "artificially" introduced after each sequence of 1 and a 0 after each string of 0s. As a consequence, k oscillations of a given polarity represents $b - 1$ bits of information.

Let us now introduce the control strategy that allows us to impose to the system dynamics to follow a desired string of bits. Say the system state point passes through branch 0 of the surface of section at $x = x_a$, and next crosses the surface of section at $x = x_b$, on either branch 0 or 1. Because we have previously determined the function $r(\chi)$, we can use the stored values to find the symbolic state $r(x_a)$. We then convert the number $r(x_a)$ to its corresponding binary sequence truncated at some chosen length N, and store this finite-length symbol sequence in a code register. As the system state trajectories move toward its next encounter with the surface of section $x = x_b$, we shift the sequence in the code register left, discarding the most significant bit, and insert the first desired information code bit in the now empty least significant slot of the code register. We then concert this new symbol sequence to its corresponding symbolic state r'_b. Now, when the system state point

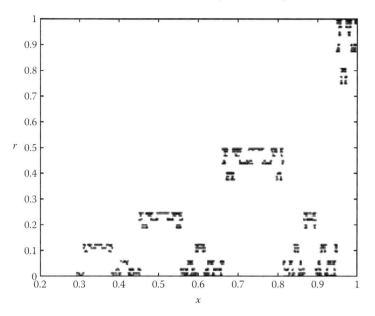

FIGURE 1.3
Binary codification function $r(\chi)$ for the double scroll system.

crosses the surface of section at $x = x_b$, we use a search algorithm to find the nearest value of the coordinate x that corresponds to the desired symbolic state r'_b. Let us call this value x'_b. By construction, $\left| r(x_b) - r\left(x'_b\right) \right| \leq 2^{-N}$. Now let $\delta x = x_b - x'_b$. Because we have chosen the branches of the surface of the section at constant values of the inductor current i_L, the deviation δx corresponds to a small deviation in the voltages v_{C_1} and v_{C_2} [9, 10, 17, 19].

1.3 Chaotic synchronization and information transmission

An impressive result related to the chaos theory was the discovery that two chaotic systems may be synchronized with a scalar signal from the full system [25]. The chaotic subsystem may be cascaded so that the driving signal is reproduced. Subsequent works showed that this result could be applied to develop chaos-based communication systems [6, 7, 29]. This approach has proven to be effective even when a large amount of noise is added to the driving signal or when the communication signal is not effective. Let us here see how this synchronization phenomenon can take place in chaotic systems.

Let us consider a dynamical system, the sender, that may be described by the following equation:

$$\dot{\mathbf{u}}(t) = \mathbf{f}(\mathbf{u}). \tag{1.3}$$

This system is then divided into two subsystems, $\mathbf{u} = (\mathbf{v}, \mathbf{w})$, so that

$$\begin{cases} \dot{\mathbf{v}} = \mathbf{g}(\mathbf{v}, \mathbf{w}) \\ \dot{\mathbf{w}} = \mathbf{h}(\mathbf{v}, \mathbf{w}) \end{cases} \tag{1.4}$$

where $\mathbf{u} = (u_1, \cdots, u_m)$, $\mathbf{g} = (f_1(u), \cdots, f_m(u))$, $\mathbf{w} = (u_{m+1}, \cdots, u_n)$, and $\mathbf{h} = (f_{m+1}(u), \cdots, f_n(u))$. It is important to know this division is truly arbitrary.

The receiver system may be created by duplicating a new subsystem $\mathbf{w}'$ identical to the $\mathbf{w}$ system, substituting the set of variables $\mathbf{v}$ for the corresponding $\mathbf{v}'$ in the function $\mathbf{h}$, and augmenting Equation (1.4) with this new system, giving the following full system, that describes the dynamics in both sender and receiver:

$$\begin{cases} \dot{\mathbf{v}} = \mathbf{g}(\mathbf{v}, \mathbf{w}) \\ \dot{\mathbf{w}} = \mathbf{h}(\mathbf{v}, \mathbf{w}) \\ \dot{\mathbf{w}}' = \mathbf{h}(\mathbf{v}, \mathbf{w}'). \end{cases} \tag{1.5}$$

This construction is called *complete replacement* [26]. Note that in this scheme the sender's signal $\mathbf{v}$ acts as a driver for the receiver system. The main result here is that if all the Lyapunov exponents of the $\mathbf{w}$ system (as well as $\mathbf{w}'$) are less than zero, then $\left| \mathbf{w}' - \mathbf{w} \right| \to 0$ as $t \to \infty$, which means that receiver and sender systems synchronize with each other.

Let us apply those ideas for the Lorenz system [14] with parameters $\sigma = 10$, $r = 28$, and $b = 8/3$. The equation for the sender is the following:

$$\begin{cases} \dot{x} = -\sigma\,(y - x) \\ \dot{y} = -xz + rx - y \\ \dot{z} = xy - bz. \end{cases} \tag{1.6}$$

Following the previously discussed ideas, the "receiver" system is as follows:

$$\begin{cases} \dot{y}' = -xz' + rx - y' \\ \dot{z}' = xy' - bz'. \end{cases} \tag{1.7}$$

Note that we have replaced x' by x in the second system and eliminated the equation for $\dot{x}'$ in the receiver, since it is superfluous. For this configuration, x actually acts as the driving signal for the second Lorenz system. This idea is schematically shown in Figure 1.4. As so, the first system can be regarded as the *driver*, and the second one as the *response*.

For this configuration, if the parameters used imply that the two Lyapunov exponents for System (1.7) are less than zero, then starting both systems from

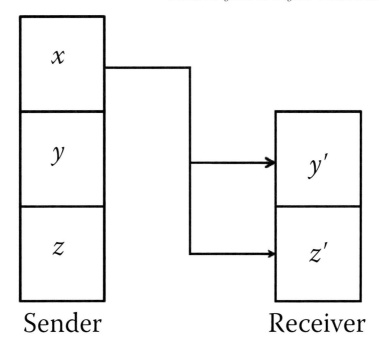

FIGURE 1.4
Drive-response scheme for complete replacement synchronization.

arbitrary initial conditions, after some time, we see that $|y' - y| \to 0$ and $|z' - z| \to 0$, which means that the chaotic evolution of both systems synchronizes [25, 26]. For our example, the Lyapunov exponents of the subsystem composed by variables (y', z') are equal to $(-1.81, -1.86)$, which implies in synchronization, as can be seen in Figure 1.5.

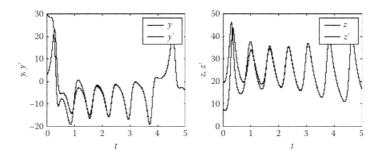

FIGURE 1.5
Synchronization between two Lorenz systems using the variable x as the transmitter variable.

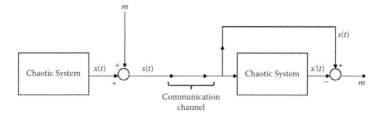

FIGURE 1.6
First approach to a chaos-based communication system.

Several methods using chaotic carriers and based on chaotic synchronization have been studied over the years [11, 19, 23, 24, 26, 27, 29]. In the first approach by Cuomo and Oppenheim [7], the information message to be transmitted, assumed small in magnitude, is added to the chaotic signal, assumed to be much larger, of one of the variables. This variable is transmitted to the receiver, allowing it to synchronize with the sender. Then, the information message can be recovered by subtraction (see Figure 1.6).

In this communication architecture, the signal x is transmitted from the sender and a small speech signal added to it. At the receiver, the difference $x - x'$ is taken and so the speech signal is recovered. This strategy opens the way for a chaos-based communication system.

1.4 A chaos-based communication system

The architecture proposed by Cuomo and Oppenheim [6, 7] works properly. In another better-elaborated and less sensitive to noise approach, the information message actually drives the chaotic transmitter system and so is incorporated into the system dynamics [19, 26]. A schematic representation of this approach appears in Figure 1.7.

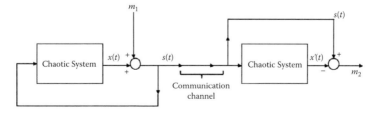

FIGURE 1.7
Chaos-based communication system in which the transmitted message is incorporated into the system dynamics.

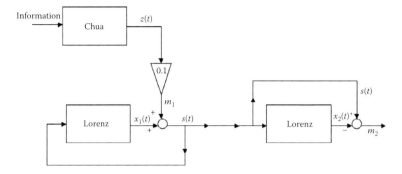

FIGURE 1.8
Chaos-based communication system with chaotic codification strategy.

Let us now show how those previously discussed topics can be combined to result in a very efficient chaos-based communication system. In this approach, a chaotic waveform is used to transmit messages that are codified in a chaotic invariant set [19]. This architecture is depicted in Figure 1.8.

In this approach, the information to be transmitted is codified in a Chua circuit [20] using the method of codification by feedback control of symbolic dynamics previously discussed. The codified information (m_1) is added just after the system chaotic carrier output (x_1) and the resulted transmitted signal (s) is injected into the feedback path and so incorporated into the system dynamics that provides the chaotic carrier. The receiver can recover the codified information (m_2) by subtracting the output of the synchronized chaotic carrier (x_2) from the transmitted signal (s). We use a Lorenz system to generate the chaotic carrier. Note that this architecture yields exact recovery of the message, i.e., $m_2 \to m_1$ after an initial transient.

Let us now present the results of our numerical simulation on this architecture. For all the cases, our goal is to send the chaotic message over the communication signal. Figure 1.9 shows the result of using the method of codification by feedback control of symbolic dynamics to codify this message on the Chua chaotic dynamical system. Just the z state of this system is represented. The arrows indicate the time location in which the signal crosses the Poincaré section, while the bold digits were artificially introduced in the message to overcome the problem of forbidden sequence on the Chua system.

In Figure 1.10 we exhibit the power spectrum of z state variable of the controlled Chua system together with the spectrum of the x state variable of the Lorenz systems, which provides the chaotic carrier that is used to encompass the codified information. Note that the relevant parts of both signals are concentrated on the low frequency part of the signals.

Next, Figure 1.11 shows the power spectrum of the transmitted signal. Comparing this figure with the last one, we can see that our communication

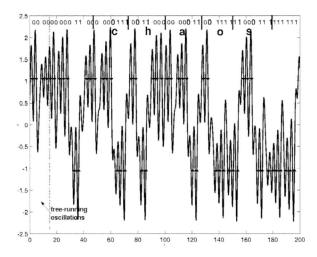

FIGURE 1.9
Message "chaos" codified in the Chua chaotic dynamics.

strategy preserves the shape of the chaotic carrier signal, which results in a very robust communication strategy.

The spectrum of the recovered codified information (m_2) can be compared with the spectrum of the original codified message (m_1) in Figure 1.12. Note that all the relevant parts of the signal are effectively recovered.

1.5 A chaos-based communication system — main advantages and further developments

Since the introduction of the concept to the present day, research effort that aims to explore communication systems based on chaos is only increasing. From it, today the perception is clear that these systems are potentially much more efficient than one might initially expect [3]. From these efforts, considerably effective and innovative methods have been developed related to synchronization, noisy filtering from noisy chaotic trajectories, chaotic error correcting codes, chaos coding, and cryptography with chaos [24]. However, the major difference in efficiency was the realization that all these operations that are as essential to any communication network commercially used, can be performed in an integrated and simple way by exploiting a single chaotic

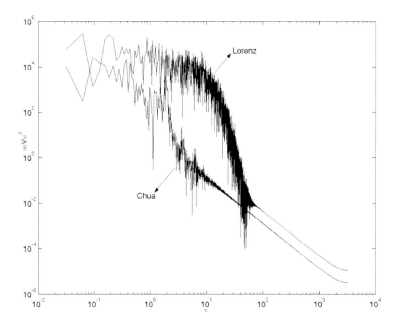

FIGURE 1.10
Power spectrums of the Chua and Lorenz systems in normalized coordinate units.

subsystem. This feature of excellence is unparalleled in traditional systems in use today.

Let us see, for example, the two fundamental operations that a standard digital communication system must perform before sending information through a communication channel: (i) source encoding, which compacts, compresses, and encrypts the source message; (ii) channel encoding, which guarantees that the encoded message is robust against the presence of noise in the channel. Both operations encode one bit stream into another. In a standard digital communication scheme, each of these functions is not only accomplished by different subsystems, but, cumbersomely, the final sequence of bits must be modulated a posteriori into a wave form signal that can be adequately transmitted over a channel. This complex and involved scenario can be radically simplified if the communication system is based on chaos instead. All those functions and operations can be performed with the use of only one subsystem, the same one that performs the proper modulation of the signal for transmission over the communication channel [3]. Furthermore, a nonlinear chaotic oscillator that generates a wave form for transmission can be easily built, while all the electronics that are necessary for encoding the information

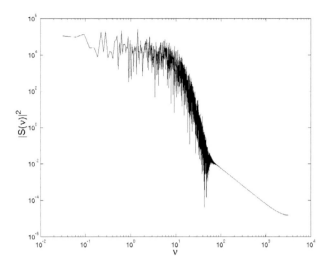

FIGURE 1.11
Power spectrum of the transmitted signal in a normalized coordinate unit.

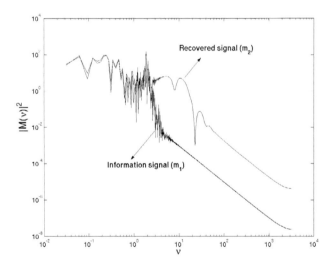

FIGURE 1.12
Power spectrum of the transmitted signal in a normalized coordinate unit.

in the chaotic signal remain as a low-power and inexpensive microelectronic circuit. The final scenario of this efficient, simple, and low-power consumption process is a chaotic waveform that encompasses compressed and encrypted information with robustness to the presence noise. As so the resulting signal is not only the carrier but also the message itself [3, 19].

In summary, one can say that this approach has the advantages listed below, which will undoubtedly make it as a solution of excellence to be considered for the next generation of communication systems:

1. Efficiency: The ongoing communication systems require a considerable amount of energy to operate complex circuits to accomplish the modulation and conditioning of the information to be transmitted. Chaotic communication systems operate using very small amounts of energy because they just use small perturbations to accomplish their work.

2. Compactness: In communication with chaos, nonlinear devices, amplification, and circuit for adjusting input to the proper range are eliminated. Consequently the number of components is reduced, as well as the weight and the volume of the circuits involved.

3. Great information-bearing capacity: Chaotic signals are broad-band and may have a spatial dimension in addition to a temporal dimension. As a consequence, they present a potential great capacity to bear information without a counterpart in non-chaotic signals.

4. Low-cost manufacturing: Many of the components used to implement communication systems with chaos are electronic/optical components currently used in conventional communication devices. Thus, costs for development of new components will be limited. Furthermore, the devices needed use few components and simpler circuits, which imply low cost of manufacturing.

Bibliography

[1] K. T. Alligood, T. Sauer, and J. A. Yorke. *Chaos: An Introduction to Dynamical Systems*. Textbooks in Mathematical Sciences. Springer, New York, 1996.

[2] A. Argyris, D. Syvridis, L. Larger, V. Annovazzi-Lodi, P. Colet, I. Fischer, J. Garcia-Ojalvo, C.R. Mirasso, L. Pesquera, and K.A. Shore. Chaos-based communications at high bit rates using commercial fibre-optic links. *Nature*, 438(7066):343–346, November 2005.

[3] M. S. Batista, E. E. N. Macau, C. Grebogi, Y. Lai, and E. Rosa. Integrated chaotic ommunication scheme. *Phys. Rev. E*, 62:4835–4845, 2000.

[4] E. M. Bollt. Review of chaos communication by feedback control of symbolic dynamics. *International Journal of Bifurcation and Chaos*, 13(02):269–285, 2003.

[5] S. Chen and H. Leung. Ergodic chaotic parameter modulation with application to digital image watermarking. *IEEE Transactions on Image Processing*, 14:1590–1602, 2005.

[6] K. M. Cuomo and A. V. Oppenheim. Circuit implementation of synchronized chaos with applications to communications. *Phys. Rev. Lett.*, 71(1):65–68, July 1993.

[7] K. M. Cuomo, A. V. Oppenheim, and S. H. Strogatz. Synchronization of lorenz-based chaotic circuits with applications to communications. *IEEE Transactions on Circuits and Systems II*, 40:626–633, October 1993.

[8] R. L. Devaney. *An Introduction to Chaotic Dynamical Systems*. Addison Wesley, Redwood City, CA, 2nd edition, 1989.

[9] S. Hayes, C. Grebogi, and E. Ott. Communicating with chaos. *Phys. Rev. Lett.*, 70:3031–3034, May 1993.

[10] S. Hayes, C. Grebogi, E. Ott, and A. Mark. Experimental control of chaos for communication. *Phys. Rev. Lett.*, 73:1781–1784, 1994.

[11] L. Illing. Digital communication using chaos and nonlinear dynamics. *Nonlinear Analysis: Theory, Methods & Applications*, 71(12):E2958–E2964, December 2009.

[12] B. P. Lathi and Z. Ding. *Modern Digital and Analog Communication Systems*. Oxford University Press, Oxford, 4th edition, 2009.

[13] F. C. M. Lau and C. K. Tse. *Chaos-Based Digital Communication Systems*. Springer, Berlin 2003.

[14] E. N. Lorenz. Deterministic nonperiodic flow. *J. Atmosf. Sci.*, 20:130–141, 1963.

[15] E. E. N. Macau and C. Grebogi. Driving trajectories in chaotic systems. *International Journal of Bifurcation and Chaos*, 11(5):1423–1442, 2001.

[16] E. E. N. Macau and C. Grebogi. Control of chaos and its relevancy to spacecraft steering. *Phil. Trans. R. Soc. A*, 364:2463–2481, 2006.

[17] E. E. N. Macau and C. M. P. Marinho. Communication with chaos over band-limited channels. *Acta Astronautica*, 53:465–475, 2003.

[18] G. Maral, M. Bousquet, and Z. Sun. *Satellite Communications Systems: Systems, Techniques and Technology*. Wiley, Chichester, West Sussex, 5th edition, 2010.

[19] C. M. P. Marinho, E. E. N. Macau, and T. Yoneyama. Chaos over chaos: A new approach for satellite communication. *Acta Astronautica*, 57:230–238, 2005.

[20] T. Matsumoto, L. Chua, and M. Komuro. The double scroll. *IEEE Transactions on Circuits and Systems*, 32:797–818, 1985.

[21] E. Ott. *Chaos in Dynamical Systems*. Cambridge, 2002.

[22] E. Ott, C. Grebogi, and J. A. Yorke. Controlling chaos. *Phys. Rev. Lett.*, 64:1196–1199, March 1990.

[23] U. Parlitz, L. Kokarev, T. Stojanovski, and H. Preckel. Encoding messages using chaotic synchronization. *Phys. Rev. E*, 53:4351–4361, 1996.

[24] L. Pecora, editor. *Chaos in Communications*. SPIE - the International Society for Optical Engineering, Europan, 2012.

[25] L. M. Pecora and T. L. Carroll. Synchronization in chaotic systems. *Phys. Rev. Lett.*, 64:821–824, February 1990.

[26] L. M. Pecora, T. L Carroll, G. A. Johnson, and D. J. Mar. Fundamentals of synchronization in chaotic systems. *Chaos*, 7:520–543, 1997.

[27] S. Sengupta and M. K. Kasotiya. A chaotic circuit demo for communications. *IEEE Potentials*, 23:28–31, 2004.

[28] S. Smale. Differentiable dynamical systems I. Diffeomorphisms. *Bull. Am. Math. Soc.*, 73:747–817, 1967.

[29] G. D. Van Wiggeren and R. Roy. Optical communication with chaotic waveform. *Phys. Rev. Lett.*, 81:3547–3550, 1998.

[30] T. Yamada and H. Fujisaka. Stability theory of synchronized motion in coupled-oscillator systems ii. *Progress of Theoretical Physics*, 70:1240, 1983.

[31] T. Yamada and H. Fujisaka. Stability theory of synchronized motion in coupled-oscillator systems iii. *Progress of Theoretical Physics*, 72:885, 1984.

2

Overview of digital communications

Ivan R. S. Casella, Aline de O. N. Panazio, and Murilo B. Loiola

Centro de Engenharia, Modelagem e Ciências Sociais Aplicadas (CECS)
Universidade Federal do ABC (UFABC)

CONTENTS

2.1   Introduction

Digital communication is the foundation of modern telecommunications. Basically, it is an evolution of analog communication that explores the advantages of the information in a digital format. For this reason, differently from analog communication systems, digital systems have to identify, at the receiver, only a finite number of possible transmit waveforms from the received signal and, consequently, they are much more powerful than any analog system.

The major benefits of digital communication systems are:

(a) High immunity to noise and channel distortions (accomplished by error correction coding, equalization, diversity techniques, etc.)

(b) Long distance communication in low quality channels

(c) Easy to encrypt the information in the digital domain

(d) Easy to store the information in the digital domain

(e) Easy to compress the information in the digital domain

(f) High dynamic range of information signal

(g) Multimedia capability

(h) Modern digital signal processing (DSP) techniques (e.g., software defined radio)

A digital communication system can be generically described by the processes of conversion, at the transmitter side, of the digital information symbols into signal waveforms more suitable to the communication channel, transmission of these waveforms through the channel, and, finally, recovery of the transmitted information symbols at the receiver side.

Usually, for short distance and some long distance (e.g., optical communications) wireline transmissions, baseband communication systems are employed. In these cases, a sequence of digital symbols generates variations in the amplitude, width, or position of a pulse waveform to represent each different symbol without any ambiguity so that they can be recovered at the receiver. This conversion process is called digital baseband modulation or, more often, digital coding and will be presented in Section 2.4. On the other hand, for long distance wireless transmissions, bandpass communication systems are generally used. In this case, a sequence of digital symbols varies the amplitude, phase, or frequency of a sinusoidal carrier waveform to represent each different symbol without ambiguity. This process, denoted as digital bandpass modulation, will be covered in Section 2.5.

2.1.1 Communication system model

Any digital communication system can be represented by the simplified model presented in Figure 2.1, composed by the following three main blocks:

(a) The transmitter, that is responsible for converting the digital information symbols into signal waveforms suitable for transmission over the communication channels.

(b) The communication channel, that is the medium through which information propagates.

(c) The receiver, that is responsible to recover the transmitted digital symbols from the received signal waveforms.

For these systems, the information signal can be represented by a time sequence of symbols of T_s seconds, each one encompassing n_b bits of T_b seconds.

Considering that a possible transmit symbol is part of a finite alphabet of M different symbols (i.e., the signal set), the system can be denoted as an M-ary digital communication system [66, 93]. In this case, the symbol rate, also known as *baud rate*, can be defined as

$$R_s = \frac{1}{T_s}. \tag{2.1}$$

As $n_b = \log_2(M)$ bits per symbol, the bit rate is easily obtained by

FIGURE 2.1
Simple M-ary digital communication system model.

$$R_b = R_s \cdot \log_2(M). \tag{2.2}$$

For baseband systems, the transmitter encompasses the process of baseband modulation (i.e., digital encoding) and the receiver includes the process of baseband demodulation (i.e., digital decoding). On the other hand, for bandpass systems, the transmitter encompasses the process of bandpass modulation and the receiver, of bandpass demodulation. An example comparing these processes is presented in Figure 2.2.

2.1.2   Communication channels

As stated in Section 2.1.1, the communication channel corresponds to the medium through which the information signals propagate. Examples of communication channels are the air in cellular and wi-fi systems, the space in satellite transmissions, copper wires in telephony and cable television, and optical fibers in optical communications. Although very different in nature, all communication channels exhibit, at least, attenuation and some kind of noise, thus changing the transmitted signal waveform.

Attenuation designates the energy spread a signal suffers when propagating through a dispersive (nonideal) medium. Hence, the channel attenuation is a function of the distance between transmitter and receiver, since the greater this distance, the higher the attenuation. Consequently, the energy of the

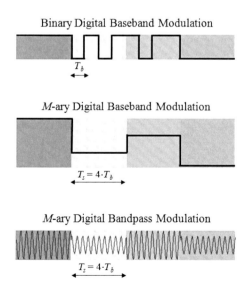

FIGURE 2.2
Example of baseband and bandpass digital modulation waveforms.

signal at the receiver side is always less than that at the transmitter side. Attenuation is also a function of frequency. For instance, copper wires used for telephony and Internet access through DSL (Digital Subscriber Line) systems present attenuations that decay approximately exponentially with frequency [42, 88, 89].

Noise is the term generally used to indicate any random signal that corrupts the transmitted waveforms. The most common kind of noise is *thermal noise*, present in any communication channel. This noise is produced by the random motion of electrons in a medium and its intensity increases with increasing temperature, being zero only at the absolute zero [71]. Because thermal noise is the result of the random motion of many independent electrons in the medium, it may be modeled as a Gaussian random process by the central limit theorem. Another characteristic of this Gaussian noise is that its power spectral density (PSD)[1] is approximately flat up to frequencies on the order of 10^{13} Hz [71]. Thus, thermal noise may also be considered a white noise in this frequency range. As mathematical modeling of communication channels considers that thermal noise has an additive effect on the transmitted signals, this noise is most often known as Additive White Gaussian Noise (AWGN). Besides AWGN, different channels may possess other specific noises, such as the impulsive noise present in wireless [7, 12] and Power-Line Communication (PLC) [25, 55] channels, and the shot noise encountered in optical fiber channels [1, 6].

Communication channels may also distort the transmitted waveforms as they propagate toward the receiver. Since the channels cannot pass infinite frequencies, any sharp corners of the waves are rounded [46, 71]. In fact, all communication channels have a cutoff frequency beyond which the transmitted signals are almost entirely attenuated. There are also many channels that exhibit a low-frequency cutoff. Thus, channels possessing just high-frequency cutoffs are usually modeled as lowpass filters, while bandlimited channels are modeled as bandpass filters [40].

The filtering effect may cause the transmitted waveforms to widen, possibly resulting in an overlap between pulses sent in different time instants if these pulses are not sufficiently apart. This overlapping, known as *intersymbol interference* (ISI), is one of the main factors limiting the performance of a digital communication system.

The ISI can also be caused by multipath propagation [6, 25, 33, 46, 66, 67]. In wireless communications, for instance, a signal may reach the receive antenna through many different paths due to atmospheric scattering and refraction, or reflections from building and other objects. The signals arriving along different paths will present different attenuations and delays, and might add at the receiver either constructively or destructively [33, 40, 66, 67], thus generating ISI. In PLC systems, multipath propagation is due to reflections produced by impedance mismatch between elements of the lines [25].

---

[1]The PSD shows how the power of a signal distributes along its frequency components.

In general, ISI channels are modeled by linear finite impulse response (FIR) filters [25,40]. Hence, the received signal, except from the AWGN, can be modeled by the convolution of the transmitted signal with the impulse response of the channel. Some communication channels, such as optical fibers, however, may present severe nonlinear effects [1]. In these cases, linear FIR filters are not well suited and nonlinear channel models must be employed.

It is important to highlight that communication channels may also be either time-invariant or time-varying. In time-invariant channels, such as the wire cables used for telephony and DSL systems, the impulse response or, equivalently, the frequency response, remains practically unchanged with time. On the other hand, time-varying channels have their impulse and frequency responses changing with time. This can occur, for example, in wireless communications, where transmitters and receivers might be mobile. The relative motion between transmitter and receiver leads to Doppler effect, which reflects on a time variation of the channel [33,46,66,67,87].

2.2 Review of fundamentals

This section will briefly present some fundamental concepts to the design and analysis of any digital communication system. At first, it is worth providing a definition for continuous- and discrete-time signals.

A continuous-time signal $s(t)$ is a signal defined on the continuum of time values, i.e., $s(t)$ is a function of a continuous independent variable.[2] On the other hand, a discrete-time signal $s(n)$ is defined only for specific time values, i.e., $s(n)$ is a function of a discrete independent variable [45,58].

Besides this classification, signals may also be classified according to the nature of their amplitudes. Signals whose amplitudes may assume any values in a continuous range are called analog, while signals whose amplitudes may assume just a finite number of values are called digital. Hence, the terms continuous-time and discrete-time refer to the nature of the independent variable, while analog and digital, to the nature of the dependent variable. The process of converting an analog continuous-time signal to a digital discrete-time signal (a bitstream) will be described in the sequel.

2.2.1 Sampling theorem

The simple communication system model presented in Section 2.1.1 considers the transmission of discrete-time digital information symbols. Although

---

[2]Although the independent variable is usually represented by time, a continuous-time signal may in fact represent a function of any continuous independent variable. For instance, a picture could be described as a "continuous-time" function of two variables representing a two-dimensional space.

digital information naturally arises in computer-to-computer communication, many other signals, such as voice, music, pictures, and video, are inherently continuous in time (or space) and amplitude. Therefore, to benefit from the advantages of digital communications, analog continuous-time signals must be accurately represented by a sequence of bits. This conversion from the analog domain to the digital domain is performed by an analog-to-digital converter (ADC or A/D), whose basic block diagram is shown in Figure 2.3.

The function of the sampling block in Figure 2.3 is to generate an analog discrete-time signal by taking samples of an analog continuous-time signal at regular time intervals. However, how often must these samples be taken in order that the resulting analog discrete-time signal correctly represents the original continuous-time signal? To answer this question, consider the frequency-domain representation of a bandlimited analog continuous-time signal shown in Figure 2.4(a), where f_m represents the maximum frequency component of this signal. The sampling operation is represented by

$$s(n) = s(nT_{sa}),\tag{2.3}$$

where n is the time index, T_{sa} is the sampling period, and $f_{sa} = 1/T_{sa}$ is the sampling frequency. The resulting discrete-time signal can then be written as [6]

$$\hat{s}(t) = \sum_{n=-\infty}^{\infty} s(n)\delta(t - nT_{sa}) = s(t) \sum_{n=-\infty}^{\infty} \delta(t - nT_{sa}),\tag{2.4}$$

where $\delta(t)$ is the Dirac delta function [45, 58]. The spectrum shown in Figure 2.4(b) is the result of sampling the signal in Figure 2.4(a) at a sampling frequency f_{sa}.

As can be seen from these figures, the sampled signal will correctly represent the continuous-time signal provided that they have the same spectrum between $-f_m$ and f_m. This is only possible if

$$f_{sa} - f_m > f_m \Rightarrow f_{sa} > 2f_m.\tag{2.5}$$

If the sampling frequency f_{sa} is lower than $2f_m$, which is known as the *Nyquist frequency*, portions of the spectrum will overlap and it will not be possible to

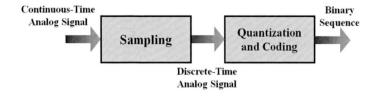

FIGURE 2.3
Basic block diagram of an analog-to-digital converter.

(a) Spectrum of a bandlimited continuous-time analog signal.

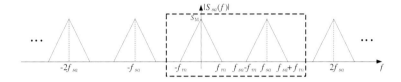

(b) Spectrum of the signal sampled at a frequency f_{sa}.

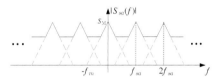

(c) Spectrum of the signal sampled at a frequency lower than $2f_m$.

FIGURE 2.4
Frequency-domain analysis of the sampling process.

recover the original signal from the sampled one, as illustrated in Figure 2.4(c). This phenomenon is called *aliasing* [6, 45, 58, 71] and its name comes from the fact that higher frequency components of a signal disguise themselves in the form of lower frequencies when sampling is too slow [71]. To see this "disguise," consider a 0.5 Hz sinusoid, represented by the dashed line in Figure 2.5, sampled at a frequency $f_{sa} = 0.4$ Hz. In this case, $f_m = 0.5$ Hz, $f_{sa} < 2f_m$ and these same samples could also be obtained from a sinusoid at 0.1 Hz, shown by the dotted line in Figure 2.5. Hence, regarding the samples, the 0.5 Hz signal disguises itself as a 0.1 Hz signal.

The inequality in Equation (2.5), referred to as the *Nyquist sampling theorem* or simply the *sampling theorem*, establishes a lower bound on the sampling frequency in order that the sampled discrete-time signal correctly represents the original continuous-time signal.

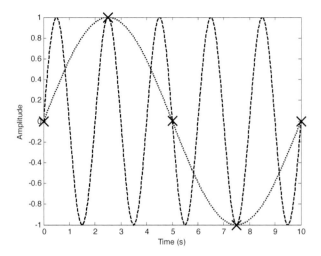

FIGURE 2.5
Effect of aliasing in time.

> *Sampling theorem*: To accurately represent a bandlimited analog continuous-time signal by its samples, the sampling frequency (f_{sa}) must be greater than two times the maximum frequency component (f_m) of the signal, i.e., $f_{sa} > 2f_m$.

It is worth noting that the sampling theorem assumes the existence of a bandlimited signal at the input of the sampler. However, physical signals, such as voice and music, are limited in time and thus unlimited in frequency. To ensure that an analog continuous-time signal is properly bandlimited prior to sampling, a lowpass filter called *anti-aliasing filter* is normally used [45, 58].

Once the signal is properly sampled, the resulting samples must be converted to a sequence of bits. This task is carried out by the block named quantization and coding in Figure 2.3. The goal of the quantization process is to transform an analog discrete-time signal into a digital discrete-time signal (a bitstream). To that end, the quantizer rounds off the amplitudes of the samples to several (finite) possible levels. Each level is then labeled, i.e., coded, with binary numbers [46, 71, 81]. For instance, for eight quantization levels, three-bit binary numbers are required. This can be visualized in the example shown in Figure 2.6. In this figure, the digital signal resulting from the ADC is 100111001101101. It is important to note that the quantization levels determine the *resolution* of the ADC. The greater the number of quan-

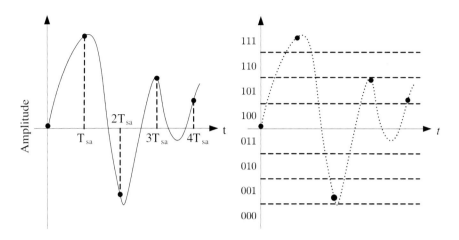

(a) Sampling a continuous-time analog signal.

(b) Example of quantization and coding.

FIGURE 2.6
Basic steps of an analog-to-digital converter.

tization levels, the greater the number of bits needed to represent them, and the smaller the spacing between the levels. Further details on quantization and coding techniques, as well as on practical realizations of ADCs, can be found in [46, 71, 81].

2.2.2 Bandwidth

A bandpass digital modulated signal can be obtained by translating, in the frequency domain, a baseband digital modulated signal to a given frequency f_o. This process, usually denoted as heterodyning, is illustrated in Figure 2.7. If the bandpass system is linear, a baseband signal with maximum frequency component f_m generates a bandpass signal with bandwidth of $2f_m$, i.e., twice the bandwidth of the baseband signal.

   Considering the Nyquist sampling theorem presented in Section 2.2.1 and the use of ideal pulse shaping, which will be discussed in Section 2.6.1, the baseband signal bandwidth can be defined as

$$B = \frac{R_s}{2}. \tag{2.6}$$

Consequently, for a linear bandpass communication system, the bandpass signal bandwidth can be estimated by

$$W = R_s. \tag{2.7}$$

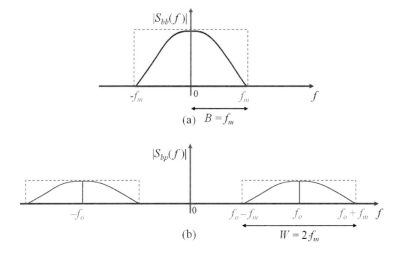

FIGURE 2.7
Spectrum of baseband (a) and bandpass (b) signals.

In a later section, a more realistic pulse shaping will be introduced and a new evaluation of the signal bandwidth will be presented.

2.2.3 Measuring efficiency of communication systems

A desirable digital communication system should (i) provide low Bit Error Rate (BER)[3] at low receiver Signal-to-Noise Ratio (SNR), which is defined by the ratio of signal power (P_s) to noise power (P_n), i.e.,

$$\text{SNR} = \frac{P_s}{P_n}, \tag{2.8}$$

(ii) occupy a minimum of bandwidth, and (iii) be easy and cost-effective to implement. In general, existing systems do not simultaneously satisfy all of these requirements [67].

The performance of digital communication systems can often be measured in terms of its power or energy efficiency and bandwidth efficiency. Power efficiency η_p and energy efficiency η_e measure the ability of a communication system to preserve digital information (quantified, for example, by means of BER) at low power or energy levels, respectively. η_p is often represented by a normalized version of the SNR defined by the ratio of signal energy per bit (E_b) to noise power spectral density (N_0) required at the receiver input

---

[3]The BER is an estimate of the bit error probability at the receiver for a given SNR and is determined by the ratio of the number of bit errors to the number of transmitted information bits.

for a certain probability of error [66, 81]. The relationship between E_b/N_0 and SNR can be expressed as

$$\frac{E_b}{N_0} = \text{SNR} \, \frac{B_w}{R_b}, \qquad (2.9)$$

where B_w is the channel bandwidth. For baseband systems, $B_w = B$, and for bandpass systems, $B_w = W$.

Bandwidth efficiency η_b, also known as spectral efficiency, measures the ability of a digital communication system to transmit digital information within a specific bandwidth. It is usually defined as the ratio of the bit rate per unit bandwidth.

The capacity of a digital communication system is directly related to its bandwidth efficiency [67]. The fundamental Shannon's channel coding theorem states that for an arbitrarily small probability of error, the maximum possible bandwidth efficiency is limited by the noise in the communication channel and can be obtained by the channel capacity formula [80]

$$\frac{C}{B_w} = \log_2 \left(1 + \text{SNR}\right), \qquad (2.10)$$

where C is the channel capacity.

In the design of a digital communication system, very often there is a trade-off between bandwidth efficiency and power efficiency. For example, by adding error control coding to the transmitted information, the bandwidth occupancy is increased, thus reducing the bandwidth efficiency. At the same time, it reduces the required received power for a particular BER. On the other hand, the use of M-ary schemes decreases bandwidth occupancy but increases the required received power, and hence trade power efficiency for bandwidth efficiency [67, 81].

2.3 Vector signal space

Digital communication encompasses the transmission of waveforms $s_m(t)$, belonging to a finite signal set S, through the communication channel and the recovery of the transmitted bits by choosing the most likely waveforms from the received signal. Each waveform $s_m(t)$ in S has a duration T_s and represents a group of n_b information bits.

Thereby, in binary digital communication systems, the signal set S is composed by only two signal waveforms and each one represents just one bit of information ($n_b = 1$). On the other hand, in M-ary digital communication systems, the signal set S is composed by M signal waveforms and each one represents $n_b = \log_2(M)$ information bits.

A fundamental approach proposed in [43, 90] represents the elements of S as points in a vector space. This widely used representation is particularly general and can simplify the performance analysis of different digital communication systems by converting a continuous-time detection problem into a discrete-time finite-dimensional detection problem.

2.3.1 Definition of the signal space

Based on an Euclidean geometry point of view, any finite set of signal wave-forms can be represented by a linear combination of N orthonormal wave-forms, which form a basis to a N-dimensional vector space. Thus, when the signal waveforms are replaced by vectors of appropriate dimension N, a signal communication system can be fully described by an equivalent vector communication system [90].

Definition 1 *Let Φ be the set composed by N different waveforms in the time interval $t_0 \leq t \leq t_0 + T_s$, given by*

$$\Phi = \{\phi_1(t), \cdots, \phi_N(t)\}. \tag{2.11}$$

Definition 2 *Let* SPAN $\{\Phi\}$ *be the set of all signal waveforms $s_m(t)$ formed by the linear combination of the elements of Φ in the time interval $t_0 \leq t \leq t_0 + T_s$, such that*

$$s_m(t) \in \text{SPAN}\{\Phi\} \Leftrightarrow s_m(t) = \{s_{m,1} \cdot \phi_1(t), \cdots, s_{m,N} \cdot \phi_N(t)\}, \tag{2.12}$$

where $s_{m,1}, \ldots, s_{m,N}$ are weighting coefficients.

Thus, the set Φ is linearly independent **if and only if** just the trivial combination can result in

$$s_{m,1} \cdot \phi_1(t) + \cdots + s_{m,N} \cdot \phi_N(t) = 0. \tag{2.13}$$

If Φ is *Linearly Independent*, then it forms a *basis* for $S = \text{SPAN}\{\Phi\}$ and the dimension of S is the number of elements of Φ.

Definition 3 *Let the inner product between two elements of Φ be given by*

$$\langle \phi_j(t), \phi_k(t) \rangle = \int_{t_0}^{t_0+T_s} \phi_j(t) \cdot \phi_k^*(t) \, dt, \tag{2.14}$$

where $\phi_k^*(t)$ represents the complex conjugate of $\phi_k(t)$.

Definition 4 *If Φ is an Orthogonal Basis, all pairs of elements of Φ satisfy*

$$\langle \phi_j(t), \phi_k(t) \rangle = 0, \ j \neq k. \tag{2.15}$$

Definition 5 *If Φ is an Orthonormal Basis, each of its elements $\phi_n(t)$ is normalized to have unit energy, i.e.,*

$$E_n = \langle \phi_n(t), \phi_n(t) \rangle = \int_{t_0}^{t_0+T_s} |\phi_n(t)|^2 dt = 1. \tag{2.16}$$

In a similar way, any two signal waveforms $s_j(t)$ and $s_k(t)$ of the SPAN$\{\Phi\}$ are orthogonal if they satisfy

$$\langle s_j(t), s_k(t) \rangle = 0, \; j \neq k. \tag{2.17}$$

The projection of a given signal waveform $s_m(t)$ of the SPAN$\{\Phi\}$ in one of the components $\phi_n(t)$ of the set Φ is given by

$$s_{m,n} = \langle s_m(t), \phi_n(t) \rangle = \int_{t_0}^{t_0+T_s} s_m(t) \cdot \phi_n^*(t) dt. \tag{2.18}$$

Thus, if a set of M signal waveforms $s_m(t)$, $m = 1, \cdots, M$, can be decomposed into an N-dimensional orthonormal basis (i.e., a *signal space*), where $N \leq M$, then $s_m(t)$ can be represented by the following vector notation:

$$\mathbf{s}_m = \begin{bmatrix} s_{m,1} & \cdots & s_{m,N} \end{bmatrix}^T, \tag{2.19}$$

where each element of $\mathbf{s}_m$ corresponds to the projection of $s_m(t)$ on each of the components $\phi_n(t)$ of the signal space.

In this way, the energy of $s_m(t)$ can be computed as

$$E_{s_m} = |\mathbf{s}_m|^2 = \sum_{n=1}^{N} |s_{m,n}|^2. \tag{2.20}$$

Also, the correlation between two different signals $\mathbf{s}_j$ and $\mathbf{s}_k$ can be obtained by

$$E_{s_j,s_k} = \langle \mathbf{s}_j, \mathbf{s}_k \rangle = \sum_{n=1}^{N} s_{j,n} \cdot s_{k,n}^*, \tag{2.21}$$

while the distance between $\mathbf{s}_j$ and $\mathbf{s}_k$ is given by

$$d_{s_j,s_k} = |\mathbf{s}_j - \mathbf{s}_k| = \sqrt{\sum_{n=1}^{N} |s_{j,n} - s_{k,n}|^2}. \tag{2.22}$$

Hence, the waveforms of a basis can be considered as a coordinate system for the vector space. The Gram-Schmidt procedure, described in the sequel, can provide a systematic way of obtaining the vector space for a given set of signal waveforms.

2.3.2 Gram-Schmidt and the geometric representation of signals

Given a signal set S composed by M signal waveforms $s_m(t)$ with energies E_{s_m}, respectively, Gram-Schmidt (GS) orthogonalization technique can be employed to define an N-dimensional *orthonormal signal space*, composed by N waveforms $\phi_n(t)$, $N \leq M$, to represent each element of S in a vector space, according to the following procedure [46, 66, 93].

GS Procedure

(a) To determine $\phi_1(t)$, consider without any loss of generality, that $\phi_1(t)$ is equal to $s_1(t)$ normalized to unit energy:

$$\phi_1(t) = \frac{s_1(t)}{\sqrt{E_{s_1}}}. \tag{2.23}$$

(b) To determine $\phi_2(t)$, consider that

$$g_2(t) = s_2(t) - s_{2,1} \cdot \phi_1(t), \tag{2.24}$$

$$\phi_2(t) = \frac{g_2(t)}{\sqrt{E_{g_2}}}, \tag{2.25}$$

where

$$s_{2,1} = \langle s_2(t), \phi_1(t) \rangle. \tag{2.26}$$

(c) To determine a generic $\phi_n(t)$, consider that

$$g_n(t) = s_n(t) - \sum_{k=1}^{n-1} s_{n,k} \cdot \phi_k(t), \tag{2.27}$$

$$\phi_n(t) = \frac{g_n(t)}{\sqrt{E_{g_n}}}, \tag{2.28}$$

where all coefficients $s_{n,k}$ can be determine by

$$s_{n,k} = \langle s_n(t), \phi_k(t) \rangle. \tag{2.29}$$

After defining the set of the N orthonormal waveforms $\phi_n(t)$, each of the M signal waveforms $s_m(t)$ can be represented by the corresponding signal vector $\mathbf{s}_m$, whose elements are the coefficients $s_{m,n}$ described in the GS procedure:

$$\mathbf{s}_m = \begin{bmatrix} s_{m,1} & \cdots & s_{m,N} \end{bmatrix}^T, \quad m = 1, \cdots, M. \tag{2.30}$$

2.3.3   Karhunen-Loeve and the geometric representation of noise

Unfortunately, the GS procedure cannot be directly applied to random signals. In this case, the Karhunen-Loeve (KL) expansion offers a powerful tool to obtain an orthonormal basis for a random process $w(t)$ through the solution of the following integral equation [46, 63]:

$$\lambda_n \cdot \phi_n(t) = \int_{t_0}^{t_0+T_s} R_w(t,\tau) \cdot \phi_n^*(\tau) d\tau, \tag{2.31}$$

where $R_w(t,\tau)$ is the autocorrelation function of $w(t)$, and λ_n and $\phi_n(t)$ are the eigenvalues and eigenfunctions of Equation (2.31), respectively.

One of the most important random processes employed in the study of communication systems is the AWGN [46, 66]. The AWGN is an uncorrelated, zero mean Gaussian random process with variance $\sigma_w^2 = N_0/2$. All its Z-th order probability density functions (PDF) are Z-variate Gaussian random variables given by [63]

$$f_{\mathbf{W}_z}(\mathbf{w}_z) = \frac{1}{(2\pi)^{\frac{Z}{2}} \cdot |\det(\mathbf{\Sigma}_w)|^{\frac{1}{2}}} \cdot e^{-\frac{1}{2}[\mathbf{w}_z - \overline{\mathbf{w}}_z]^T \cdot \mathbf{\Sigma}_w^{-1} \cdot [\mathbf{w}_z - \overline{\mathbf{w}}_z^*]}, \tag{2.32}$$

where $\mathbf{W}_z = [w(t_1), \cdots, w(t_Z)]$, $\mathbf{w}_z = [\mathrm{w}_1, \cdots, \mathrm{w}_Z]$, and $\overline{\mathbf{w}}_z = 0$ is the mean of $\mathbf{w}_z$. Due to the properties of the AWGN process, the autocovariance matrix $\mathbf{\Sigma}_w$ converges to the following diagonal autocorrelation matrix:

$$\mathbf{\Sigma}_w = \begin{bmatrix} \sigma_w^2 & & 0 \\ & \ddots & \\ 0 & & \sigma_w^2 \end{bmatrix}. \tag{2.33}$$

Since the AWGN random process is considered white, its PSD can be represented by [46, 63]

$$S_w(f) = \frac{N_0}{2}, \quad -\infty \leq f \leq \infty. \tag{2.34}$$

Also, in accordance with the Wiener-Khinchin theorem, the autocorrelation function can be obtained by the inverse Fourier transform of the PSD, resulting in [46, 63, 66]

$$R_w(t,\tau) = \frac{N_0}{2} \cdot \delta(t-\tau). \tag{2.35}$$

In this way, Equation (2.31) can be reduced to

$$\lambda_n \cdot \phi_n = \frac{N_0}{2} \cdot \phi_n. \tag{2.36}$$

This result implies that any orthonormal basis can be used to represent an

AWGN random process as, for example, the one obtained by the GS procedure. However, in accordance with the KL expansion, to perfectly represent an AWGN process, it is necessary to employ an infinite-dimensional orthonormal basis, i.e.,

$$w(t) = \sum_{n=1}^{\infty} w_n \cdot \phi_n(t), \ t_0 \leq t \leq t_0 + T_s, \tag{2.37}$$

where $w_n = \langle w(t), \phi_n(t) \rangle$ is the projection of $w(t)$ over the component of the basis $\phi_n(t)$.

One very important and useful result in the analysis of random signals is established by the *Theorem of the Irrelevancy* [90].

Theorem 1 *Only the components of a white Gaussian random process that are projected on the signal space affect the decision process.*

Proof 1 *The proof can be viewed in [90].*

Therefore, from the signal detection point of view, the noise representation presented in Equation (2.37) can be reduced to a more tractable finite N-dimensional signal space by using

$$w(t) = \sum_{n=1}^{N} w_n \cdot \phi_n(t), \ t_0 \leq t \leq t_0 + T_s. \tag{2.38}$$

Note that the joint PDF of the components w_n of the AWGN can be obtained by [63]

$$f_{\mathbf{W}}(\mathbf{w}) = \frac{1}{(\pi \cdot N_0)^{\frac{N}{2}}} \cdot e^{-\frac{|\mathbf{w}|^2}{N_0}}, \tag{2.39}$$

where $\mathbf{W} = [w_1, \cdots, w_N]$ and $\mathbf{w} = [\mathrm{w}_1, \cdots, \mathrm{w}_N]$. The elements of $\mathbf{W}$ are independent and identically distributed (i.i.d) zero mean Gaussian random variables with $\sigma_w^2 = N_0/2$ and the elements of $\mathbf{w}$ are the values assumed by the corresponding random variables.

In the next section, an optimum receiver for M-ary systems in AWGN channels will be introduced, considering the geometric representation of signals and noise discussed previously.

2.3.4 Optimum receiver structure (MAP/ML criteria)

An M-ary communication system transmits digital information from a transmitter to a receiver through a communication channel. Usually, the channel causes some different kind of undesirable degradations in the transmitted information. Due to the random nature of the degradations, they are usually described by their statistical distributions.

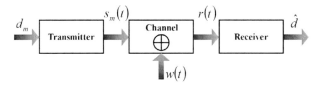

FIGURE 2.8
M-ary digital communication system model for AWGN channels.

One of the most common sources of degradation presented in a communication system is the thermal noise. This kind of undesirable signal, as already mentioned in Section 2.1.2, is generated by the random motion of electrons in a conductive material (e.g., electronic devices) at temperatures higher than zero Kelvin and is normally added to the information signal at the receiver. Often, the thermal noise is modeled as an AWGN random process [46,66].

A generic communication system model is depicted in Figure 2.8. In this model, d_m denotes one of the M possible symbols that compose the symbol alphabet and that is transmitted to the receiver by means of the corresponding signal waveform $s_m(t)$. Each symbol carries n_b different information bits.

Considering, without any loss of generality, that the symbol d_i is transmitted through the corresponding signal waveform $s_i(t)$ and assuming that the system is memoryless, as the corresponding transmitted signal is just corrupted by an AWGN $w(t)$, the received signal $r(t)$ can be represented by

$$r(t) = s_i(t) + w(t). \tag{2.40}$$

The geometric approach can be used to simplify the design of the optimum receiver for AWGN channels. As presented previously, the received signal can be represented in an N-dimensional vector signal space by

$$\mathbf{r} = \mathbf{s}_i + \mathbf{w}, \tag{2.41}$$

where

$$\mathbf{r} = \begin{bmatrix} r_1 & \cdots & r_N \end{bmatrix}^T, \tag{2.42}$$

$$\mathbf{s}_i = \begin{bmatrix} s_{i,1} & \cdots & s_{i,N} \end{bmatrix}^T, \tag{2.43}$$

$$\mathbf{w} = \begin{bmatrix} w_1 & \cdots & w_N \end{bmatrix}^T. \tag{2.44}$$

The optimum receiver has to decide which is the most likely transmitted symbol, given that $\mathbf{r}$ was received. Considering that $\mathbf{d}_i$ was transmitted, the conditional probability of making the correct decision $\hat{\mathbf{d}} = \mathbf{d}_i$, given that $\mathbf{r}$ was received, is commonly denoted as the *a posteriori* probability (APP) of $\mathbf{d}_i$ and can be represented by [46]

$$P(\text{Correct decision} = \mathfrak{C} \,|\mathbf{r}) = P(\mathbf{d}_i \text{ was transmitted} \,|\mathbf{r} \text{ was received}), \tag{2.45}$$

while the unconditional probability of making a correct decision, independently on the received signal $\mathbf{r}$, is given by

$$P(\mathfrak{C}) = \int_{\mathbf{r}} P(\mathfrak{C}\,|\mathbf{r}) \cdot f_R(\mathbf{r})d\mathbf{r}. \tag{2.46}$$

As $f_R(\mathbf{r}) \geq 0$, the integral is maximum when $P(\mathfrak{C}\,|\mathbf{r})$ is maximum. Therefore, given that the decision is to $\mathbf{d}_i$, the decision error probability can be minimized if $P(\mathbf{d}_i\,|\mathbf{r})$ is maximized. This maximization procedure, known as the maximum a *posteriori* probability (MAP) criteria, is described as follows.

MAP Criteria

(a) Receive $\mathbf{r}$

(b) Evaluate all M APP of $\mathbf{d}_m$

(c) Decide to the symbol $\mathbf{d}_i$ that presents the largest APP, according to

$$P(\mathbf{d}_i\,|\mathbf{r}) > P(\mathbf{d}_m\,|\mathbf{r})\,, \text{ for all } m \neq i \tag{2.47}$$

Thus, the optimum receiver, in the sense of minimizing the decision error probability, is the MAP receiver.

Using Bayes' rule, the MAP criteria can be formally represented by

$$\operatorname*{argmax}_{\mathbf{d}_m} \left[\frac{P(\mathbf{d}_m) \cdot f_R(\mathbf{r}\,|\mathbf{d}_m)}{f_R(\mathbf{r})} \right]. \tag{2.48}$$

The decision will be in favor of symbol $\mathbf{d}_i$ if Equation (2.48) is maximum for $m = i$. As $f_R(\mathbf{r})$ is common to all $\mathbf{d}_m$, the MAP criteria simplify to

$$\operatorname*{argmax}_{\mathbf{d}_m} \left[P(\mathbf{d}_m) \cdot f_R(\mathbf{r}\,|\mathbf{d}_m) \right], \tag{2.49}$$

where $P(\mathbf{d}_m)$ is the a *priori* probability of sending a specific symbol $\mathbf{d}_m$ and $f_R(\mathbf{r}\,|\mathbf{d}_m)$ is the PDF of $\mathbf{r}$ to be received conditioned to the transmission of $\mathbf{d}_m$.

Under these conditions, the corresponding transmit vector $\mathbf{s}_m$ is considered constant during the symbol interval T_s and the conditional probability density function is given by

$$f_R(\mathbf{r}\,|\mathbf{d}_m) = \frac{1}{(\pi \cdot N_0)^{\frac{N}{2}}} \cdot e^{\frac{|\mathbf{r}-\mathbf{s}_m|^2}{N_0}}. \tag{2.50}$$

Therefore, the decision function is given by [46, 66]

$$L_m = P(\mathbf{d}_m) \cdot f_R(\mathbf{r} | \mathbf{d}_m) = \frac{P(\mathbf{d}_m)}{(\pi \cdot N_0)^{\frac{N}{2}}} \cdot e^{\frac{|\mathbf{r} - \mathbf{s}_m|^2}{N_0}}. \tag{2.51}$$

As all the terms of the decision function are positive, it can be expressed by means of the natural logarithm. Making some simplifications, the resulting logarithmic decision function can be represented by

$$\ell_m = \ln\left[P(\mathbf{d}_m)\right] - \frac{|\mathbf{r} - \mathbf{s}_m|^2}{N_0}. \tag{2.52}$$

Note that $|\mathbf{r} - \mathbf{s}_m|^2$ is the square of the distance between $\mathbf{r}$ and $\mathbf{s}_m$.

Expanding the terms of Equation (2.52) and making some additional simplifications, the decision function can finally be expressed as [46, 66]

$$\ell_m = \xi_m + \langle \mathbf{r}, \mathbf{s}_m \rangle, \tag{2.53}$$

where $\langle \mathbf{r}, \mathbf{s}_m \rangle$ is the correlation between $\mathbf{r}$ and $\mathbf{s}_m$, and ξ_m is given by

$$\xi_m = \frac{N_0}{2} \cdot \ln\left[P(\mathbf{d}_m)\right] - E_{s_m}, \tag{2.54}$$

with E_{s_m} representing the energy of $\mathbf{s}_m$.

Thus, the optimum receiver based on the MAP criteria, illustrated in Figure 2.9, consists in calculating ℓ_m, $m = 1, \cdots, M$, and deciding to $\hat{\mathbf{d}} = \mathbf{d}_i$ if the decision function is maximum to $m = i$.

> One interesting result obtained in this analysis is that the optimum MAP receiver for AWGN channels, that minimizes the error probability, is a linear system.

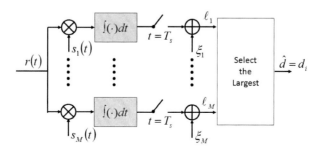

FIGURE 2.9
Optimum MAP receiver based on correlators.

The optimum receiver can also be implemented through a different approach based on a special filter designed to perform equivalently to the operation of correlation $\langle \mathbf{r}, \mathbf{s}_m \rangle$.

If $r(t)$ is applied to a filter with impulse response $h_m(t)$, such that

$$h_m(t) = s_m(T_s - t), \tag{2.55}$$

the output of the filter at instant T_s will be

$$y_m(T_s) = \int_0^{T_s} r(\tau) \cdot s_m(\tau) d\tau. \tag{2.56}$$

This output is exactly the same obtained by a correlator at time instant T_s and this filter is usually called Matched Filter (MF).

Thus, the bank of correlators of the optimum MAP receiver in Figure 2.9 can be replaced by a bank of MF, as shown in Figure 2.10, without any performance degradation.

Another possible implementation of the optimum MAP receiver is shown in Figure 2.11. In this case, the terms $\langle \mathbf{r}, \mathbf{s}_m \rangle$ can be obtained by first projecting the received signal $\mathbf{r}$ in each one of the N components of the signal space ϕ_n and then calculating the sum of the product of each projection r_n with all the components $s_{m,n}$ of the signal vector $\mathbf{s}_m$, as shown by

$$\langle \mathbf{r}, \mathbf{s}_m \rangle = \sum_{n=1}^{N} r_n \cdot s_{m,n}. \tag{2.57}$$

In the same way, the bank of correlators of the implementation of the MAP

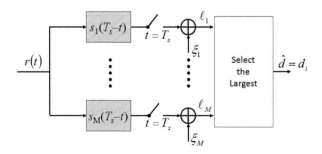

FIGURE 2.10
Optimum MAP receiver based on MF.

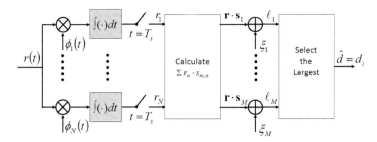

FIGURE 2.11
Optimum MAP receiver based on orthonormal basis with correlators.

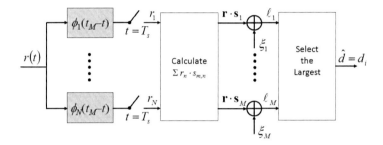

FIGURE 2.12
Optimum MAP receiver based on orthonormal basis with MF.

receiver based on orthogonal basis can also be replaced by a bank of MF, as show in Figure 2.12.

> If $N < M$ and the components of the basis are easily generated, then the optimum receiver implementation based on orthonormal basis should be chosen.

If the symbols $\mathbf{d}_m$ are equiprobable, all *a priori* probabilities are equal and the MAP criteria presented in Equation (2.49) simplifies to the *Maximum Likelihood* (ML) criteria given by

$$\operatorname*{argmax}_{\mathbf{d}_m} \left[f_R(\mathbf{r} \,|\, \mathbf{d}_m) \right]. \tag{2.58}$$

In this case, Equation (2.52) can be simplified to

$$\ell_m = - \left| \mathbf{r} - \mathbf{s}_m \right|^2, \tag{2.59}$$

and the resulting decision function for the ML criteria is given by

$$\ell_m = \langle \mathbf{r}, \mathbf{s}_m \rangle - E_{s_m}. \tag{2.60}$$

This means that the optimum receiver in the ML sense is simply a bank of correlators or MF, followed by a bank of subtractors with the symbol energy E_{s_m}. The ML criterion is summarized below.

ML Criteria

   (a)   Receive $\mathbf{r}$

   (b)   Evaluate all M correlations $\langle \mathbf{r}, \mathbf{s}_m \rangle$

   (b)   Subtract E_{s_m} from each correlation output

   (c)   Decide to the symbol $\mathbf{d}_i$ that presents the largest value

---

2.3.5   Decision region and error probability

A very important figure of performance in digital communications is the symbol error probability (P_e), usually measured in practice by the Symbol Error Rate (SER). To determine P_e for the MAP receiver, the N-dimensional signal space is split in M disjoint regions $\Psi_1, \Psi_2, \cdots, \Psi_M$, $N \leq M$, that represent all possible transmit symbols, as illustrated in Figure 2.13 for a hypothetical system with $N = 2$ and $M = 4$.

    If a signal waveform $\mathbf{s}_i$ is transmitted and the received signal $\mathbf{r}$ falls in the

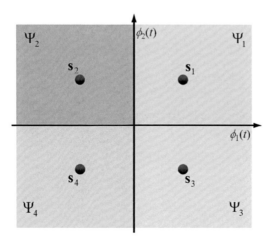

FIGURE 2.13
Decision regions for $N = 2$ and $M = 4$.

region Ψ_j, $i \neq j$, then the decision will be wrongly made in favor of $\mathbf{d}_j$ and a decision error will occur.

> The decision regions should be chosen to minimize P_e in accordance with the MAP decision function presented in Equation (2.52).

Considering for simplicity that all possible transmit symbols are equiprobable, the MAP receiver simplifies to the ML receiver, meaning that the optimum receiver has just to choose the symbol $\mathbf{d}_m$ that maximizes $-|\mathbf{r} - \mathbf{s}_m|^2$. In fact, it is equivalent to choose the symbol $\mathbf{d}_m$ that minimizes the distance between $\mathbf{r}$ and $\mathbf{s}_m$, given by $|\mathbf{r} - \mathbf{s}_m|$.

> The decision process based on the ML criteria has a very nice interpretation in the vector signal space: the decision is made in favor of symbol $\mathbf{d}_i$, if the signal vector closest to the received vector $\mathbf{r}$ is the signal vector $\mathbf{s}_i$.

Therefore the probability of making the right decision, given that $\mathbf{s}_i$ was sent, is obtained by

$$P(\mathfrak{C}\,|\mathbf{d}_i) = P(\mathbf{r} \in \Psi_i\,|\mathbf{d}_i). \tag{2.61}$$

On the other hand, the probability of making the right decision is given by

$$P(\mathfrak{C}) = \sum_{m=1}^{M} P(\mathbf{d}_m) \cdot P(\mathfrak{C}\,|\mathbf{d}_m) = \frac{1}{M} \sum_{m=1}^{M} P(\mathfrak{C}\,|\mathbf{d}_m) \tag{2.62}$$

and the error probability is easily obtained by

$$P_e = 1 - \frac{1}{M} \sum_{m=1}^{M} P(\mathfrak{C}\,|\mathbf{d}_m). \tag{2.63}$$

In the vector space, this probability can also be obtained geometrically by

$$P_e = 1 - \frac{1}{M} \sum_{m=1}^{M} P(\mathbf{r} \in \Psi_m\,|\mathbf{d}_m). \tag{2.64}$$

Bit error probability (P_b), usually estimated in practice by BER, is another very important figure of merit of digital communication systems and can be easily derived from P_e. For linear systems, such as M-ASK, M-PSK, and M-QAM (discussed in Section 2.5.1), employing gray encoding [46, 93], P_b can be obtained by

$$P_b = \frac{P_e}{\log_2(M)}. \tag{2.65}$$

For nonlinear systems, such as M-FSK, P_b can be computed as

$$P_b = \frac{M}{2 \cdot (M-1)} \cdot P_e. \tag{2.66}$$

2.3.6 Error probability bounds

Depending on the geometric representation of a given digital communication system, the determination of P_e can be very difficult. In this case, the use of bounds may be very useful.

One of the most well-known bounds for obtaining an approximation of P_e is the Union Bound (UB). The UB is an upper bound based on the following probability identity [33, 93]:

$$P\left(\bigcup_{m=1}^{M} A_m\right) \leq \sum_{m=1}^{M} P(A_m), \tag{2.67}$$

where A_m is the m-th event of the sample space and $\bigcup$ represents the union operator.

Considering that $A_{m|i}$ corresponds to the error event $|\mathbf{r} - \mathbf{s}_m| < |\mathbf{r} - \mathbf{s}_i|$, given that $\mathbf{s}_i$, $i \neq m$, was sent, the UB for equally likely transmit symbols can be expressed as [46, 66]

$$P_e \leq \frac{1}{M} \sum_{i=1}^{M} \sum_{\substack{m=1 \\ m \neq i}}^{M} Q\left(\frac{D_{m,i}}{\sqrt{2 \cdot N_0}}\right), \tag{2.68}$$

where $Q(\cdot)$ is the Q-function[4] [46, 66] and $D_{m,i}$ is the distance between $\mathbf{s}_m$ and $\mathbf{s}_i$.

Another very useful bound is the Nearest Neighbors Bound (NNB). The NBB determines an approximation of P_e by considering just the neighboring signals that are at the minimun distance D_{min} of a given signal $\mathbf{s}_m$. Assuming equally likely transmit symbols, the NBB is given by [33]

$$P_e \approx N_{min} \cdot Q\left(\frac{D_{min}}{\sqrt{2 \cdot N_0}}\right), \tag{2.69}$$

where $D_{min} = \min_{m \neq i}[D_{m,i}]$ is the minimum distance between all possible pairs of signals $\mathbf{s}_m$ and $\mathbf{s}_i$ and N_{min} is given by

$$N_{min} = \frac{1}{M} \cdot \sum_{m=1}^{M} N_m, \tag{2.70}$$

with N_m indicating the number of neighbors at a distance D_{min} from $\mathbf{s}_m$.

---

[4]The Q-function is defined as

$$Q(x) = \frac{1}{\sqrt{2\pi}} \int_x^\infty e^{-t^2/2} dt, \quad x \geq 0.$$

2.4   Baseband communication systems

A digital baseband communication system, represented in Figure 2.14, converts digital information symbols into baseband pulse waveforms that are suitable to be transmitted directly over lowpass channels. As mentioned before, this process is denoted as baseband digital modulation or simply digital coding.

A variety of digital baseband communication systems have been proposed to reach some desirable properties such as good bandwidth and power efficiencies, adequate timing information, and error detection capability.

2.4.1   Line coding

Line coding was developed in the past for digital transmission over telephone cables and digital recording on magnetic medias. Recent developments are primarily concentrated on applications in local area networks (LAN), including transmissions over unshielded twisted pairs (UTP) and fiber optic cables (FOC).

In general, line coding is a baseband scheme employed to represent digital information by means of different pulse waveforms of an M-dimensional signal set, according to some specific rules [46].

Any line coding waveform can be represented by the following M-ary Pulse

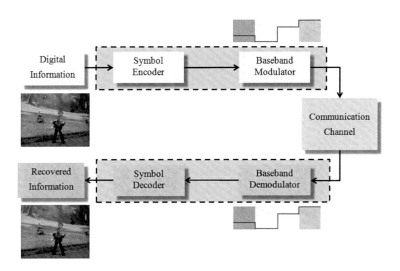

FIGURE 2.14   (SEE COLOR INSERT)
Baseband digital communication system diagram.

Amplitude Modulation (PAM) signal:

$$s_{\text{pam}}(t) = \sum_i d_i' \cdot p(t - i \cdot T_s), \qquad (2.71)$$

where d_i' is the i-th real-valued information symbol obtained by a specific encoding rule and $p(t)$ is the pulse shape.

2.4.2 Complex-valued M-ary PAM

A more general and powerful representation of M-ary PAM can be obtained by considering that any information symbol d_i may have complex values [66]:

$$s_{lp}(t) = \sum_i d_i \cdot p(t - i \cdot T_s), \qquad (2.72)$$

where d_i is the i-th complex-valued information symbol.

This representation, also called Lowpass Equivalent (LPE), is a very useful tool to simplify simulations in computational environments and theoretical analysis of bandpass digital communication signals and systems.

2.5 Bandpass communication systems

Digital communication also encompasses the transmission of digital information through wireless channels. Wireless channels are essentially analog bandpass systems and, in this case, digital communication systems are usually denoted as bandpass digital communication systems or digital modulation systems.

As shown in the previous section, a digital baseband communication system converts digital information symbols into baseband pulse waveforms that are suitable to be transmitted directly over a lowpass channel. On the other hand, a digital bandpass communication system, represented in Figure 2.15, converts digital information symbols into bandpass sinusoidal waveforms that are suitable to be transmitted through wireless channels. Thus, bandpass communication systems can be viewed as systems that map baseband signals into bandpass signals. This operation encompasses the shift of the spectrum of a baseband signal to a higher frequency [6, 46, 66, 81].

Digital bandpass communication systems or simply digital modulation systems are obtained by switching (keying) amplitude, frequency, phase, or a combination of amplitude and phase of a high frequency sinusoidal carrier in accordance with the digital information. Thus, the digital modulation techniques could be grouped into the following four major categories:

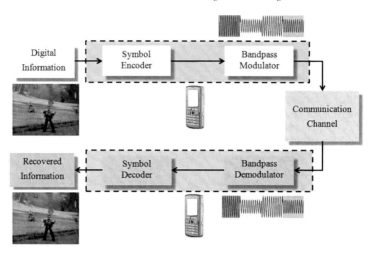

FIGURE 2.15 (SEE COLOR INSERT)
Bandpass digital communication system diagram.

(a) Amplitude Shift Keying (ASK) results in amplitude variations of the carrier waveform in accordance with the information symbols.

(b) Phase Shift Keying (PSK) results in phase variations of the carrier in accordance with the information symbols.

(c) Frequency Shift Keying (FSK) results in frequency variations of the carrier in accordance with the information symbols.

(d) Quadrature Amplitude Modulation (QAM) results in amplitude and phase variations of the carrier in accordance with the information symbols.

Other modulation categories can be obtained by combining different parameters of the carrier, but they are not commonly found in practice and are out of the scope of this analysis.

Digital bandpass systems can be classified as linear and nonlinear. Linear communication systems follow the superposition theorem and encompass ASK, PSK, and QAM schemes. Typically, they show a good bandwidth efficiency. Nonlinear communication systems do not satisfy the superposition theorem and are represented basically by different FSK schemes. Usually, they show a good energy efficiency.

Any bandpass communication system can be represented by

$$s(t) = \Re\left\{ s_{lp}(t)e^{j2\pi f_o t}\right\}, \tag{2.73}$$

where f_o is the carrier frequency and $s_{lp}(t)$ is the LPE representation given by Equation (2.72). In the receiver, the MAP detector can be employed to recover the transmitted information.

There are several factors that influence the selection of a digital modulation technique including bandwidth efficiency, energy efficiency required for detection, BER and SER at reception and circuitry complexity. Many of these factors are correlated and an improvement in one of them generally causes a degradation in another [67, 93]. Thus, an appropriate choice depends on the communication system requirements.

2.5.1 Some important digital communication schemes

2.5.1.1 Binary amplitude shift keying

As discussed before, in Binary Amplitude Shift Keying (BASK), the amplitude of the carrier is modified in accordance with the digital information symbols. One of the most important and simple BASK schemes is On-Off Keying (OOK). The OOK is widely used in FOC applications mainly because of its simple optical implementation (i.e., light/nonlight).

The OOK signal waveform can be represented by

$$s_{ook}(t) = \sum_i s_{ook}^i(t), \tag{2.74}$$

where the i-th OOK transmitted waveform, corresponding to information symbol d_i of duration T_s, is given by

$$s_{ook}^i(t) = \begin{cases} 0 & \text{for } d_i = 0 \\ \sqrt{\frac{2 \cdot E_s}{T_s}} \cdot \cos(2\pi f_o t), \ i \cdot T_s \leq t \leq (i+1) \cdot T_s & \text{for } d_i = 1 \end{cases}. \tag{2.75}$$

Employing GS procedure, the following orthonormal basis can be definied for an OOK scheme:

$$\phi_1^i(t) = \sqrt{\frac{2}{T_s}} \cdot \cos(2\pi f_o t), \ i \cdot T_s \leq t \leq (i+1) \cdot T_s. \tag{2.76}$$

Thus, the LPE representation of an OOK signal can be written as

$$s_{lp}(t) = \sum_i d_i \cdot \Pi\left(\frac{t - i \cdot T_s}{T_s}\right), \tag{2.77}$$

where $\Pi(\cdot)$ is the unit gate pulse and $d_i \in \{0, \sqrt{E_s}\}$.

Therefore, the OOK signal waveform could also be expressed by

$$s_{ook}(t) = \Re\left\{ \sum_i d_i \cdot \Pi\left(\frac{t - i \cdot T_s}{T_s}\right) \cdot e^{j2\pi f_o t} \right\}. \tag{2.78}$$

Figure 2.16 presents an example of the scattering diagram for an OOK scheme and the corresponding received signal corrupted by AWGN for an $E_b/N_0 = 10\,\text{dB}$.

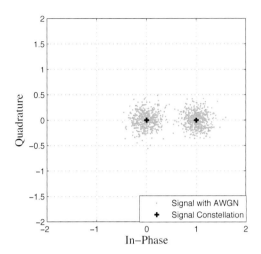

FIGURE 2.16
OOK constellation diagram.

One interesting observation is that ASK is also widely used in some wireless applications (e.g., gate control), even requiring low efficiency linear amplifiers and presenting low immunity to interferences. The main reason for this is its extremely low cost and low complexity, besides its ability to use simple non-coherent detection.

2.5.1.2 Binary phase shift keying

As discussed before, in Binary Phase Shift Keying (BPSK), the phase of the carrier is modified in accordance with the digital information symbols. The BPSK scheme is widely used in wireless applications that require low P_e at low E_b/N_0. However, BPSK presents low bandwidth efficiency, as any binary scheme.

The i-th BPSK transmitted waveform, corresponding to information symbol d_i, can be represented by

$$
s_{bpsk}^i(t) = \begin{cases} -\sqrt{\frac{2 \cdot E_s}{T_s}} \cdot \cos(2\pi f_o t), \ i \cdot T_s \leq t \leq (i+1) \cdot T_s & \text{for } d_i = 0 \\[3mm] \sqrt{\frac{2 \cdot E_s}{T_s}} \cdot \cos(2\pi f_o t), \ i \cdot T_s \leq t \leq (i+1) \cdot T_s & \text{for } d_i = 1 \end{cases}
$$
$$(2.79)$$

Using GS procedure, an orthonormal basis for a BPSK scheme can be defined as

$$
\phi_1^i(t) = \sqrt{\frac{2}{T_s}} \cdot \cos(2\pi f_o t), \ i \cdot T_s \leq t \leq (i+1) \cdot T_s. \tag{2.80}
$$

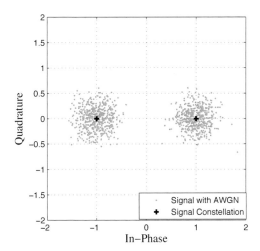

FIGURE 2.17
BPSK constellation diagram.

Thus, the LPE representation of a BPSK signal can be written as

$$s_{lp}(t) = \sum_i d_i \cdot \Pi\left(\frac{t - i \cdot T_s}{T_s}\right), \tag{2.81}$$

where $d_i \in \left\{-\sqrt{E_s}, \sqrt{E_s}\right\}$.

In Figure 2.17, an example of the scattering diagram for a BPSK scheme is presented, as well as the corresponding received signal corrupted by AWGN for an $E_b/N_0 = 10\,\mathrm{dB}$.

2.5.1.3 Quartenary phase shift keying

The i-th Quartenary Phase Shift Keying (QPSK) transmitted waveform, corresponding to information symbol d_i of duration T_s, can be written as

$$s_{qpsk}^i(t) = \sqrt{\frac{E_s}{T_s}} \cdot \cos(2\pi f_o t - \theta_i), \; i \cdot T_s \le t \le (i+1) \cdot T_s, \tag{2.82}$$

where $\theta_i \in \left\{\frac{(2 \cdot m - 1) \cdot \pi}{4}\right\}$, $m = 1, \cdots, 4$.

Using GS procedure, an orthonormal basis for a QPSK scheme can be computed as

$$\phi_1^i(t) = \sqrt{\frac{2}{T_s}} \cdot \cos(2\pi f_o t), \; i \cdot T_s \le t \le (i+1) \cdot T_s$$
$$\phi_2^i(t) = \sqrt{\frac{2}{T_s}} \cdot \sin(2\pi f_o t), \; i \cdot T_s \le t \le (i+1) \cdot T_s \tag{2.83}$$

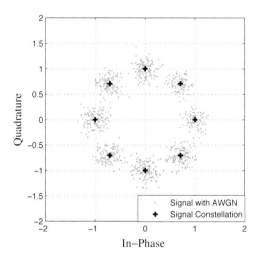

FIGURE 2.18
QPSK constellation diagram.

Thus, the LPE representation of a QPSK signal can be obtained by

$$s_{lp}(t) = \sum_i d_i \Pi \left(\frac{t - i \cdot T_s}{T_s} \right), \tag{2.84}$$

with $d_i \in \left\{ \sqrt{\frac{E_s}{2}} \cdot (1+j), \sqrt{\frac{E_s}{2}} \cdot (1-j), \sqrt{\frac{E_s}{2}} \cdot (-1+j), \sqrt{\frac{E_s}{2}} \cdot (-1+j) \right\}$. Figure 2.18 shows an example of the scattering diagram for a QPSK scheme and the corresponding received signal corrupted by AWGN for an $E_b/N_0 = 10$ dB.

2.5.1.4 *M*-ary phase shift keying

The i-th M-ary Phase Shift Keying (M-PSK) transmitted waveform, corresponding to information symbol d_i, can be expressed as

$$s_{mpsk}^i(t) = \sqrt{\frac{E_s}{T_s}} \cdot \cos(2\pi f_o t - \theta_i), \ i \cdot T_s \leq t \leq (i+1) \cdot T_s, \tag{2.85}$$

with $\theta_i \in \left\{ \frac{2 \cdot \pi \cdot (m-1)}{M} \right\}$, $m = 1, \cdots, M$.

An orthonormal basis for an M-PSK signal can be defined as

$$\phi_1^i(t) = \sqrt{\frac{2}{T_s}} \cdot \cos(2\pi f_o t), \ i \cdot T_s \leq t \leq (i+1) \cdot T_s$$

$$\phi_2^i(t) = \sqrt{\frac{2}{T_s}} \cdot \sin(2\pi f_o t), \ i \cdot T_s \leq t \leq (i+1) \cdot T_s \tag{2.86}$$

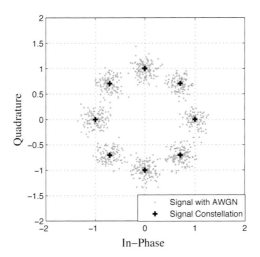

FIGURE 2.19
8-PSK constellation diagram.

Thus, the LPE representation of an M-PSK signal is given by

$$s_{lp}(t) = \sum_i d_i \cdot \Pi\left(\frac{t - i \cdot T_s}{T_s}\right), \qquad (2.87)$$

where $d_i \in \{a_m \cdot \sqrt{E_s} + jb_m \cdot \sqrt{E_s}\}$, $\sqrt{a_m^2 + b_m^2} = 1$, and the pair $(a_m, b_m) = \left(\cos(2\pi\frac{(m-1)}{M}), \sin(2\pi\frac{(m-1)}{M})\right)$, $m = 1, \cdots, M$. It is worth noting that the i-th symbol d_i could also be represented in a two-dimensional vector space by the signal vector $\mathbf{d}_i \in \{[a_m \cdot \sqrt{E_s}, b_m \cdot \sqrt{E_s}]\}$, $m = 1, \cdots, M$.

In Figure 2.19, an example of the scattering diagram for an 8-PSK scheme and the corresponding received signal corrupted by AWGN for an $E_b/N_0 = 10\,\mathrm{dB}$ is shown.

2.5.1.5 *M*-ary quadrature amplitude modulation

The i-th transmitted waveform, corresponding to information symbol d_i of duration T_s, of a generic M-ary Quadrature Amplitude Modulation (M-QAM) scheme, can be represented by

$$s_{mqam}^i(t) = \sqrt{\frac{2 \cdot E_i}{T_s}} \cdot \cos(2\pi f_o t + \theta_i), \ i \leq t \leq i \cdot T_s, \qquad (2.88)$$

with $E_i \in \{E_m\}$ and $\theta_i \in \{\theta_m\}$, $m = 1, \cdots, M$.

An orthonormal basis for an M-QAM signal can be written as

$$\phi_1^i(t) = \sqrt{\tfrac{2}{T_s}} \cdot \cos(2\pi f_o t), \ i \cdot T_s \leq t \leq (i+1) \cdot T_s$$
$$\phi_2^i(t) = \sqrt{\tfrac{2}{T_s}} \cdot \sin(2\pi f_o t), \ i \cdot T_s \leq t \leq (i+1) \cdot T_s \qquad (2.89)$$

Specifically for a square M-QAM signal, the LPE representation is given by

$$s_{lp}(t) = \sum_i d_i \cdot \Pi \left(\frac{t - i \cdot T_s}{T_s} \right), \qquad (2.90)$$

where $d_i \in \left\{ a_m \cdot \sqrt{E_{min}} + j b_m \cdot \sqrt{E_{min}} \right\}$, a_m and $b_m \in \left\{ -\sqrt{M} + 1, -\sqrt{M} + 3, \cdots, \sqrt{M} - 1 \right\}$, $m = 1, \cdots, \sqrt{M}$, and E_{min} is the energy of the symbol with minimum energy.

The i-th symbol d_i can also be represented in a two-dimensional space by the signal vector $\mathbf{d}_i \in \left\{ \left[a_m \cdot \sqrt{E_{min}}, b_m \cdot \sqrt{E_{min}} \right] \right\}$, $m = 1, \cdots, M$. Figure 2.20 presents an example of the scattering diagram for a 16-QAM scheme and the corresponding received signal corrupted by AWGN for an $E_b/N_0 = 10\,\mathrm{dB}$.

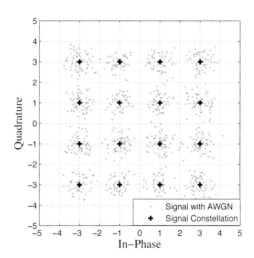

FIGURE 2.20
16-QAM constellation diagram.

2.5.1.6 M-ary frequency shift keying

The i-th orthogonal M-ary Frequency Shift Keying (M-FSK) transmitted waveform, corresponding to information symbol d_i, can be described by

$$s^i_{mfsk}(t) = \sqrt{\frac{2 \cdot E_s}{T_s}} \cdot \cos(2\pi f_i t),\ i \leq t \leq i \cdot T_s, \tag{2.91}$$

where $f_i \in \left\{ f_o + \frac{m}{2 \cdot T_s} \right\}$, $f_o = \frac{N_{fo}}{2 \cdot T_s}$, $m = 1, \cdots, M$, and $N_{fo} \in \mathbb{N}$ to keep the orthogonality of the symbol waveforms.

Since the set of possible M-FSK transmitted waveforms is orthogonal, an orthonormal basis can be simply obtained by normalizing the own set of M-FSK signal waveforms, i.e.,

$$\phi^i_m(t) = \sqrt{\frac{2}{T_s}} \cdot \cos(2\pi f_i t),\ 0 \leq t \leq T_s. \tag{2.92}$$

Thus, the LPE representation of an M-FSK signal can be written as

$$s_{lp}(t) = \sum_i d_i \cdot \Pi \left(\frac{t - i \cdot T_s}{T_s} \right), \tag{2.93}$$

with $d_i \in \left\{ \sqrt{E_s} \cdot \cos \left(\frac{\pi \cdot m \cdot t}{2 \cdot T_s} \right) + j\sqrt{E_s} \cdot \sin \left(\frac{\pi \cdot m \cdot t}{2 \cdot T_s} \right) \right\}$, $m = 1, \cdots, M$.

The i-th symbol d_i can also be represented in an M-dimensional vector space by the signal vector $\mathbf{d}_i \in \{[0, \cdots, d_{i,m}, \cdots, 0]\}$ and $d_{i,m} = \sqrt{E_s}$ is the m-th vector element, with $m = 1, \cdots, M$. Therefore, an M-FSK scheme requires an M-dimensional signal space for its correct representation.

2.5.2 Performance of digital communication schemes in AWGN

As discussed in the beginning of this chapter, the performance of a digital communication system can be measured in terms of its bandwidth efficiency η_b and power efficiency η_p.

In the design of digital bandpass communication systems, there is a trade-off between η_b and η_p that must be taken into account in the choice of the best scheme for a given application, since hardly a digital bandpass communication system will present good bandwidth and power efficiencies at the same time.

In Figure 2.21, the null-to-null bandwidth expressions of some bandpass modulation schemes [46, 93] employing, for simplicity, rectangular pulse shaping instead of Nyquist pulse shaping[5] are presented. In a complementary manner, the error probability expressions of some bandpass modulation schemes [46, 47] are shown in Figure 2.22.

Hence, with the presented specifications of bandwidth occupancy and error

---

[5]The definition of the Nyquist pulse is presented in Section 2.6.1.

Bandpass Modulation	Null-to-Null Bandwidth
M-ASK, M-PSK, M-QAM	$W = \dfrac{2 \cdot R_b}{\log_2(M)}$
M-FSK (Coherent)	$W = \dfrac{(M+3) \cdot R_b}{2 \cdot \log_2(M)}$
M-FSK (Non-Coherent)	$W = \dfrac{(M+1) \cdot R_b}{\log_2(M)}$

FIGURE 2.21
Bandpass modulation bandwidth occupancies.

probability for a given E_b/N_0, the choice of the most suitable scheme for a given application can be made based on η_p, represented by

$$\eta_p = \frac{E_b}{N_0}, \text{for a given BER} \tag{2.94}$$

and on η_b, expressed as

$$\eta_b = \frac{R_b}{W}. \tag{2.95}$$

Systems that present lower η_p are considered more power efficient, since they require lower E_b/N_0 to achieve a given BER. On the other hand, systems that present larger η_b are more bandwidth efficient, since they transmit higher bit rates per unit bandwidth. Ideally, a good system is the one that offers the highest η_b at a given η_p or the one that requires the lowest η_p at a given η_b [66].

2.6 Band-limited transmission

In the discussion so far, time-limited unit gate pulses $\Pi(t)$ have been used as the pulse shaping filter $p(t)$ in the LPE representation of transmitted PAM waveforms. As the gate pulses are not limited in frequency, part of their spectra will be suppressed by a bandlimited channel, thus resulting in pulse distortion and, possibly, ISI at the receiver, as already discussed in Section 2.1.2. To solve this problem, one can try to replace $\Pi(t)$ with bandlimited pulses. However, bandlimited pulses are not time-limited. Hence, they will overlap in time, causing ISI, too.

To get out of this apparently dead end, most digital communication systems employ pulse shaping and equalization techniques to mitigate ISI and recover the transmitted signals. While pulse shaping concerns the design of

Bandpass Modulation	Symbol Error Probability (P_e)	Bit Error Probability (P_b)
BPSK	$P_e = Q\left(\sqrt{\dfrac{2 \cdot E_s}{N_0}}\right)$	$P_b = Q\left(\sqrt{\dfrac{2 \cdot E_b}{N_0}}\right)$
QPSK	$P_e \approx 2 \cdot Q\left(\sqrt{\dfrac{2 \cdot E_s}{N_0}}\right)$	$P_b \approx Q\left(\sqrt{\dfrac{2 \cdot E_b}{N_0}}\right)$
M-PSK	$P_e \approx 2 \cdot Q\left(\sqrt{\dfrac{2 \cdot E_s}{N_0}} \cdot \sin\left(\dfrac{\pi}{M}\right)\right)$	$P_b \approx \dfrac{2}{n_b} \cdot Q\left(\sqrt{\dfrac{2 \cdot n_b \cdot E_b}{N_0}} \cdot \sin\left(\dfrac{\pi}{M}\right)\right)$
M-QAM	$P_e \approx \dfrac{4 \cdot \left(\sqrt{M}-1\right)}{\sqrt{M}} \cdot Q\left(\sqrt{\dfrac{3}{(M-1)} \cdot \dfrac{E_s}{N_0}}\right)$	$P_b \approx \dfrac{4 \cdot \left(\sqrt{M}-1\right)}{n_b \cdot \sqrt{M}} \cdot Q\left(\sqrt{\dfrac{3 \cdot n_b}{(M-1)} \cdot \dfrac{E_b}{N_0}}\right)$
M-FSK (Coherent)	$P_e \approx (M-1) \cdot Q\left(\sqrt{\dfrac{E_s}{N_0}}\right)$	$P_b \approx \dfrac{M}{2} \cdot Q\left(\sqrt{\dfrac{n_b \cdot E_b}{N_0}}\right)$

FIGURE 2.22
Bandpass modulation error probabilities.

LPE signals such that, despite pulse spreading, there is no ISI at the decision-making instants [46], equalization deals with the design of filters to compensate for the distorsions imposed by the channel on the transmitted waveforms [6, 46]. These two techniques will be briefly described in the next sections.

2.6.1 Nyquist criteria for zero ISI

In Section 2.5, it was shown that the most used digital modulation schemes can be represented by the LPE waveform given by Equation (2.72), repeated here for convenience:

$$s_{lp}(t) = \sum_i d_i \cdot p(t - iT_s), \qquad (2.96)$$

where d_i is the i-th complex-valued digital information symbol, $p(t)$ is the pulse shape, and $1/T_s$ is the symbol rate.

Consider for a moment the transmission of the LPE waveform $s_{lp}(t)$ over an ideal (distortion-free and noiseless) channel. Even in this case, ISI might occur if the pulse shape is not chosen appropriately. To see this, assume the receiver will try to recover the symbols d_i by sampling $s_{lp}(t)$ at multiples of the symbol period. Then, the n-th sample is given by [6]

$$s_{lp}(nT_s) = \sum_i d_i \cdot p(nT_s - iT_s) = d_n * p(nT_s), \qquad (2.97)$$

where the $*$ operator represents the discrete-time convolution of the symbol sequence with a sampled version of the pulse shape. This convolution sum can be further decomposed into two parts as [6]

$$s_{lp}(nT_s) = p(0)d_n + \sum_{i \neq n} d_i \cdot p(nT_s - iT_s), \tag{2.98}$$

where the first term on the right-hand side contains the desired symbol and the second term represents the ISI, i.e., the interference from neighboring symbols.

From Equation (2.98), it is clear that there will be no ISI only if the second term on the right-hand side is zero. This happens if the sampled pulse shape reduces to a Kronecker delta function [45, 57] in the sampling instants, i.e.,

$$p(nT_s) = \delta_n. \tag{2.99}$$

Remembering that $p(nT_s)$ is the sampled version of $p(t)$ and using the properties of Fourier transform [45, 58], it is possible to show that the Fourier transform of Equation (2.99) is given by [6, 46, 66]

$$\frac{1}{T_s} \sum_i P\left(f - \frac{i}{T_s}\right) = 1, \tag{2.100}$$

where $P(f)$ is the Fourier transform of $p(t)$. Equation (2.100) is the *Nyquist criterion for zero ISI* or simply *Nyquist criterion*. A pulse $p(t)$ that satisfies Equation (2.100) is said to be a *Nyquist pulse*, meaning that it does not induce ISI when sampled properly [6, 66].

Figure 2.23 shows an example of a Nyquist pulse. Specifically, this figure shows sinc[6] pulses used to shape the transmission of two successive digital information symbols with values $d_0 = 2$ and $d_1 = 1$.

It is clear from this figure that the transmitted digital symbols can be correctly recovered by sampling the received waveform, which is composed by the sum of the two sinc pulses corresponding to d_0 and d_1, at multiples of the symbol period. This is due to the fact that Nyquist pulses are designed such that only one pulse is nonzero at each multiple of T_s.

> Therefore, even though neighboring pulses overlap, they do not interfere with one another at proper sampling times if Equation (2.100) is satisfied.

There are an infinite number of pulses that satisfy the Nyquist criterion. The sinc pulse is shown to be the minimum-bandwidth pulse that satisfies Equation (2.100) [6, 46, 66].

---

[6]The sinc pulse is defined as $\mathrm{sinc}(t) = \frac{\sin(\pi t/T_s)}{\pi t/T_s}$.

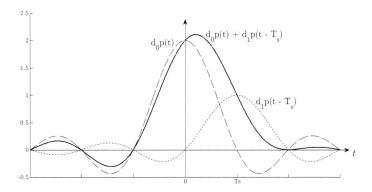

FIGURE 2.23
Transmission of two successive symbols with values $d_0 = 2$ and $d_1 = 1$ using sinc pulses.

> The minimum-bandwidth (or Nyquist bandwidth) required to avoid ISI is half the symbol rate, i.e., $\dfrac{1}{2T_s}$.

Although a minimum-bandwidth pulse is desirable, it is not realizable. A practical pulse will have a bandwidth larger than the minimum value by a factor $1 + \alpha$, where α is known as *excess-bandwidth parameter* or *roll-off factor*. For instance, a pulse with $\alpha = 1$ has 100% excess bandwidth, that is, it occupies twice the band of a sinc pulse. The most commonly used Nyquist pulses are, perhaps, the *raised-cosine pulses*, defined as [6, 46, 66]

$$p(t) = \left(\frac{\sin\left(\pi t/T_s\right)}{\pi t/T_s} \right) \left(\frac{\cos\left(\alpha\pi t/T_s\right)}{1 - \left(2\alpha t/T_s\right)^2} \right), \qquad (2.101)$$

which have Fourier transform

$$P(f) = \begin{cases} T_s, & |f| \le \frac{1-\alpha}{2T_s} \\ T_s \cos^2\left[\frac{\pi T_s}{2\alpha} \left(|f| - \frac{1-\alpha}{2T_s} \right) \right], & \frac{1-\alpha}{2T_s} < |f| \le \frac{1+\alpha}{2T_s} \\ 0, & |f| > \frac{1+\alpha}{2T_s} \end{cases} . \qquad (2.102)$$

Figure 2.24 shows raised-cosine pulses for some values of α.

In this way, using raised-cosine pulses, the resulting baseband signal bandwidth can be represented by

$$B = (1 + \alpha) \cdot \frac{R_s}{2}, \qquad (2.103)$$

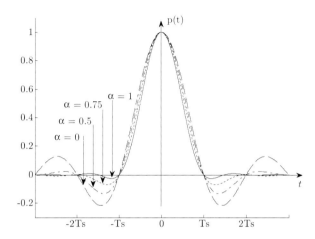

(a) Time-domain.

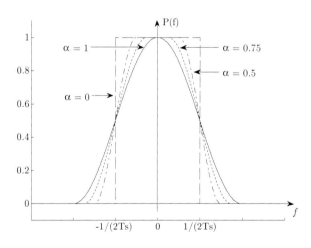

(b) Frequency-domain.

FIGURE 2.24
Raised-cosine pulses.

and the resulting bandpass signal bandwidth, for linear bandpass systems, can be estimated by

$$W = (1 + \alpha) \cdot R_s. \tag{2.104}$$

2.6.2 Equalization

In the previous section, the Nyquist criterion for zero ISI was derived for a distortion-free and noiseless channel. However, usual communication channels will distort the transmitted waveforms. In this case, it is possible to apply the same theory presented in Section 2.6.1 to design pulse shaping filters $p(t)$ such that the overall pulse shape at the detector input satisfies the Nyquist criterion. The problem with this approach is that the channel is not generally known prior to transmission. Also, the channel may be time-varying, thus forcing the design of new pulses $p(t)$ before the transmission of each new symbol. Clearly, this is not practical. Hence, to solve this problem, digital communication systems often employ a filter called equalizer before the detection process.

The role of the equalizer is to compensate for the channel distortion. By carefully designing this filter, it is possible to use a Nyquist pulse, such as the raised-cosine, to obtain at the detector input a pulse shape that presents no ISI at the sampling instants, no matter how the channel conditions. If the channel is modeled by a known linear time-invariant system, linear time-invariant equalizers can be derived, as will be shown in Sections 2.6.2.1 and 2.6.2.2. However, if the channel is unknown and/or time-varying, adaptive equalizers, discussed in Sections 2.6.2.3 and 2.6.2.4, are used instead.

Before presenting some equalizer design criteria, it is convenient to define an equivalent discrete-time model for a channel with ISI. To that end, consider a transmitter sending digital symbols d_i at a rate $1/T_s$ symbols/s. As the sampled output of the matched filter at the receiver is also a discrete-time signal generated at a rate $1/T_s$ samples/s, it is possible to show that the concatenation of the pulse shaping filter $p(t)$, the analog channel, the matched filter, and the sampler can be represented by an equivalent discrete-time FIR filter [6, 66]. The input to this discrete-time channel model is the sequence of information symbols $s(n)$ and the output is the sequence of discrete-time signals $y(n)$ defined in Section 2.3.4. Hereinafter, the equivalent discrete-time channel model will be represented by its transfer function $F(z)$, that is, the z-transform of the channel impulse response.

2.6.2.1 ZF equalization

As stated previously, the goal of any equalizer is to eliminate the undesired effects the channel might impose on the transmitted signals. To achieve this, the *zero-forcing* (ZF) equalizer is designed to invert the channel transfer function $F(z)$. Thus, the z-transform of the impulse response of a ZF equalizer is given by

$$H_{ZF}(z) = \frac{1}{F(z)}. \tag{2.105}$$

The term "zero-forcing" refers to the fact that the equalizer forces the ISI to zero. The major drawback of this equalizer is its high susceptibility to the phenomenon known as *noise enhancement*, i.e., an excessive noise amplification when the channel presents spectral nulls [36].

2.6.2.2   MMSE equalization

In general, to reduce noise enhancement the zero-ISI constraint must be relaxed, thus allowing equalizers to have residual-ISI at their outputs. Equalizers based on the *mean-squared error* (MSE) criterion are examples of filters that follow this approach.

In the MSE criterion, the coefficients of the equalizer are adjusted to minimize the variance (i.e., the energy) of the error between the transmitted symbol $s(n)$ and the detected symbol $y(n)$ [6,36,66]. In mathematical terms, the MSE is defined as

$$J_{MSE} = \mathrm{E}\left\{|s(n) - y(n)|^2\right\}, \tag{2.106}$$

where $\mathrm{E}\{.\}$ is the expectation operator.

Equalizers that minimize the MSE are called *minimum mean-squared error* (MMSE) equalizers. It is possible to show that these filters are computed by the well-known *Wiener solution* [6,36,66]

$$\mathbf{h} = \mathbf{R}_r^{-1}\mathbf{p}_{rs}, \tag{2.107}$$

where $\mathbf{h}$ is the vector of equalizer's coefficients, $\mathbf{R}_r$ is the autocorrelation matrix of the received symbols, and $\mathbf{p}_{rs}$ is the cross-correlation vector between the transmitted and received symbols [36,66].

2.6.2.3   Adaptive filtering: supervised methods

In the previous section, we discussed the MMSE equalizers and Wiener solution. Such criterion presents some advantages: it is rather simple, since the cost function J_{MSE} presents a unique global minimum. Nonetheless, a major drawback in applying Equation (2.107) is the need of knowing the matrix $\mathbf{R}_r$ and the vector $\mathbf{p}_{rs}$. Since the statistics of the environment are usually not known, those of $r(n)$ are also not available and difficult to estimate. Thus, an adaptive method to attain or approximate the optimum solution given by Equation (2.107) is needed.

Let us start by considering filters given by discrete-time linear structures, modeled as transversal *finite impulse response* (FIR) filters. When considering adaptive filters, *infinite impulse response* (IIR) filters may present stability problems, and we also consider that they may be approximated by a long FIR

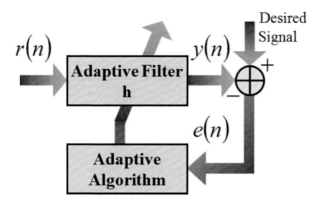

FIGURE 2.25
System model for adaptive filtering.

filter [65]. The output of an FIR filter may be written as

$$y(n) = \sum_{k=0}^{L} h_k^* r(n-k) = \mathbf{h}^H \mathbf{r}(n), \tag{2.108}$$

i.e., the output signal $y(n)$ is given by a linear combination of delayed samples from the input signal, $r(n)$, multiplied by the filter coefficients h_k. In Equation (2.108), L is the filter order and the superscript $*$ denotes complex conjugation, while $\mathbf{h} = [h_0\ h_1\ ...\ h_L]^T$, $\mathbf{r}(n) = [r(n)\ r(n-1)\ ...\ r(n-L)]^T$, and the superscript H denotes the Hermitian transpose.

The problem that arises is, as illustrated in Figure 2.25, how to adjust the values of h_i, so that $y(n)$ will be as close as possible to a desired signal $s(n)$ (transmitted signal). In other words, how could we find $\mathbf{h}$ in order to make the error signal $e(n)$ as small as possible, where $e(n) = s(n) - y(n)$? From such demand came the development of the *Least-Mean Squared* (LMS) algorithm.

The LMS is based on Equation (2.107), avoiding the need of knowing $\mathbf{R}_r$ and $\mathbf{p}_{rs}$ *a priori* by using instantaneous estimates based directly on $r(n)$ and $s(n-\tau)$, i.e., $\hat{\mathbf{R}}(n) = \mathbf{r}(n)\mathbf{r}^H(n)$ and $\hat{\mathbf{p}}(n) = \mathbf{r}(n)s(n-\tau)$, where τ is a delay. Doing so, a recursive relation to the adaptation of the filter coefficients $\mathbf{h}$ may be stated:

$$\mathbf{h}(n+1) = \mathbf{h}(n) + \mu\mathbf{r}(n)e^*(n), \tag{2.109}$$

where μ is the algorithm's step-size and $e(n) = (s(n-\tau) - y(n))$. The LMS is one of the most important algorithms among stochastic gradient algorithms, due to its features: it is extremely simple and can be applied to a large number of problems. It is also an algorithm that converges easily, given that the step-size is correctly chosen. Its solution is usually close to the one obtained by the Wiener solution. A detailed analysis of its conditions for convergence and comparison with Wiener solution can be found in [36].

An alternative approach to Wiener theory and the LMS algorithm is the *least squares* filters. In this case, a finite set of observed data is considered and the objective is to minimize the following cost function:

$$J_{LS} = \sum_{i=1}^{n} \lambda^{n-i} |e(i)|^2, \qquad (2.110)$$

where $e(i) = s(i) - y(i)$ and λ is a weighting factor or a forgetting factor. With such approach, at each time instant n, a new cost function is defined and, as a consequence, a new optimum filter $\mathbf{h}$ is obtained. Minimizing Equation (2.110) with respect to $\mathbf{h}$ results in

$$\mathbf{h}(n) = \mathbf{R}_{LS}^{-1}(n)\mathbf{p}_{LS}(n), \qquad (2.111)$$

where $\mathbf{R}_{LS}(n) = \sum_{i=1}^{n} \lambda^{n-i}\mathbf{r}(i)\mathbf{r}^H(i)$ and $\mathbf{p}_{LS}(n) = \sum_{i=1}^{n} \lambda^{n-i}\mathbf{r}(i)s(i)$. Again, it would be interesting to find a recursive method to attain the solution given by Equation (2.111). Since at each new observed data a new optimum solution may be found, it would be interesting to write $\mathbf{h}(n + 1)$ as a function of $\mathbf{h}(n)$, $s(n + 1)$, and $\mathbf{r}(n + 1)$. In fact, following directly from Equation (2.111) and the definitions of $\mathbf{R}_{LS}(n)$ and $\mathbf{p}_{LS}(n)$, we may write

$$\mathbf{h}(n + 1) = \mathbf{h}(n) + \mathbf{R}_{LS}^{-1}(n + 1)\mathbf{r}(n + 1)e_a(n + 1), \qquad (2.112)$$

where $e_a(n + 1)$ is the *a priori* estimation error defined as $e_a(n + 1) = s(n + 1) - \mathbf{r}^H(n + 1)\mathbf{h}(n)$. The matrix inversion may be avoided by using the matrix inversion lemma. The final algorithm is presented below and is called the *Recursive Least Squares algorithm* (RLS).

Summary of the RLS Algorithm

Initialization: $\begin{cases} \mathbf{R}_{LS}(0) = \delta\mathbf{I}, \delta = \text{small positive constant} \\ \mathbf{h}(0) = \mathbf{0} \end{cases}$

For each time instant n:

$$\mathbf{k}(n + 1) = \frac{\lambda^{-1}\mathbf{R}_{LS}(n)\mathbf{r}(n + 1)}{1 + \lambda^{-1}\mathbf{r}^H(n + 1)\mathbf{R}_{LS}(n)\mathbf{r}(n + 1)}$$

$$e_a(n + 1) = s(n + 1) - \mathbf{h}^H(n)\mathbf{r}(n)$$

$$\mathbf{h}(n + 1) = \mathbf{h}(n) + \mathbf{k}(n + 1)e_a^*(n + 1)$$

$$\mathbf{R}_{LS}(n + 1) = \lambda^{-1}\mathbf{R}_{LS}(n) - \lambda^{-1}\mathbf{k}(n + 1)\mathbf{r}^H(n + 1)\mathbf{R}_{LS}(n)$$

Even though RLS has a higher computational cost when compared to LMS, it presents a few advantages:

- RLS converges much faster than LMS.

- LMS convergence depends on the characteristics of the input signal $r(n)$ and, consequently, those of $\mathbf{R}_r$. The higher the eigenvalue spread (defined as the ratio of the largest to the smallest eigenvalues of $\mathbf{R}_r$), the slower will be the convergence of the algorithm. Since RLS includes the inverse of $\mathbf{R}_{LS}$ in its formulation, it does not suffer from such effect.

- RLS does not present excess mean-squared error, defined as the difference between the final value obtained for J_{MSE} using LMS and $J_{MSE_{min}}$ obtained by attaining the optimum Wiener solution.

On the other hand, the RLS algorithm presented suffers from numerical instability, due to the recursive estimation of $\mathbf{R}_{LS}$, that does not assure its symmetry. Several methods were proposed to cope with this problem. Namely, the so-called square-root adaptive filters are robust to such problem. In this class of algorithms, we may find the QR-decomposition-based RLS algorithm, its extended and inverse versions, and also lattice-based implementations of RLS. For more details, please check [36, 73].

In addition, several versions of LMS have also been proposed, always with the objective of improving its performance or aiming at particular applications of the algorithm. A few interesting references are [36, 37].

Since in a large number of applications the environment statistics are unknown and difficult to estimate, adaptive filtering is an important tool in statistical signal processing. It has been applied to such diverse fields as communications, control, radar, sonar, seismology, and biomedical engineering, among others. Further references on this subject include [24, 36, 51, 65, 73].

Up to this point, we have only considered the classical methods that depend on the knowledge of a desired signal, $s(n)$. In several applications, however, such signal is not known or only a few of its samples are known. In these cases, unsupervised methods have to be adopted. The following section considers a few classical techniques in such context.

2.6.2.4 Adaptive filtering: unsupervised methods

First, it is interesting to present an unsupervised method, that is, a method that does not depend on the knowledge of $s(n)$, that follows directly from what was discussed in the previous section: the decision directed (DD) method. Usually, such method is applied after a first stage known as training period, in which the desired signal is known. Thus, a supervised method like LMS may be applied to adjust the adaptive filter weights and, at the end of such period, the adaptation is switched to using the following equation:

$$\mathbf{h}(n+1) = \mathbf{h}(n) - \mu \mathbf{r}(n) e_{DD}^*(n), \tag{2.113}$$

where $e_{DD} = (\varphi(y(n)) - y(n))$ with $\varphi(.)$ being a zero-memory nonlinearity [36]. Comparing Equations (2.113) and (2.109), we may note that the only difference is that the later uses $s(n)$ while the former replaces its values by those of $\varphi(y(n))$. We may understand this new approach as using an error signal obtained from comparing $\varphi(y(n))$ with the equalizer output, $y(n)$. In an equalization context we may consider that the first training stage is sufficient to obtain a good solution for the equalizer coefficients. Thus, when switching to DD mode, the equalizer output is already sufficiently close to the transmitted signal. A decision device (nonlinear function) is then used at the equalizer output to obtain an estimate of $s(n)$, which becomes the reference for the error computation. Such approach is useful to track possible channel variations after the training period. It may also be used in other applications as system identification and spatial filtering, among others.

When not even a training sequence is available, different kinds of criteria, alternatives to minimizing the mean-squared error, have been proposed. In this context, only the signal $r(n)$ is known. Considering the classical theory involving unsupervised or blind methods, two approaches are possible:

- Methods based on the use of higher-order statistics (HOS); such methods may also be divided into methods that use HOS implicitly or methods that use it explicitly.

- Methods based on exploiting the second-order cyclostationary statistics of $r(n)$.

The first question that arises is how to propose a criterion able to adapt the coefficients of a filter in order to obtain a good equalization or system identification, if a desired signal is not available. Two important results are able to answer this question: the Benveniste-Goursat-Ruget (BGR) [10] and the Shalvi-Weinstein (SW) [77] theorems.

The BGR theorem states that considering independent and identically distributed (i.i.d.) signals and linear filters, if the probability density function of the transmitted signal (non-Gaussian) is equal to that of the equalizer output, then a perfect equalization will be achieved (assuming the equalizer is capable of perfectly inverting the channel) [10]. Such theorem was the first result showing that a blind criterion would be able to achieve equalization. The SW theorem simplifies the above statement. In [77], the authors show that if the fourth-order cumulant and second-order moment of the transmitted signal and of the equalizer output are equal, perfect equalization will be achieved. Such result was also of major importance, since it greatly simplifies what was previously stated by BGR. Every method proposed for blind equalization depends, implicitly or explicitly, on the results discussed above.

One of the most studied and used techniques in the context of blind methods is the *constant modulus criterion* (CMC) and its stochastic gradient algorithm, the *constant modulus algorithm* (CMA). Such technique was first proposed by Godard in [32,86] and searches to penalize deviations of the equalizer output, $y(n)$, from a constant modulus obtained from the transmitted signal, $s(n)$:

$$J_{CM} = E\left\{ \left(|y(n)|^2 - R_2 \right)^2 \right\},\qquad(2.114)$$

where $R_2 = E\{|s(n)|^4\}/E\{|s(n)|^2\}$. Comparing Equations (2.114) and

(2.106), we may observe that the former is not so simple. Obtaining the optimum **h** is not a direct operation [82]. In addition, J_{CM} presents local minima [23].

The resulting stochastic gradient algorithm, CMA, is given by:

$$\mathbf{h}(n+1) = \mathbf{h}(n) - \mu \mathbf{r}^*(n)y(n)\left(|y(n)|^2 - R_2\right). \tag{2.115}$$

The CMA is a particular case of a larger family of algorithms called the Bussgang algorithms [9, 36], whose general form is given by:

$$\mathbf{h}(n+1) = \mathbf{h}(n) - \mu \mathbf{r}^*(n)\left(y(n) - \varphi(y(n))\right), \tag{2.116}$$

where $\varphi(y(n))$ is a zero-memory nonlinearity. All of such algorithms implicitly satisfy the BGR theorem. CMA is the most popular due to its simplicity and robustness to phase errors from the carrier, since it only considers the modulus of the signals involved. The DD algorithm, discussed in the beginning of this section, is another example of a Bussgang algorithm. This class of algorithms uses HOS implicitly. In the following, we will discuss another family of criteria that uses HOS explicitly.

An alternative approach that leads to a family of criteria and algorithms was proposed by Shalvi and Weinstein, following directly the theorem proposed by them [77, 79]. Several versions of such criteria have been considered. Since they are all equivalent, we will only discuss the following:

$$\max \left|C_{2,2}^y\right|, \quad \text{subject to } C_{1,1}^y = C_{1,1}^s, \tag{2.117}$$

where $C_{p,q}^y$ is the p,q-order cumulant of $y(n)$, $C_{1,1}^y$ is equivalent to the power of $y(n)$ (second-order moment), and $C_{1,1}^s$ is equivalent to the power of $s(n)$. Doing $p = q = 2$ results in the fourth-order cumulant, also known as kurtosis. The above criterion takes into account that, under the power restriction, $|C_{2,2}^y| \leq |C_{2,2}^s|$, with equality if and only if perfect equalization is achieved.

The corresponding algorithm is named the *super-exponential algorithm* and was first proposed and deduced in [78]. It has the interesting property of presenting the optimum step size in the sense of convergence speed [52, 53]. On the other hand, it has not been so studied and used as CMA mainly due to its complexity. Equations (2.118) represent the super-exponential algorithm, where β is a constant, $\gamma = C_{2,2}^s/C_{1,1}^s$, and the matrix $\mathbf{Q}(n)$ is proportional to $\mathbf{R}_r^{-1}$.

$$\mathbf{h}(n+1) = \mathbf{h}(n) + \frac{\beta}{\gamma}\mathbf{Q}(n)\mathbf{r}^*(n)y(n)\left(|y(n)|^2 - R_2\right),$$

$$\mathbf{Q}(n+1) = \frac{1}{1-\beta}\left(\mathbf{Q}(n) - \frac{\beta\mathbf{Q}(n)\mathbf{r}^*(n)\mathbf{r}^T(n)\mathbf{Q}(n)}{1 - \beta + \beta\mathbf{r}^T(n)\mathbf{Q}(n)\mathbf{r}^*(n)}\right). \tag{2.118}$$

Several works have been done comparing the constant modulus (CM), Shalvi-Weinstein (SW), and Wiener filters. In [68], the author finds a direct relation between CM and SW cost functions, showing that they present the same stationary points. In addition, [35,82,92] show that CM, SW, and Wiener solutions are collinear and uniquely related.

As for methods based on exploiting the second-order cyclostationary statistics of $r(n)$, one of the most used is the subspace decomposition. In such technique, the received signal $r(n)$ has to be oversampled, that is, it must be sampled several times in each symbol duration. This results in the so-called fractionally spaced equalizers, since the equalizer coefficients are spaced closer than the incoming symbol rate. Computing the autocorrelation matrix of the oversampled received signal, $r(n)$, and analyzing its eigenvalues and eigenvectors, a *signal subspace* and a *noise subspace* may be defined. The first one is given by the eigenvectors related to eigenvalues that are larger than the noise variance. The second one is given by eigenvectors related to eigenvalues that are equal to the noise variance. Since such approach differs from what will be addressed in the rest of this book, we suggest the interested reader to obtain more details in [9,36].

So far, we have considered only the problem of a system having one signal as input, $s(n)$, and one signal as output, $r(n)$. Such problem may be generalized to represent a more realistic scenario, in which we have several users transmitting at the same time. Such signals will interfere with each other and, besides eliminating ISI, the equalizer will also have to separate such signals. This problem may also be viewed as a *blind source separation* problem, and will be discussed in the next section.

2.6.2.5   Blind source separation

Extending the problem discussed in the previous section, let us consider several users transmitting at the same time. At the receiver, such signals will have to be equalized and separated. Figure 2.26 illustrates the system model being considered.

Such problem may also be viewed as a *blind source separation* (BSS) problem. In such context, the original signals, named sources, will not be restrained to transmitted communication signals. They may also be speech,

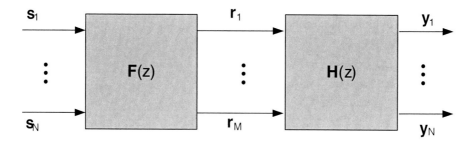

FIGURE 2.26
Multi-user and blind source separation system model.

image, biomedical, geophysical signals, among others. Nonetheless, independently of the signals at hand, the problem remains the same: the objective is to recover the original signals without having access to them and without knowing how they were mixed, i.e., without knowing matrix $\mathbf{F}$ in Figure 2.26.

In Figure 2.26, we consider I different sources s_i, $i = 1, ..., I$, mixed by a matrix $\mathbf{F}$, resulting in J different observed or received signals r_j, $j = 1, ..., J$. The BSS problem consists in estimating the demixing matrix $\mathbf{H}$, so that the outputs y_i are as close as possible to the original sources s_i.

Two main scenarios have been studied:

- Instantaneous mixtures: In this case, r_j is obtained as a linear combination of the sources s_i, that is, matrix $\mathbf{F}$ has scalar coefficients.

- Convolutive mixtures: In this case, $r_j(n)$ also depends on delayed samples of s_i, $i = 1, ..., I$. Thus, the coefficients of $\mathbf{F}$ are given by filters. An example would be:

$$\mathbf{F}[z] = \begin{bmatrix} z^{-1} & -1 - 3z^{-1} \\ 1 - z^{-1} & -1 + z^{-1} \\ z^{-1} & -1 \end{bmatrix}. \qquad (2.119)$$

In this context, the inversibility of $\mathbf{F}[z]$ is only possible if certain conditions are satisfied [16, 41]. This case encompasses a communication scenario, in which, at the receiver, we have multiple delayed copies of the transmitted signals interfering with each other (see Section 2.1.2). The antennas at the receiver will capture a mixture of the transmitted signals and delayed versions of them, which may be modeled exactly as a convolutive mixture.

In addition, cases in which $I \leq J$, that is, there are at least as many sensors as sources, are easier to treat. Nevertheless, several methods have also been proposed to solve the problem when there are not enough sensors [20].

The structure of the adaptive filter $\mathbf{H}[z]$ depends largely on the optimization criterion chosen, on the nature and statistical properties of the sources and noise (random, deterministic, Gaussian, non-Gaussian, stationary, among others), and also on the kind of mixture (instantaneous or convolutive) [19]. For example, the optimal filtering structure for non-Gaussian noise and interferences is non-linear. When considering non-stationary sources, time-varying filtering structures are necessary. Nevertheless, linear, time-invariant structures have been the most widely studied in the literature and, even in cases where they become sub-optimal, it does not necessarily mean lack of practical usefulness [19]. Thus, in the following, we will only consider this kind of structure.

One of the major results that prompted researches in this field was the paper of Pierre Comon on independent component analysis (ICA) [18]. In this paper, the author shows that the hypotheses of having independent sources is crucial to achieve a blind separation. Thus, recovering decorrelated signals is not sufficient and, as a consequence, principal component analysis (PCA) will not correctly separate the mixed sources. From such analysis, the use of criteria based on concepts able to measure independence between different signals became clear. Such concepts are easily found and well understood in information theory. For example, Kullback-Leibler divergence measures the deviation between two probability distributions. Consider $f_y(\mathbf{y}; \mathbf{H})$ as being the distribution of $\mathbf{y} = [y_1\ y_2\ ...\ y_N]^T$ obtained using $\mathbf{H}$. Consider also $q(\mathbf{y})$ as being a reference distribution. The Kullback-Leibler divergence between the two may be defined as [3]

$$
\begin{aligned}
D(\mathbf{H}) &= \int f_y(\mathbf{y}; \mathbf{H}) log \frac{f_y(\mathbf{y}; \mathbf{H})}{q(\mathbf{y})} d\mathbf{y} \\
&= -\mathcal{H}(\mathbf{H}) - \sum_{i=1}^{N} E\{\log\ q_i(y_i)\},
\end{aligned} \tag{2.120}
$$

where $\mathcal{H}(\mathbf{H})$ is the entropy of $f_y(\mathbf{y}; \mathbf{H})$. The above equation is equivalent to obtaining the expectation of $l_q(\mathbf{y}, \mathbf{H})$ defined as

$$
l_q(\mathbf{y}, \mathbf{H}) = -\log|\det(\mathbf{H})| - \sum_{i=1}^{N} \log\ q_i(y_i), \tag{2.121}
$$

except for a constant term. Minimizing Equation (2.120) or the expectation of Equation (2.121) enables obtaining a demixing matrix $\mathbf{H}$ that leads to having independent y_i. The Kullbak-Leibler divergence also leads to the concept of mutual information. Several criteria may be understood under this framework, with different choices for the reference function $q(\mathbf{y})$ [3]. This is the case of the entropy maximization [8, 91], that uses nonlinear transformations $v_i = \varphi_i(y_i)$ for q_i; ICA [4, 18], that uses the marginalized independent distribution of $f_y(\mathbf{y}; \mathbf{H})$; nonlinear PCA [56] and the maximum-likelihood approach [14,54,91] that uses the true distribution of the sources as $q(\mathbf{y})$. Usually, these have to be estimated since the true distribution is not known.

Another interesting approach to achieve separation is maximizing the non-Gaussianity of the estimated signals y_i. Following the Central Limit Theorem [62], the probability density function of a sum of random variables tends to a Gaussian variable, under certain conditions. Thus, we would like to search for a matrix $\mathbf{H}$ that maximizes the non-Gaussianity of $\mathbf{y} = \mathbf{Hr}$ [39].

A classical measure of Gaussianity is the kurtosis (fourth-order cumulant). Signals that have a positive kurtosis are called *super-Gaussian*, while signals that present a negative kurtosis are called *sub-Gaussian*. The Gaussian distribution is one of the few exceptions that presents a kurtosis equal to zero. Thus, a possible criterion would be to maximize the modulus of the kurtosis.

A different approach would be to consider that Gaussian distribution presents the largest entropy among all distributions with the same variance [62]. Such result may also be used to measure the Gaussianity of a signal and is behind the definition of a measure named *negentropy*:

$$
\mathcal{N}(\mathbf{y}) = \mathcal{H}(\mathbf{y}_{gauss}) - \mathcal{H}(\mathbf{y}), \tag{2.122}
$$

where $\mathcal{H}(\cdot)$ is Shannon entropy and $\mathbf{y}_{gauss}$ is a random vector with Gaussian distribution and the same mean and variance of $\mathbf{y}$. It is possible to show that such measure is directly related to the Kullback-Leibler divergence and to mutual information [3, 20, 39]. Minimizing mutual information is equivalent to maximizing negentropy.

It is important to mention that, in several applications, it is interesting to apply a preprocessing on the observed signals r_j, in order, for example, to remove redundancy (in the case of more sensors than sources), to reduce additive noise or to improve the convergence properties of the adaptive algorithms used for signal separation. Commonly used preprocessing techniques include prewhitening and PCA [3].

Even though the methods discussed here seem completely different from the ones presented in Section 2.6.2.4, several similarities may be stated between the techniques used in both contexts [5]. For example, maximizing the kurtosis is equivalent to SW criteria. In fact, here we consider sources to be independent while in blind equalization, the transmitted signal is i.i.d. In addition, Bussgang techniques, negentropy, and nonlinear PCA use nonlinear functions as estimators and can also be seen through the same framework [5]. Such similarities may also be extended to algorithms. A well-known algorithm for BSS, based on negentropy, is the FastICA [38, 39]. Depending on the nonlinearity chosen, its expression is equivalent to the super-exponential algorithm given in Equation (2.118) [5].

Considering again a communication scenario, the methods viewed in Section 2.6.2.4, such as CMA, have been extended to the multiuser case [61], which is equivalent to a convolutive mixture. Basically, it is the same algorithm presented in Equation (2.115) but with an additional term that penalizes the cross-correlation between filters related to different users. Such strategy avoids multiple filters converging to the signal of the same user. Kurtosis maximization, used in SW criteria, has also been extended to the multiuser scenario, resulting in a method equivalent to that developed for BSS and discussed above [59, 60].

2.7 Synchronization

The demodulation process, performed at the receivers of digital bandpass systems, can be classified as coherent and non-coherent. In coherent demodulation, frequency and phase information of the carrier signal are required to the correct recovery of the transmitted information. On the other hand, in noncoherent demodulation, the information of the carrier signal is not required in the demodulation process.

Coherent demodulation usually has the advantage of presenting a lower P_e compared with noncoherent demodulation for the same received signal power, although at the cost of more receiver complexity [22]. Also, both coherent and noncoherent digital demodulation schemes require timing information from the received signal to perfectly recover the digital transmitted symbols.

In a real digital bandpass communication system, the output of the demodulator must be sampled periodically at a rate of at least once per symbol period, in order to recover the transmitted information. Since the propagation delay from the transmitter to the receiver is generally unknown at the receiver, symbol timing must be derived from the received signal in order to synchronously sample the output of the demodulator. The propagation delays of the communication channel also result in offsets in both frequency and phase of the carrier, which must be estimated at coherent receivers [30].

Therefore, to reach optimal carrier and timing estimation, synchronization schemes are employed at the receiver. Maximum likelihood estimation is one of the most employed approaches since it can provide a unified framework for developing optimal synchronization algorithms for digital communication receivers [30].

In the following, carrier and timing synchronization are discussed for coherent receivers.

2.7.1   Carrier synchronization

In general, coherent digital demodulation requires knowledge of the sinusoidal basis functions at the receiver. It means that the receiver has to estimate the frequency and the phase of the carrier from the corresponding received signal. This estimation process is called carrier synchronization.

There are many possibilities for implementing carrier synchronization in a digital bandpass communication system, but the core of most of them is a Phase-Locked Loop (PLL).

A generic PLL, represented by the block diagram of Figure 2.27, is a dynamic control system whose controlled parameter is the phase of a locally generated replica of the incoming carrier signal.

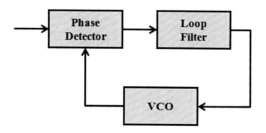

FIGURE 2.27
Generic phase-locked loop.

The main components of a generic PLL are:

(a) The phase detector (PD) – measures the phase difference between the local carrier generated by the VCO and the PLL input signal;

(b) The voltage controlled oscillator (VCO) – a variable-frequency oscillator whose output signal frequency is controlled by an input voltage signal (voltage-frequency converter). In normal operation, the free running frequency of the VCO, corresponding to an input voltage of zero, should be equal or close to the frequency of the PLL input signal;

(c) The loop filter (LF) – a lowpass filter. Usually, first or second order filters are employed.

The phase detector generates an error signal resulting from the phase difference between the local carrier and the PLL input signal. This error signal is then fed to a loop filter, which is designed to track the changes in the error signal and to reduce the effects of noise. The filtered error signal controls the frequency of the VCO so that the local generated carrier signal reaches the same phase of the PLL input signal. As the frequency is the derivative of the phase, keeping the phases of the PLL input signal and of the VCO output signal locked, their frequencies will also be locked. A detailed description of the PLL operation can be found in [69].

2.7.2 Timing synchronization

Coherent and noncoherent digital demodulations require knowledge of the timing information of the received signal to perfectly recover the transmitted information. This estimation process is called timing synchronization. The main purpose of the timing recovery is to obtain symbol synchronization.

Symbol decisions are based on the MF output at the end of each symbol period T_s. The detector samples the MF output and uses this information to decide which symbol is most likely to have been sent. In order to make these decisions, the detector must know when the symbols begin and end.

Basically, there are three main techniques to determine the optimum sampling point [22]:

(a) The first method finds the point where the slope of the MF output is zero by considering ML criterion. If the current timing estimate is too early, then the slope of the MF output is positive indicating that the timing phase should be advanced. If the current timing estimate is too late, then the slope of the MF output is negative indicating that the timing phase should be retarded.

(b) The second method, commonly denoted Gardner loop, uses the zero crossings in the oversampled MF output, with oversampling rate of two samples per symbol, to estimate the times in-between the optimum sampling points. Zero crossings are found by searching for sign changes between the previous y_{i-1} and following MF y_{i+1} outputs with respect to current analyzed output y_i. A sign change means that the current MF output resides on a zero crossing trajectory, whose error can be estimated by $e_i = [y_{i+1} - y_{i-1}] \cdot y_i$. This method has the advantage of being insensitive to carrier offsets and timing recovery can be locked first, simplifying the task of carrier recovery.

(c) The third method, usually called dither loop, estimates the optimum sampling point by finding the position that minimizes the variance of the MF output. Typically, the search is performed by estimating the variance at the next interpolation point. If the variance increases, then the timing estimate is not advanced. If the variance decreases, then the timing estimate is advanced.

2.8   Recent trends in digital communications

New communication systems promise to offer a wide range of multimedia applications, which demand high data rates and low error probabilities. These requirements, generally characterized by a demanded quality of service (QoS), combined with the ever increasing demand for high speed Internet access, motivate the development and the use of new technologies.

In wireless communication systems, for instance, several new structures and techniques have been proposed in recent years to mitigate the deleterious effects of fading channels and the different kinds of interferences, thus enabling an increase in system capacity and wireless speed access.

Among these techniques, OFDM (Orthogonal Frequency Division Multiplexing) has risen as a very efficient, flexible, and cost-effective modulation technique [6, 87]. In OFDM, digital information can be first mapped to a high-order M-ary scheme to improve system throughput. Then, Inverse Fast Fourier Transform (IFFT) and Fast Fourier Transform (FFT) operations are performed in the transmission and reception processes, respectively. By combining IFFT/FFT and Cyclic Prefix (CP), OFDM can mitigate ISI, converting a frequency selective fading channel into multiple orthogonal frequency flat fading channels. Consequently, the fading can be compensated by a simple one-tap frequency domain equalizer (FDE) at the receiver [15, 46].

Antenna arrays can also be used to improve system performance [13, 87]. The interest in communication systems with multiple transmit and receive antennas is continuously increasing since theoretical studies [28, 29, 85] demonstrated a significant increase in system capacity without sacrificing precious bandwidth and power resources. Communication systems with multiple transmit (multiple input) and multiple receive (multiple output) antennas are called

MIMO (multiple input/multiple output) systems. The idea of MIMO systems is to exploit not only the time dimension of the information signals but also the space dimension provided by the use of multiple antennas. By jointly processing the signals from the multiple antennas, the interference can be effectively suppressed, thus increasing the system capacity.

One of the most effective and practical ways to improve the quality of the receive signals in MIMO systems is to mitigate the fading effects by using space-time codes [26, 44, 87]. As the name suggests, the encoding of space-time codes is done in both temporal and spatial domains, thus introducing correlation between signals transmitted by different antennas in different time instants. Among the existing space-time coding schemes, orthogonal space-time block codes (OSTBC) are of particular interest because their ML receiver consists of a simple linear combiner followed by a symbol-by-symbol decoder [2, 26, 44, 83]. MIMO systems can also be used to increase the data rate of a digital communication system by the use of spatial multiplexing techniques [13, 26, 87].

It is important to highlight that, although nearly associated with wireless communications, recently MIMO systems have been employed in other systems, such as fiber optic [34, 76], PLC [74, 75], and digital subscriber line (DSL) [17, 27, 31, 84] communications.

To increase the immunity of the transmitted signals to noise and to protect the information from interference, error correcting codes, or channel codes, are normally used. These codes add a controlled redundancy to the transmitted message in order to allow the receiver to detect and/or correct errors originated through the transmission. Although channel codes have been used for a long time, a variety of powerful channel codes have been designed during the last decade. Among these, the most important classes are that of the turbo codes [11, 48, 72] and that of the low-density parity-check (LDPC) codes [48–50, 70, 72]. The main characteristic of these codes is that they can achieve a capacity near to the Shannon limit [6, 48, 66, 72].

By combining turbo or LDPC codes with OFDM modulation in MIMO systems employing OSTBC, fourth generation (4G) cellular systems will improve 3G services, offering higher data rates and new high quality multimedia services.

2.8.1   Recent trends in signal processing for communications

As a direct consequence of the information theory framework used to obtain blind source separation, several methods using concepts as entropy and mutual information are being developed for the problem of supervised or unsupervised deconvolution. The main motivation is that such measures consider more information about the signals involved than considering only a few moments of such signals. Such methods are being called *information theoretic learning* (ITL) and have shown some interesting results [64]. For example, such techniques perform better than CMA or super-exponential algorithms when equalizing correlated signals or in the presence of impulsive noise [64]. In addition, in a supervised scenario, they perform better than MSE when treating a signal with probability density having multiple peaks [64].

In its early days, signal processing for communications focused on physical layer technologies specially for wireless communications. Nowadays, such interest shifted to networking [21]. Equalization, precoding, beamforming, and channel estimation algorithms have been developed aiming a decentralized operation, so that each node of the network is able to process its data and exchange information with its neighbors [94]. Cognitive networks may perform a variety of tasks as environmental or resource monitoring, event detection, target tracking, communications over cognitive radio channels, and design of multiagent systems, among others [94].

In addition, the processing of sparse signals has also called the attention of researchers in several fields. In communications, compressive sensing, i.e., the study of compression and reconstruction of analog signals that are sparse in some domain (space, time, frequency), has been applied to radar and spectrum sensing, among others [21]. In BSS, such techniques are called *sparse component analysis* and have been applied, for example, when less sensors than sources are available [20]. Several signals may be considered in this framework such as speech, image (fMRI), and electroencephalograms.

Bibliography

[1] G. P. Agrawal. *Fiber-Optic Communication Systems*. Wiley Interscience, New York, 2002.

[2] S. M. Alamouti. A simple transmit diversity technique for wireless communications. *IEEE Transactions on Information Theory*, pages 1451–1458, October 1998.

[3] S. Amari, A. Cichocki, and H. Yang. *Unsupervised Adaptive Filtering, Volume I: Blind Source Separation*, volume 1, chapter Blind Signal Separation and Extraction: Neural and Information-Theoretic Approaches, pages 63–138. John Wiley and Sons, New York, 2000.

[4] S. Amari, A. Cichocki, and H. H. Yang. A new learning algorithm for blind signal separation. *Adv. in Neural Inform. Processing Systems*, 8:752–763, 1996.

[5] R. Attux, A. Neves, L. T. Duarte, R. Suyama, C. C. Junqueira, L. Rangel, T. Dias, and J. M. Romano. On the relationships between blind equalization and blind source separation - part ii: Relationships. *Journal of Communication and Information Systems*, 22:53–61, 2007.

[6] J. R. Barry, E. A. Lee, and D. G. Messerschmitt. *Digital Communications.* Springer, New York, 2003.

[7] O. Z. Batur, M. Koca, and G. Dundar. Measurements of impulsive noise in broad-band wireless communication channels. In *Research in Microelectronics and Electronics, 2008. PRIME 2008. Ph.D.*, pages 233–236, April 2008.

[8] A. J. Bell and T. J. Sejnowski. An information-maximization approach to blind separation and blind deconvolution. *Neural Computation*, 7:1129–1159, 1995.

[9] S. Bellini. *Blind Deconvolution*, chapter Bussgang Techniques for Blind Deconvolution and Equalization, pages 8–52. Prentice-Hall, Upper Saddle River, NJ, 1994.

[10] A. Benveniste, M. Goursat, and G. Ruget. Robust identification of a nonminimum phase system: Blind adjustment of a linear equalizer in data communications. *IEEE Transactions on Automatic Control*, AC-25(3):385–399, 1980.

[11] C. Berrou, A. Glavieux, and P. Thitimajshima. Near Shannon limit error-correcting coding and decoding: Turbo-codes. *IEEE International Communications Conference*, pages 1064–1070, 1993.

[12] K. L. Blackard, T. S. Rappaport, and C. W. Bostian. Measurements and models of radio frequency impulsive noise for indoor wireless communications. *IEEE Journal on Selected Areas in Communications*, 11(7):991–1001, September 1993.

[13] H. Bölcskei, D. Gesbert, C. B. Papadias, and A. J. van der Veen. *Space-Time Wireless Systems: From Array Processing to MIMO Communications.* Cambridge University Press, Cambridge, 2006.

[14] J.-F. Cardoso. Informax and maximum likelihood for blind source separation. *IEEE Signal Processing Letters*, 4:112–114, April 1997.

[15] I. R. S. Casella. Analysis of turbo coded OFDM systems employing space-frequency block code in double selective fading channels. *IEEE International Microwave and Optoelectronics Conference*, pages 516–520, November 2007.

[16] R. T. Causey. *Blind Multiuser Detection Based on Second-Order Statistics.* PhD thesis, School of Electrical and Computer Engineering - Georgia Institute of Technology, Atlanta, Georgia, 1999.

[17] R. Cendrillon. *Multi-User Signal and Spectra Co-ordination for Digital Subscriber Lines.* PhD thesis, Katholieke Universiteit Leuven, December 2004.

[18] P. Comon. Independent component analysis, a new concept? *Signal Processing*, 36:287–314, 1994.

[19] P. Comon and P. Chevalier. *Unsupervised Adaptive Filtering, Volume I: Blind Source Separation*, volume 1, chapter Blind Source Separation: Models, Concepts, Algorithms and Performance, pages 191–235. John-Wiley and Sons, New York, 2000.

[20] P. Comon and C. Jutten. *Handbook of Blind Source Separation - Independent Component Analysis and Applications*. Elsevier, Oxford, 2010.

[21] S. Cui, R. Heath Jr., and G. Leus. Signal processing for networking and communications. *IEEE Signal Processing Magazine*, 28(5):151–152, 2011.

[22] C. Dick, F. Harris, and M. Rice. Synchronization in software radios - carrier and timing recovery using fpgas. *IEEE Symposium on Field-Programmable Custom Computing Machines*, pages 195–204, 2000.

[23] Z. Ding, R. Kennedy, B. Anderson, and C. Jonhson. Ill- convergence of Godard blind equalizers in data communication systems. *IEEE Transactions on Communications*, 39(9):1313–1327, 1991.

[24] P. Diniz. *Adaptive Filtering: Algorithms and Practical Implementation*. Kluwer Academic Publishers, New York, 1997.

[25] K. Dostert. *Powerline Communications*. Prentice-Hall, Upper Saddle River, NJ, 2001.

[26] T. M. Duman and A. Ghrayeb. *Coding for MIMO Communication Systems*. Wiley, Chichester, 2007.

[27] D. Zanatta Filho, R. R. Lopes, R. Ferrari, M. B. Loiola, R. Suyama, G. C. C. P. Simões, and B. Dortschy. Achievable rates of dsl with crosstalk cancellation. *European Transactions on Telecommunications*, 20:81–86, January 2009.

[28] G. J. Foshini. Layered space-time architecture for wireless communication in a fading environment when using multi-element antennas. *Bell Labs Technical Journal*, 1:41–59, October 1996.

[29] G. J. Foshini and M. J. Gans. On limits of wireless communications in a fading environment when using multi-element antennas. *Wireless Personnal Communications*, pages 311–335, June 1998.

[30] L. E. Franks. Carrier and bit synchronization in data communication — a tutorial review. *IEEE Transactions on Communications*, 28:1107–1121, 2000.

[31] G. Ginis and J. M. Cioffi. Vectored transmission for digital subscriber line systems. *IEEE Journal on Selected Areas in Communications*, 20(5):1085–1104, June 2002.

[32] D. Godard. Self-recovering equalization and carrier tracking in two-dimensional data communication systems. *IEEE Transactions on Communications*, 28(11):1867–1875, 1980.

[33] A. Goldsmith. *Wireless Communications*. Cambridge, 2005.

[34] M. Greenberg, M. Nazarathy, and M. Orenstein. Multimode fiber as random code generator – application to massively parallel mimo transmission. *Journal of Lightwave Technology*, 26(8):882–890, April 2008.

[35] M. Gu and L. Tong. Geometrical characterizations of constant modulus receivers. *IEEE Transactions on Signal Processing*, 47(10):2745–2756, 1999.

[36] S. Haykin. *Adaptive Filter Theory*. Prentice-Hall, Upper Saddle River, NJ, 4th edition, 2002.

[37] S. Haykin and B. Widrow. *Least-Mean-Square Adaptive Filters*. Adaptive and Learning Systems for Signal Processing, Communications and Control Series, Wiley-Interscience, Hoboken, NJ, 2003.

[38] A. Hyvärinen. *Independent Component Analysis - Principles and Practice*, chapter Fast ICA by a Fixed-Point Algorithm That Maximizes Non-Gaussianity, pages 71–94. Cambridge University Press, Cambridge, 2001.

[39] A. Hyvärinen, J. Karhunen, and E. Oja. *Independent Component Analysis*. Adaptive and Learning Systems for Signal Processing, Communications, and Control. Wiley-Interscience, New York, 2001.

[40] M. C. Jeruchim, P. Balaban, and K. S. Shanmugan. *Simulation of Communication Systems*. Kluwer Academic Press, New York, 2nd edition, 2000.

[41] T. Kailath. *Linear Systems*. Prentice-Hall, Upper Saddle River, NJ, 1980.

[42] E. Karipidis, N. Sidiropoulos, A. Leshem, L. Youming, R. Tarafi, and M. Ouzzif. Crosstalk models for short VDSL2 lines from measured 30 MHz data. *EURASIP Journal on Applied Signal Processing*, 2006(85859):785–800, 2006.

[43] V. A. Kotelnikov. *The theory of optimum noise immunity*. Molotov Energy Institute, Moscow, 1947.

[44] E. G. Larsson and P. Stoica. *Space-Time Block Coding for Wireless Communications*. Cambridge University Press, Cambridge, 2003.

[45] B. P. Lathi. *Linear Systems and Signals*. Oxford University Press, Oxford, 2004.

[46] B. P. Lathi and Z. Ding. *Modern Analog and Digital Communication Systems*. Oxford, 2010.

[47] K. B. Letaief, J. C.-I. Chuang, and M. L. Liou. M-PSK and m-QAM ber computation using signal-space concepts. *IEEE Transactions on Communications*, 47:181–184, February 1999.

[48] S. Lin and D. J. Costello Jr. *Error Control Coding*. Prentice-Hall, Upper Saddle River, NJ, 2nd edition, 2004.

[49] D. J. C. MacKay. Good error correcting codes based on very sparse matrices. *IEEE Transactions on Information Theory*, 45:399–431, March 1999.

[50] D. J. C. MacKay and R. M. Neal. Near Shannon limit performance of low-density parity-check codes. *IET Electronics Letters*, 32:1645–1646, 1996.

[51] D. Manolakis, V. Ingle, and S. Kogon. *Statistical and Adaptive Signal Processing*. Artech House, Norwood, MA, 2005.

[52] M. Mboup and P. Regalia. On the equivalence between the super-exponential algorithm and a gradient search method. In *Proc. of IEEE Int. Conference on Acoustic, Speech and Signal Processing*, volume 5, pages 2643–2646, Phoenix, 1999.

[53] M. Mboup and P. Regalia. A gradient search interpretation of the super-exponential algorithm. *IEEE Transactions on Information Theory*, 46(7):2731–2734, 2000.

[54] E. Moulines, J.-F. Cardoso, and E. Gassiat. Maximum likelihood for blind separation and deconvolution of noisy signals using mixture models. In *Proc. Int. Conf. Acoust., Speech, Signal Processing, ICASSP-97, Munich, Germany*, pages 3617–3620, 1997.

[55] M. Nassar, K. Gulati, Y. Mortazavi, and B.L. Evans. Statistical modeling of asynchronous impulsive noise in powerline communication networks. In *Global Telecommunications Conference (GLOBECOM 2011), 2011 IEEE*, pages 1–6, December 2011.

[56] E. Oja. The nonlinear PCA learning rule in independent component analysis. *Neurocomputing*, 17:25–45, 1997.

[57] A. V. Oppenheim and R. W. Schafer. *Discrete-Time Signal Processing*. Prentice-Hall signal processing series. Prentice-Hall, Upper Saddle River, NJ, 2010.

[58] A. V. Oppenheim, A. S. Willsky, and S. Hamid. *Signals and Systems*. Prentice-Hall, Upper Saddle River, NJ, 2nd edition, 1996.

[59] C. B. Papadias. Globally convergent blind source separation based on a multiuser kurtosis maximization criterion. *IEEE Transactions on Signal Processing*, 48(12):3508–3519, 2000.

[60] C. B. Papadias. *Unsupervised Adaptive Filtering*, volume 2, chapter Blind Separation of Independent Sources Based on Multiuser Kurtosis Optimization Criteria, pages 147–149. John Wiley and Sons, New York, 2000.

[61] C. B. Papadias and A. J. Paulraj. A constant modulus algorithm for multiuser signal separation in presence of delay spread using antenna array. *IEEE Signal Processing Letters*, 4(6):178–181, 1997.

[62] A. Papoulis. *Probability, random variables and stochastic processes*. MC, Singapore, 3rd edition, 1991.

[63] A. Papoulis and S. U. Pillai. *Probability, Random Variables and Stochastic Processes*. McGraw-Hill, New York, 2002.

[64] J. C. Principe. *Information Theoretic Learning*. Springer, New York, 2010.

[65] J. Proakis and D. Manolakis. *Digital Signal Processing - Principles, Algorithms and Applications*. Pearson Education, Prentice-Hall, Upper Saddle River, NJ, 4th edition, 2007.

[66] J. G. Proakis. *Digital Communications*. McGraw-Hill, New York, 2008.

[67] T. S. Rappaport. *Wireless Communications Principles and Practice*. Prentice-Hall, Upper Saddle River, NJ, 2002.

[68] P. Regalia. On the equivalence between the Godard and Shalvi Weinstein schemes of blind equalization. *Signal Processing*, 73(1-2):185–190, 1999.

[69] M. Rice. *Digital Communications - A Discrete-Time Approach*. Prentice-Hall, Upper Saddle River, NJ, 2009.

[70] T. Richardson, A. Shokrollahi, and R. Urbanke. Design of capacity-approaching low-density parity-check codes. *IEEE Transactions on Information Theory*, 47:619–637, February 2001.

[71] M. S. Roden. *Analog and Digital Communication Systems*. Prentice-Hall, Upper Saddle River, NJ, 4th edition, 1996.

[72] W. Ryan and S. Lin. *Channel Codes: Classical and Modern*. Cambridge University Press, Cambridge, 2009.

[73] A. Sayed. *Fundamentals of Adaptive Filtering*. John Wiley and Sons, Hoboken, NJ, 2003.

[74] D. Schneider, A. Schwager, J. Speidel, and A. Dilly. Implementation and results of a mimo plc feasibility study. In *Power Line Communications and Its Applications (ISPLC), 2011 IEEE International Symposium on*, pages 54 –59, April 2011.

[75] A. Schwager, D. Schneider, W. Baschlin, A. Dilly, and J. Speidel. Mimo PLC: Theory, measurements and system setup. In *Power Line Communications and Its Applications (ISPLC), 2011 IEEE International Symposium*, pages 48–53, April 2011.

[76] A. R. Shah, R. C. J. Hsu, A. Tarighat, A. H. Sayed, and B. Jalali. Coherent optical mimo (comimo). *Journal of Lightwave Technology*, 23(8):2410–2419, August 2005.

[77] O. Shalvi and E. Weinstein. New criteria for blind deconvolution of non-minimum phase systems (channels). *IEEE Transactions on Information Theory*, 36(2):312–321, 1990.

[78] O. Shalvi and E. Weinstein. Super-exponential methods for blind deconvolution. *IEEE Transactions on Information Theory*, 39(2):504–519, 1993.

[79] O. Shalvi and E. Weinstein. *Blind Deconvolution*, chapter Universal Methods for Blind Deconvolution. Prentice-Hall, Upper Saddle River, NJ, 1994.

[80] E. C. Shannon. A mathematical theory of communication. *Bell Syst. Tech. Journal*, 27:379–423, 1948.

[81] B. Sklar. *Digital Communications*. Prentice-Hall, Upper Saddle River, NJ, 2nd edition, 2004.

[82] R. Suyama, R. Attux, J. M. T. Romano, and M. Bellanger. Relations entre les critères du module constant et de wiener. *19e Colloque GRETSI - Paris - França*, 2003.

[83] V. Tarokh, N. Seshadri, and A. R. Calderbank. Space-time block codes from orthogonal designs. *IEEE Transactions on Information Theory*, 45:1456–1467, July 1999.

[84] G. Tauböck and W. Henkel. Mimo systems in the subscriber-line network. In *Proc. 5th Int. OFDM-Workshop*, pages 18.1–18.3, August 2000.

[85] I. E. Telatar. Capacity of multi-antenna gaussian channels. *European Transactions on Telecommunications*, 10:585–595, November–December 1999.

[86] J. Treichler and B. Agee. New approach to multipath correction of constant modulus signals. *IEEE Transactions on Acoustics, Speech and Signal Processing*, ASSP-31(2):459–472, 1983.

[87] D. Tse and P. Viswanath. *Fundamentals of Wireless Communication*. Cambridge University Press, Cambridge, 2005.

[88] R. F. M. van den Brink. Cable reference models for simulating metallic access networks. Technical Report STC TM6(97)02, ETSI, 1998.

[89] J. J. Werner. The HDSL environment. *IEEE Journal on Selected Areas in Communications*, 9(6):785–800, August 1991.

[90] J. M. Wonzencraft and I. M. Jacobs. *Principles of Communication Engineering*. Waveland, 1990.

[91] H. H. Yang and S. Amari. Adaptive on-line learning algorithms for blind separation-maximum entropy and minimum mutual information. *Neural Computation*, 9:1457–1482, 1997.

[92] H. H. Zeng, L. Tong, and C. R. Johnson. Relationships between the Constant Modulus and Wiener receivers. *IEEE Transactions on Information Theory*, 44:1523–1538, July 1998.

[93] R. E. Ziemer and R. L. Peterson. *Introduction to Digital Communication*. Prentice-Hall, Upper Saddle River, NJ, 2001.

[94] A. Zoubir, V. Krishnamurthy, and A. Sayed. Signal processing theory and methods. *IEEE Signal Processing Magazine*, 28(5):152–156, 2011.

3

Overview of dynamical systems and chaos

Luiz H. A. Monteiro

Escola de Engenharia da Universidade Presbiteriana Mackenzie
Escola Politécnica da Universidade de São Paulo

CONTENTS

Dynamical Systems Theory (DST) is the branch of applied mathematics dedicated to qualitatively characterizing the long-term behavior of systems evolving according to difference and differential equations. DST has its roots in classical mechanics, a major area in physics dealing with dynamics, kinematics, and statics of solids and fluids. Classical mechanics is based on the premise that the future can be predicted from an accurate knowledge of the present; effects can be determined from its causes. Thus, predictability would be a logical consequence of causal determinism. Remarkable works on these subjects were done, for instance, by (e.g., [9,10]) Thales of Miletus ($\pm$624-547 BC), who was the first to forecast a solar eclipse; Aristotle (384-322 BC), who proposed qualitative laws based on circles, spheres, and "natural places" that would rule the movements of earthly and heavenly objects; Archimedes of Syracuse ($\pm$287-212 BC), who conceived the method of integration for calculating areas and a principle related to the equilibrium of floating bodies; Claudius Ptolemy ($\pm$85-165), who elaborated a geocentric model of the universe considered precise during 14 centuries; Nicolaus Copernicus (1473-1543), who proposed a heliocentric model in which the center of the universe would be near the motionless Sun; Galileo Galilei (1564-1642), who concluded from experiments that all bodies (no matter the value of their masses) fall according to the same acceleration in the absence of air resistance[1]; Johannes Kepler (1571-1630), who, based on Copernicus' model and observational data, empirically found out three laws governing the planetary motion; René Descartes (1596-1650), who conjectured about conservation of quantities of motion and wrote about the foundations of the scientific method; Pierre de Fermat (1601-1665), who proposed a method for determining minima and maxima of functions from tangent lines; Evangelista Torricelli (1608-1647), who studied the kinematics of projectiles that are obliquely thrown and asserted that the momentum of a body should be determined from its quantity of matter and its velocity; Christiaan Huygens (1629-1695), who analyzed periodic motions of pendulums and investigated elastic impacts; Isaac Barrow (1630-1677), who recognized differentiation and integration as inverse operations, which is the fundamental theorem of infinitesimal calculus; Isaac Newton (1642-1727), Barrow's student, who wrote the work called "Philosophiae Naturalis Principia Mathematica" in which the concept of "mass" (quantity of matter), a law of universal gravitation, and three laws of motion are introduced, and the three Kepler's laws are deduced[2]; Gottfried W. von Leibniz (1646-1716), who developed the infinitesimal calculus independently of Newton and coined the word "dynamics" in the sense of a theory of motion and its causes; Jacob Bernoulli (1654-1705), who elaborated a method for solving separable differential equations; Johann Bernoulli (1667-1748), who proposed and solved the brachistochrone problem[3]; Pierre L. Maupertuis (1698-1759), who is credited for formulating the principle of least action, by which nature always takes the path minimizing a quantity called "action"; Leonhard Euler (1707-1783), who solved different kinds of ordinary and partial differential equations and introduced the calculus of variations related to maximize or minimize functionals[4]; Jean L.R. d'Alembert (1717-1783), who formulated the principle stating that a system is in dynamic equilibrium if the sum of applied forces and inertial forces vanishes; Joseph L. Lagrange (1736-1813), who developed an original formalism in which the behavior of a system is derived from differential equations written in terms of energy (and not in terms of forces, as in the Newtonian formulation); Pierre S. Laplace (1749-1827), who attempted to demonstrate the stability of the Solar System[5]; Johann C.F. Gauss (1777-1855), who proved the fundamental theorem of algebra about the number of roots of polynomials and supposedly created the method of least squares for solving a kinematics problem: the prediction of the trajectory of the asteroid Ceres; Siméon D.

---

[1]In 1971, during the Apollo 15 mission, the astronaut David R. Scott (1932-) simultaneously released from a same height a geological hammer and a falcon feather and confirmed Galileo's theory when observed that both objects reached the lunar surface at the same time.

[2]Thus, mechanics laws of motion were first employed to explain celestial phenomena.

[3]The aim is to find out the shape of a frictionless wire, joining two points, that minimizes the descent time, between these two points, of a bead constrained to slide due to gravity along this wire.

[4]A functional is a function whose domain is a function space.

[5]According to Newton, Solar System was kept stable due to periodic divine interventions.

Poisson (1781-1840), who investigated differential equations concerning systems of various natures; and William R. Hamilton (1805-1855), who presented an alternative formulation of classical mechanics,[6] with deep implications in quantum mechanics. This paragraph is a tribute to these great scientists, who accomplished many other relevant works (e.g., [9,10]).

The approach originally developed by Newton, Leibniz, the Bernoulli brothers, and Euler of searching for analytical exact solutions for difference and differential equations usually fails if there are nonlinearities. The main goal of DST is to qualitatively determine the dynamics in the permanent regime of systems described by discrete and continuous-time nonlinear models.

Five scientists (among others) who made important works laying the foundations of the DST are Jules H. Poincaré (1854-1912), Aleksandr M. Lyapunov (1857-1918), George D. Birkhoff (1884-1944), Aleksandr A. Andronov (1901-1952), and Andrey N. Kolmogorov (1903-1987). In fact, DST owes much to (e.g., [9,35]) the Poincaré's analysis, from a geometrical/topological perspective, of the dynamics of three mutually attracting (celestial) bodies, finished in 1890; the Lyapunov's doctoral thesis on stability theory defended in 1892; the Birkhoff's investigations on *homoclinic tangles*[7] done in 1913; a book [3] co-authored by Andronov about oscillations first published in 1937; the notion of structural stability ("robustness") introduced by Andronov and Lev S. Pontryagin (1908-1988) also in 1937; and a theorem proposed by Kolmogorov in 1954 about the effects of a small nonlinear perturbation on an integrable Hamiltonian system, concerning the preservation of quasiperiodic solutions of the unperturbed equations. In the early 1960s, this problem was rigorously investigated by Vladimir I. Arnold (1937-2010) in a continuous-time context and by Jürgen K. Moser (1928-1999) in a discrete-time system, resulting in the *KAM theory* (e.g., [9]).

Typically, an autonomous dynamical system (that is, a system in which its time evolution does not explicitly depend on the time variable t) exhibits one of the following behaviors in permanent regime (that is, when $t \to \infty$): unlimited solution, steady state, periodic or quasi-periodic behavior, or chaotic movement. *Chaos* is a limited and aperiodic behavior produced by a deterministic system exhibiting sensitive dependence on initial conditions. Thus, small changes in the initial conditions lead to aperiodic solutions that (exponentially) diverge while remaining bounded. In this sense, the word "chaos" was first employed by Tien Y. Li (1945-) and James A. Yorke (1941-), in 1975, in a paper about the dynamics of first-order difference equations [21]. In practice, long-term predictions in chaotic systems cannot be valid, because the accuracy of measurements of initial conditions is always limited.

Theoretically, chaos was supposed to appear in systems with a great num-

---

[6]In Hamiltonian mechanics, the dynamics of a system with m freedom degrees is represented by $2m$ first-order differential equations, instead of m second-order differential equations as in Newtonian and Lagrangian mechanics.

[7]Homoclinic tangle is a kind of "stochastic" or chaotic layer appearing in the neighborhood of an unstable periodic orbit of a Hamiltonian system affected by a small perturbation.

ber of variables (for instance, a gas composed of molecules) by James C. Maxwell (1831-1879) in an essay published in 1873, in which he claimed the impossibility of predicting future events when the "knowledge of the present state is only approximate, and not accurate" (e.g., [6]). Chaos was also predicted to occur in systems with a small number of variables by Poincaré, in his work about the three-body problem. In order to analyze this problem, he introduced many of the concepts concerning DST, such as bifurcation, phase portrait, and Poincaré section (explained in the next sections). By 1900, Poincaré philosophically examined the relation between long-term predictability and sensitivity to initial conditions (e.g., [9]). Such a sensitivity was also found by Jacques S. Hadamard (1865-1963) in 1898, when he investigated the movement of a free particle in a surface remembering the number eight (e.g., [35]).[8] Stephen Smale (1930-), during his postdoctoral research in Rio de Janeiro, conceived a chaotic discrete-time system called *horseshoe map*, an abstraction of the dynamics in a homoclinic tangle [38]. This map presents a strange attractor. The attractor is named "strange" because it corresponds to a set with non-integer Hausdorff dimension. Objects with this feature had already been generated by routines envisaged by (e.g., [37]) George F.L.P. Cantor (1845-1918), Niels F.H. von Koch (1870-1924), and Waclaw Sierpinski (1882-1969); however, by 1960 no one knew that a fractal could be created by the asymptotical behavior of a dynamical system, when such a behavior is plotted in the corresponding phase space. The term "fractal" was proposed by Benoit B. Mandelbrot (1924-2010) in 1975 (e.g., [37]) to designate an object with details on arbitrary small scales, some degree of self-similarity, and fractional Hausdorff dimension. The name "strange attractor" was first used in a theoretical manuscript by David P. Ruelle (1935-) and Floris Takens (1940-2010) to designate the attractor that would characterize a fluid in turbulent (chaotic) regime [36]. Usually, a strange attractor is a geometric manifestation of chaotic dynamics.[9] In 1976, Robert M. May (1938-) presented an extensive analysis of the *logistic map* $x(t + 1) = px(t)(1 - x(t))$ for $0 \leq p \leq 4$, which is considered a paradigmatic example of "simple mathematical models with very complicated dynamics" [23].

Numerically, chaos was found by John von Neumann (1903-1957) and Stanislaw M. Ulam (1909-1984) by simulating in a computer the logistic map with $p = 4$, which was used, in 1947, as a random number generator. The title of their short paper is "On Combination of Stochastic and Deterministic Processes," because apparently random sequences were obtained from a deterministic equation [44]. In time-continuous systems, numerical chaotic solutions

---

[8]In this investigation, he used *symbolic dynamics*, a way of (grossly) describing the time evolution of a dynamical system by a sequence of symbols, in which each symbol corresponds to a specific region visited in the corresponding phase space.

[9]There are exceptions. For instance [34], a damped pendulum forced by the quasiperiodic stimulus $g(t) = g_0 + g_1 \cos(l_1 t) + g_2 \cos(l_2 t)$ (l_1/l_2 is a irrational number and g_0, g_1, and g_2 are constants) can produce a strange nonchaotic attractor; that is, a strange attractor (an attractor with fractional Hausdorff dimension) without sensitivity to initial conditions (no positive Lyapunov exponent).

were reported in 1963 by Edward N. Lorenz (1917-2008) in the manuscript "Deterministic Non-periodic Flow," about a hydrodynamic model for meteorology, which was solved in a computer [22]. In this context, the *butterfly effect* is a poetic metaphor of sensitivity to initial conditions: the flight of a butterfly certainly causes tiny alterations in the atmosphere; however, if this system produces chaotic solutions, then such minor changes can result in completely different weather scenarios. Observe that, in simulations, rounding and truncation errors always occur; therefore, a numerically calculated chaotic solution will diverge from the true solution with the same initial conditions. Under certain circumstances, the *shadowing lemma* assures that there is a true solution with slightly different initial coordinates that stays near ("shadows") the numerically computed solution (e.g., [16]). Thus, the characterization of a chaotic system from data obtained via simulations can be valid. This characterization is usually based on its spatio-temporal "statistical" properties (as in truly stochastic processes). Some tools employed are analysis of *power spectrum density*, *entropy* derived from Information Theory, and *invariant measure* provided by Ergodic Theory (e.g., [11]). Other tools are Lyapunov exponent and Hausdorff dimension, which will be defined in the next sections.

Experimentally, chaos was detected by Balthasar van der Pol (1889-1959) and Jan van der Mark (1893-1961) in 1927, in an electric circuit described by a second-order nonlinear differential equation with a sinusoidal forcing. However, this detection was recognized as a chaotic signal only in the last decades. Instead of the word "chaos," they wrote to notice an "irregular noise" [45]. A chaotic electrical circuit represented by a third-order autonomous nonlinear differential equation was intentionally conceived by Leon O. Chua (1936-) in 1983; thus, chaos was observed in the laboratory by using an oscilloscope (e.g., [7]). This observation reinforced the idea that chaos is a real physical behavior and not a collective numerical hallucination produced by computers or delirious theoretical researchers. The strange chaotic attractor found by Lorenz was proved to really exist almost 40 years after his numerical simulations [43].

The possibility of modeling a n-dimensional dynamical system from experimental records of a single variable was investigated by Takens [41], Norman H. Packard (1954-) and colleagues [30], and others. These investigations strongly expanded the applications of DST in practical cases (e.g., [1]). For instance, by using the search machine ISI Web of Science in the end of 2012, the topics chaos and nonlinear dynamics were related to more than 142,000 papers (about 8,000 papers were published in 2012). Moreover, the best-selling book *Chaos: Making a New Science* [15] written by James Gleick (1954-) and the blockbuster movie "Jurassic Park" [39] directed by Steven A. Spielberg (1946-) helped to popularize these subjects to the non-specialized public.

Differential equations were developed in the 17th century to describe motions of macroscopic objects, like pendulums and planets; difference equations appeared much earlier, for instance, to represent the time evolution of the number of rabbits leading to the famous Fibonacci sequence in 1202 (known by

Indian mathematics several centuries before) and to describe an algorithm created by Babylonians for numerically computing square roots (e.g., [46]). Such kinds of equations can also be employed to model the dynamics of molecules in a chemical reaction, the spreading of a contagious disease in a host population, the electrical activity of neurons composing a nervous system, and the financial situation of a family or a country (e.g., [25]). These examples illustrate the interdisciplinary nature of DST and, in this book, applications on Communication Engineering are shown in several chapters. Other potential applications of chaos theory in the "real world" include, for instance, monitoring of physiologic aging (e.g., [20]) and modeling of financial markets (e.g., [18]).

Here, a simplistic overview of nonlinear dynamical systems and chaos is presented with a minimum of mathematical formalism. Five excellent books are the references [2,4,17,28,40]. The following five sections are based on these books and also in [25]. The electronic device called phase-locked loop (PLL) is used to exemplify several aspects of this theory in the penultimate section of this chapter.

3.1   State-variable representation and some terminology

This section is primarily based on [2, 4, 25, 40]. Let the *state* of a system in the time t be represented by the vector $\mathbf{x}(t) = (x_1(t), x_2(t), ..., x_n(t)) \in \mathbb{R}^n$, in which $x_j(t)$ ($j = 1, 2, ..., n$) are the *state variables* of this system. A *dynamical system* corresponds to a set of equations describing the time evolution of the state $\mathbf{x}(t)$ from the initial condition $\mathbf{x}(0) = \mathbf{x_0}$. Its *phase space* (or *state space*) is formed by the axis-x_1, the axis-x_2, ..., the axis-x_n. In this n-dimensional space, the time evolution from an initial condition $\mathbf{x_0}$ corresponds to a *trajectory* if this is a continuous-time system, or an *orbit* if this is a discrete-time system.[10] The *order* of the system is equal to n. In DST, qualitative information about dynamics is extracted by sketching trajectories or orbits. Such a geometric representation made in the corresponding phase space is called *phase portrait* of the system.

Assume that the dynamical evolution of a continuous-time system ($t \in \mathbb{R}_+$) is represented by the following n first-order differential equations

$$
\begin{aligned}
\dot{x}_1(t) &= f_1(x_1(t), x_2(t), ..., x_n(t), t) \\
\dot{x}_2(t) &= f_2(x_1(t), x_2(t), ..., x_n(t), t) \\
&\vdots \qquad \vdots \\
\dot{x}_n(t) &= f_n(x_1(t), x_2(t), ..., x_n(t), t)
\end{aligned}
$$

---

[10]Some authors use "trajectory" and "orbit" as synonyms.

in which $\dot{x}_j(t) \equiv dx_j(t)/dt$ $(j = 1, 2, ..., n)$. These equations can be rewritten in terms of the *vector field* $\mathbf{f} = (f_1, f_2, ..., f_n)$ as $\dot{\mathbf{x}}(t) = \mathbf{f}(\mathbf{x}(t), t)$ with $f_j : D \times \mathbb{R}_+ \mapsto C$, and $D \subseteq \mathbb{R}^n$ and $C \subseteq \mathbb{R}$.

Suppose that the dynamics of a discrete-time system $(t \in \mathbb{Z}_+)$ is written in terms of the following set of n first-order difference equations

$$
\begin{aligned}
x_1(t+1) &= f_1(x_1(t), x_2(t), ..., x_n(t), t) \\
x_2(t+1) &= f_2(x_1(t), x_2(t), ..., x_n(t), t) \\
&\vdots \qquad \vdots \\
x_n(t+1) &= f_n(x_1(t), x_2(t), ..., x_n(t), t)
\end{aligned}
$$

or, in a vectorial form, as $\mathbf{x}(t+1) = \mathbf{f}(\mathbf{x}(t), t)$ with $f_j : D \times \mathbb{Z}_+ \mapsto C$, and $D \subseteq \mathbb{R}^n$ and $C \subseteq \mathbb{R}$. In time-discrete systems, $\mathbf{f}$ is also called *map*.

In both kinds of systems, if $\mathbf{f}$ explicitly depends on t, then the system is called *nonautonomous*; if $\mathbf{f} = \mathbf{f}(\mathbf{x}(t))$, it is *autonomous*. Here, only the autonomous case is considered.[11] The autonomous versions of the systems above are written as $\dot{\mathbf{x}}(t) = \mathbf{f}(\mathbf{x}(t))$ and $\mathbf{x}(t+1) = \mathbf{f}(\mathbf{x}(t))$.

The vector field of a *linear* autonomous system is given by $\mathbf{f}(\mathbf{x}(t)) = \mathbf{A}\mathbf{x}(t) + \mathbf{B}$, with $\mathbf{A} \in \mathbb{R}^{n \times n}$ and $\mathbf{B} \in \mathbb{R}^n$. If an autonomous system cannot be represented by this expression, then it is *nonlinear*, as in the three-body problem or a simple pendulum subject only to the gravitational torque. Chaos can appear only in nonlinear or piecewise linear systems.

An autonomous continuous-time dynamical system is considered *dissipative* if $\nabla \cdot \mathbf{f} < 0$, *conservative* if $\nabla \cdot \mathbf{f} = 0$, or *expansive* if $\nabla \cdot \mathbf{f} > 0$. Thus, the sign of $\nabla \cdot \mathbf{f}$ reveals if a n-dimensional volume of initial conditions in the phase space is, respectively, contracted, preserved, or increased as the time passes. Let the (i, j)-th element of the *Jacobian matrix* $\mathbf{J}(\mathbf{x}) \in \mathbb{R}^{n \times n}$ be given by $\partial f_i(\mathbf{x})/\partial x_j$. Notice that $\nabla \cdot \mathbf{f} = \mathrm{trace}(\mathbf{J})$. An autonomous discrete-time system is dissipative if $|\det(\mathbf{J})| < 1$, conservative if $|\det(\mathbf{J})| = 1$, or expansive if $|\det(\mathbf{J})| > 1$. Such a matrix $\mathbf{J}$ was named Jacobian in honor of Carl G. J. Jacobi (1804-1851).

A closed and bounded subset A of the phase space is called *attractor* if trajectories or orbits starting from initial conditions pertaining to an open neighborhood U converge to A for future times. The largest set U of initial conditions satisfying such a convergence criterion is called *basin of attraction* of A. The subset A is *invariant* in the sense that trajectories or orbits starting in A will remain confined in A forever. Attractors can exist only in dissipative systems. In the next three sections, some kinds of attractors are presented.

---

[11]By defining $x_{n+1} \equiv t$, a nonautonomous system can be represented by an autonomous vector field in a $(n+1)$-dimensional phase space.

3.2   Steady state and Lyapunov stability

This section is primarily based on [4,17,25,28]. A *steady-state solution* $\mathbf{x}(t) = \mathbf{x}^*$, represented by a point in the phase space, is called *equilibrium point* in continuous-time systems or *fixed point* in discrete-time systems.[12] The steady states of $\dot{\mathbf{x}}(t) = \mathbf{f}(\mathbf{x}(t))$ are obtained from $\mathbf{f}(\mathbf{x}^*) = 0$ (because this implies $\dot{\mathbf{x}}(t) = 0$); the steady states of $\mathbf{x}(t+1) = \mathbf{f}(\mathbf{x}(t))$ are calculated from $\mathbf{f}(\mathbf{x}^*) = \mathbf{x}^*$ (because, in this case, $\mathbf{x}(t+1) = \mathbf{x}(t) = \mathbf{x}^*$).

According to Lyapunov, $\mathbf{x}^*$ is a *locally neutrally stable* point if, in the phase space, given $\epsilon > 0$, there is an open ball (an open n-dimensional sphere) of radius $\delta(\epsilon) > 0$ centered in $\mathbf{x}^*$ such that starting from any point $\mathbf{x_0}$ pertaining to this ball, then $\mathbf{x}(t)$ remains within a ball of radius ϵ also centered in $\mathbf{x}^*$ for $t > 0$. Thus, trajectories or orbits beginning within a distance δ of $\mathbf{x}^*$ remain within a distance ϵ of $\mathbf{x}^*$ for $t > 0$. If there is no value of δ satisfying this condition, then $\mathbf{x}^*$ is an *unstable* point. If there is a ball of radius $\delta > 0$ centered in $\mathbf{x}^*$ such that, beginning in any point pertaining to this ball then $\mathbf{x}(t) \to \mathbf{x}^*$ for $t \to \infty$, the point $\mathbf{x}^*$ is considered *locally asymptotically stable*. In this case, any trajectories or orbits starting from any initial condition $\mathbf{x_0}$ within a distance δ of $\mathbf{x}^*$ eventually converges to $\mathbf{x}^*$; that is, for $\sqrt{\sum_{j=1}^{n}(x_{0j} - x_j^*)^2} < \delta$, then $\sqrt{\sum_{j=1}^{n}(x_j(t) - x_j^*)^2} \to 0$ for $t \to \infty$. Roughly speaking, neutral stability of $\mathbf{x}^*$ means bounded movements around this point; asymptotical stability implies convergence to this point; instability leads to divergence.

A way of trying to determine the local stability of $\mathbf{x}^*$ involves the use of *Lyapunov indirect method*. In this method, the vector field $\mathbf{f}(\mathbf{x})$ is linearized around $\mathbf{x}^*$. According to the indirect method, an equilibrium point $\mathbf{x}^*$ of $\dot{\mathbf{x}}(t) = \mathbf{f}(\mathbf{x}(t))$ is locally asymptotically stable if all eigenvalues $\lambda_j \in \mathbb{C}$ ($j = 1, 2, ..., n$) of the matrix $\mathbf{J}$ computed in $\mathbf{x}^*$ have negative real parts. If there is at least one eigenvalue with a positive real part, then $\mathbf{x}^*$ is unstable. Also, according to the indirect method, a fixed point $\mathbf{x}^*$ of $\mathbf{x}(t + 1) = \mathbf{f}(\mathbf{x}(t))$ is locally asymptotically stable if all eigenvalues λ_j of the matrix $\mathbf{J}$ calculated in $\mathbf{x}^*$ have moduli less than one. If there is at least one eigenvalue with modulus greater than one, then $\mathbf{x}^*$ is unstable. This method is inconclusive concerning the stability of $\mathbf{x}^*$ if there exists λ_j with zero real part in continuous-time system; or if there exists λ_j with unit modulus in discrete-time system.

An equilibrium point is *hyperbolic* if all its eigenvalues have nonzero real parts. A theorem analyzed by David M. Grobman (1922-) and Philip Hartman (1915-) states that the phase portrait near a hyperbolic equilibrium point of a nonlinear dynamical system is topologically equivalent (homeomorphic) to the phase portrait of its linearized version around such a point. *Topolog-*

---

[12]Some authors use "equilibrium point" and "fixed point" as synonyms.

ical equivalence implies the existence of a *homeomorphism*[13] preserving the time orientation. Phase portraits with the same topological structure represent similar dynamical behaviors. The *Hartman-Grobman theorem*, proved by 1960, supports the conclusions about asymptotical stability and instability derived from the Lyapunov indirect method. This theorem is also applicable to a hyperbolic fixed point; that is, a fixed point (of a map) for which all eigenvalues have moduli different from one.

In continuous-time systems, the eigenvalues λ_j obtained by linearizing $\mathbf{f}(\mathbf{x})$ in $\mathbf{x}^*$ obey the relation $n_s+n_u+n_c = n$, in which n_s is the number of eigenvalues with negative real part, n_u is the number of eigenvalues with positive real part, and n_c is the number of eigenvalues with real part equal to zero. From these eigenvalues λ_j, the corresponding eigenvectors $\mathbf{v}_j$ can be determined from $\mathbf{J}(\mathbf{x}^*)\mathbf{v}_j = \lambda_j\mathbf{v}_j$. The eigenvectors associated with those n_s eigenvalues with negative real part span the *stable subspace* E^s; the eigenvectors related to those n_u eigenvalues with positive real part span the *unstable subspace* E^u; the eigenvectors concerning those n_c eigenvalues with zero real part span the *center subspace* E^c. In discrete-time system, E^s is spanned by the eigenvectors whose eigenvalues have moduli less than one; E^u is spanned by the eigenvectors whose eigenvalues have moduli greater than one; and E^c is spanned by the eigenvectors whose eigenvalues have moduli equal to one. Notice that $E^s \oplus E^u \oplus E^c = \mathbb{R}^n$; thus, $\mathbb{R}^n$ can be decomposed as the direct sum of the invariant subspaces E^s, E^u, and E^c.

In 1967, Allen F. Kelley Jr. (1931-) proved that, in the phase space of the original nonlinear system, there is a unique n_s-dimensional *stable manifold*[14] W^s tangent to E^s in $\mathbf{x}^*$, and there is a unique n_u-dimensional *unstable manifold* W^u tangent to E^u in $\mathbf{x}^*$. If $\mathbf{x}_0 \in W^s$ (or if $\mathbf{x}_0 \in E^s$), then $\mathbf{x}(t) \to \mathbf{x}^*$ for $t \to \infty$ (there is convergence to $\mathbf{x}^*$ as the time passes); if $\mathbf{x}_0 \in W^u$ (or if $\mathbf{x}_0 \in E^u$), then $\mathbf{x}(t) \to \mathbf{x}^*$ for $t \to -\infty$ (there is divergence from $\mathbf{x}^*$ as the time progresses). These results are in agreement with the Hartman-Grobman theorem. Observe that if $n_u > 0$, $\mathbf{x}^*$ is unstable; if $n_c = 0$, $\mathbf{x}^*$ is hyperbolic; if $n_c > 0$, $\mathbf{x}^*$ is non-hyperbolic. Kelley also showed that for $n_c > 0$ there is a n_c-dimensional *center manifold* W^c tangent to E^c in $\mathbf{x}^*$; however, W^c needs not be unique. Notice that if $n_u = 0$, then $n_s + n_c = n$ and the trajectories or orbits will tend to W^c for $t \to \infty$. Thus, the analysis can be restricted to W^c, a space with n_c dimensions; that is, the analysis (near an equilibrium or fixed point) can be performed in a reduced system of equations. In 1981, Jack Carr (1948-) showed how to determine the dynamics on W^c, which discloses the Lyapunov stability of $\mathbf{x}^*$ when the indirect method fails. In fact, solutions on W^c can converge to $\mathbf{x}^*$, diverge from $\mathbf{x}^*$, oscillate around $\mathbf{x}^*$, or remain constant as the time goes by. The behavior on W^c depends on nonlinear terms of $\mathbf{f}$ (while the dynamics on W^s and W^u are ruled by the linear approxima-

---

[13]A homeomorphism is a continuous one-to-one and onto function with a continuous inverse mapping one phase portrait onto the other.

[14]Simplistically, a k-dimensional manifold can be thought as a set which near each point has the structure of ("resembles") the Euclidian space $\mathbb{R}^k$.

tion of **f**). Center manifold reduction and normal form theory simplify the characterization of local bifurcations.

Another way of analytically investigating the stability of $\mathbf{x}^*$ is via *Lyapunov direct method*. It is called "direct" because the original equations (and not their linearized versions) are directly utilized in the analysis. In this method, it is necessary to find a function $V(\mathbf{x}) : D \mapsto C$, with $D \subseteq \mathbb{R}^n$ and $C \subseteq \mathbb{R}$, such that it is null in $\mathbf{x}^*$, it is positive in a neighborhood D of $\mathbf{x}^*$, and, as the time progresses, its value does not increase. A function with these features is named *Lyapunov function*. Suppose that the value of V decreases with time; thus, $\dot{V}(\mathbf{x}(t)) < 0$ for continuous-time systems, or $V(\mathbf{x}(t+1)) - V(\mathbf{x}(t)) < 0$ for discrete-time systems. As the minimum of $V(\mathbf{x})$ coincides with $\mathbf{x}^*$, from any initial state $\mathbf{x_0} \in D$, then $\mathbf{x}(t) \to \mathbf{x}^*$ for $t \to \infty$; consequently, $\mathbf{x}^*$ is locally asymptotically stable. If $D = \mathbb{R}^n$, then $\mathbf{x}^*$ is *globally* asymptotically stable. Variations of this method can be used to prove instability or neutral stability of $\mathbf{x}^*$.

Consider the differential equation $\dot{x}(t) = f(x(t)) = -3(px(t) + x(t)^3)$, with $x \in \mathbb{R}$. If $p = 1$, this continuous-time nonlinear system presents a unique equilibrium point given by $x^* = 0$. By using the indirect method, this point is shown to be locally asymptotically stable because $\lambda = df(x)/dx|_{x=x^*} = -3 < 0$; therefore, the unique eigenvalue λ is negative. Notice that this is a first-order system; hence, the Jacobian matrix is formed by a unique element $J = df(x)/dx = -(3 + 9x^2)$. By taking $V(x) = x^2/6$, the direct method reveals that, in fact, $x^* = 0$ is globally asymptotically stable (that is, any trajectory converges to $x^* = 0$), because $V(0) = 0$ and, for $x \in \mathbb{R} - \{0\}$, then $V(x) > 0$ and $\dot{V}(x(t)) = (dV/dx)\dot{x} = -x^2 - x^4 < 0$. Such a function $V(x)$ is a Lyapunov function. Moreover, this is a dissipative system, because $\nabla \cdot \mathbf{f} = df(x)/dx < 0$ for $x \in \mathbb{R} - \{0\}$. If $p = 0$, $x^* = 0$ is non-hyperbolic and along the center manifold (the axis-x) $x(t) \to 0$ for $t \to \infty$ (because $\dot{x} < 0$ for $x_0 > 0$, and $\dot{x} > 0$ for $x_0 < 0$). Another example: the difference equation $x(t+1) = f(x(t)) = -3(x(t) + x(t)^3)$, with $x \in \mathbb{R}$, presents a unique fixed point $x^* = 0$, which is proved to be unstable by the indirect method, because $|\lambda| = |df(x)/dx|_{x=x^*} = 3 > 1$. This discrete-time system is expansive and nonlinear.

As stressed by Valery I. Oseledets (1940-) in the middle of the 1960s, the *Lyapunov exponents* $L_j \in \mathbb{R}$ ($j = 1, 2, ..., n$) are numbers expressing the behavior of the solutions $\mathbf{x(t)}$ and $\mathbf{x'(t)}$, with slightly different initial conditions, along n orthogonal directions in the phase space. Thus, $\delta_j(t) \approx \delta_j(0)e^{L_j t}$, in which $\delta_j(t) = |x_j(t) - x_j'(t)|$; hence, $L_j \approx \lim_{t \to \infty}(1/t)\ln(\delta_j(t)/\delta_j(0))$. These exponents are numerically calculated from $\mathbf{J}$ computed along a solution and they characterize sensitivity to initial conditions, measuring how rapidly solutions starting from very close initial states exponentially converge or diverge in the phase space. Along the j-th direction, $L_j < 0$ implies convergence, $L_j > 0$ implies divergence, and if such a direction is tangent to a trajectory not end-ing in an equilibrium point, then[15] $L_j = 0$. Moreover, the sum of Lyapunov exponents presents the same sign of trace($\mathbf{J}$) in continuous-time systems and the same sign of $\ln |\det(\mathbf{J})|$ in discrete-time systems. Therefore, $\sum_{j=1}^{n} L_j < 0$ in dissipative system, $\sum_{j=1}^{n} L_j = 0$ in conservative systems, and $\sum_{j=1}^{n} L_j > 0$ in expansive systems.

In spite of the Lyapunov exponents L_j and the real part of the eigenvalues λ_j giving similar qualitative information about stability, there are important differences between them: L_j are real numbers associated to mutually orthogonal directions and obtained from the linearization of $\mathbf{f}$ along trajectories or orbits; the eigenvalues λ_j are complex numbers associated to directions that are usually not orthogonal and obtained from the linearization of $\mathbf{f}$ in an equilibrium or fixed point.

In continuous-time systems with $\nabla \mathbf{f} = \text{trace}(\mathbf{J}) = \text{constant}$ and in discrete-time systems with $\ln |\det(\mathbf{J})| = \text{constant}$, then $\sum_{j=1}^{n} L_j$ will be equal to such constants in any point of the phase space. If these quantities obtained from $\mathbf{J}$ depend on $\mathbf{x}$, then they correspond to the sum of L_j in $\mathbf{x}$. Mean values of L_j can be obtained by averaging such quantities along trajectories or orbits.

All Lyapunov exponents calculated by taking a n-dimensional volume of initial conditions in the basin of attraction of an asymptotically stable equilibrium or fixed point are negative numbers, meaning contraction of this volume in all directions of the phase space.

3.3 Periodic and quasiperiodic solutions

This section is primarily based on $[2, 4, 25, 28]$. Let $\mathbf{x}(t)$ be a *periodic* solution of period T of $\dot{\mathbf{x}}(t) = \mathbf{f}(\mathbf{x}(t))$ or $\mathbf{x}(t+1) = \mathbf{f}(\mathbf{x}(t))$; thus, $\mathbf{x}(t) = \mathbf{x}(t+T)$. Such a solution $\mathbf{x}(t)$ is asymptotically stable if, in the phase space, trajectories or orbits starting in a neighborhood of $\mathbf{x}(t)$ converge to $\mathbf{x}(t)$ as $t \to \infty$. In an autonomous continuous-time system, a periodic solution corresponds to a closed trajectory in the phase space. In an autonomous discrete-time system, a periodic solution is an orbit composed by T distinct points that are consecutively visited. A *limit cycle* is an isolated closed trajectory or orbit; that is, there exists no closed trajectories or orbits in the neighborhood of a limit cycle.

For instance, $x(t) = A\sin(t)$ is a periodic solution of the differential equation $\ddot{x}(t) + x(t) = 0$, in which A is the amplitude and $T = 2\pi$ is the period of the oscillation. By defining $x_1(t) \equiv x(t)$ and $x_2(t) \equiv \dot{x}(t)$ as state variables, then $\dot{x}_1(t) = x_2(t)$ and $\dot{x}_2(t) = -x_1(t)$; thus $dx_1/dx_2 = -x_2/x_1$, consequently, $x_1(t)^2 + x_2(t)^2 = A^2 = x_1(0)^2 + x_2(0)^2$, which is a circumfer-

---

[15]Thus, the Lyapunov exponent associated with this kind of trajectory is null (indicating neither decay nor grow of the distance between the dynamical evolutions starting from two nearby initial conditions along such a trajectory).

ence (obviously, a closed trajectory). The radius A is determined from the
initial state $\mathbf{x}(0) = (x_1(0), x_2(0))$. Observe that these circumferences are not
limit cycles (because, in any neighborhood of a circumference, another circumference can be found). In fact, limit cycle can only appear in nonlinear or
piecewise linear systems and the differential equation $\ddot{x}(t) + x(t) = 0$ is linear. As an example of periodic solution of a discrete-time system, notice that
the numbers 0 and -1 form an orbit of period $T = 2$ obeying the difference
equation $x(t + 1) = f(x(t)) = x(t)^2 - 1$, because $f(0) = -1$ and $f(-1) = 0$.

Consider a discrete-time system ruled by $\mathbf{x}(t + 1) = \mathbf{f}(\mathbf{x}(t))$ with an orbit
of period $T = 2$ formed by the states $\mathbf{x}_a$ and $\mathbf{x}_b$; therefore, $\mathbf{f}(\mathbf{x}_a) = \mathbf{x}_b$ and
$\mathbf{f}(\mathbf{x}_b) = \mathbf{x}_a$. Notice that $\mathbf{f}(\mathbf{f}(\mathbf{x}_a)) = \mathbf{x}_a$ and $\mathbf{f}(\mathbf{f}(\mathbf{x}_b)) = \mathbf{x}_b$; thus, $\mathbf{x}_a$ and
$\mathbf{x}_b$ are fixed points of $\mathbf{f}(\mathbf{f}(\mathbf{x})) \equiv \mathbf{f}^{(2)}(\mathbf{x})$ and the stability of this orbit can
be derived from the Jacobian matrix $\mathbf{J}$ obtained from this vector field. For
instance, as in the last paragraph, take $f(x) = x^2 - 1$. The fixed points x^*
calculated from $f(x^*) = x^*$ are $(1 + \sqrt{5})/2$ and $(1 - \sqrt{5})/2$, which are unstable
because, respectively, $\lambda = 1 + \sqrt{5}$ and $\lambda = 1 - \sqrt{5}$ (for both points, $|\lambda| > 1$).
In order to analytically find orbits of period 2, it is necessary to solve the
equation $f(f(x)) = x$. In this case, $f(f(x)) = f^{(2)}(x) = (x^2 - 1)^2 - 1$. Two
roots of $f^{(2)}(x) = x$ are the fixed points; other two roots are $x_a = 0$ and
$x_b = -1$. The states $x_a = 0$ and $x_b = -1$ compose the orbit of period 2
mentioned in the last paragraph, which is asymptotically stable because $|\lambda| =$
$|df^{(2)}(x)/dx|_{x=x_a \text{ or } x_b} < 1$. This reasoning can be extended: for determining
an orbit of period T of $\mathbf{x}(t + 1) = \mathbf{f}(\mathbf{x}(t))$, it is necessary to find the roots of
$\mathbf{f}^{(T)}(\mathbf{x}) = \mathbf{x}$ and the corresponding stability is derived from the eigenvalues of
the Jacobian matrix obtained from $\mathbf{f}^{(T)}(\mathbf{x})$.

Because in discrete-time systems there is an analytical procedure for computing periodic solutions, Poincaré proposed in 1881 to study the dynamical
behavior of a n-dimensional continuous-time system from a $(n-1)$-dimensional
discrete-time system. Thus, continuous-time systems are transformed into
discrete-time systems and the dimension of the phase space is reduced from
n to $n - 1$. For instance, a system represented by three first-order differential
equations can be written as a two first-order difference equations by describing
the three-dimensional flow from its intersections with a two-dimensional surface. The $(n-1)$-dimensional surface called *Poincaré section* must be transversal (not tangent) to the original flow. The discrete-time system obtained this
way is denominated *Poincaré map*. For instance, the trajectories of a system
governed by two first-order differential equations can be sketched in a two-
dimensional phase space. Now, take a line in this plane as a Poincaré section.
The successive crossings of these trajectories with such a line can be represented by a first-order map $x(t + 1) = f(x(t))$. An orbit of period T of this
map corresponds to a closed trajectory of the original system and its stability
can be inferred from $f^{(T)}(x)$. Usually, it is impossible to analytically derive
the exact form of a Poincaré map and this tool is commonly employed in
numerical investigations.

In time-continuous systems, trajectories can lie on (hyper)surfaces. For instance, in a three-dimensional system, a (two-dimensional) surface of a torus can be an invariant set. A closed trajectory lying on this toroidal surface corresponds to a periodic motion; a trajectory that runs on such a surface and never closes itself implies a *quasiperiodic* motion. To clarify this issue, consider a plane perpendicularly crossing this two-dimensional torus, such that the intersection between them is a circumference. Such a plane can be viewed as a Poincaré section. Suppose that consecutive intersections of a trajectory with this vertical plane are represented by the Poincaré map $x_1(t+1) = x_1(t)$, $x_2(t+1) = x_2(t) + 2\pi l$. Here, $x_1(t)$ is the radial position and $x_2(t)$ is the angular position of an intersection in the time step t. The initial condition is given by $(x_1(0), x_2(0))$. Because the trajectories lie on the toroidal surface, their radial positions are kept constant (these intersections occur on the circumference of radius $x_1(0)$). Notice that if l is a rational number, that is, $l = l_1/l_2$ $(l_1, l_2 \in \mathbb{Z}_+)$, then the orbit of this map returns to its initial angular position $x_2(0)$ after l_2 time steps. Consequently, the trajectory of the continuous-time system closes itself and it corresponds to a periodic motion. However, if l is an irrational number (l cannot be written as $l = l_1/l_2$), then an orbit of the difference equations never returns to the initial state. Therefore, the trajectory of the continuous-time system never closes itself; in fact, it densely fills the toroidal surface, which is interpreted as a quasiperiodic behavior. In this case, any state is never exactly repeated.

If a limit cycle is attractor of an n-dimensional system of differential equations, then $(n-1)$ Lyapunov exponents are negative and the n-th exponent is null (there is contraction in all directions except the one along the trajectory). If a toroidal surface is attractor of a three-dimensional continuous-time system, then such an attractor is characterized by two exponents equal to zero and one negative exponent (there is contraction only along the direction perpendicular to that surface). In discrete-time systems, all Lyapunov exponents related to periodic orbits present nonpositive values.

3.4 Chaotic behavior

This section is primarily based on [2, 4, 25, 40]. The *Poincaré-Bendixson theorem* was originally proved by Poincaré in 1886 and a more rigorous demonstration was given by Ivar O. Bendixson (1861-1935) in 1901. Consider a second-order autonomous continuous-time system in which all the equilibrium points are isolated. A consequence of this theorem is that equilibrium points and closed trajectories can be the unique attractors of such a dynamical system. Hence, a strange chaotic attractor can only exist in an autonomous continuous-time system with at least three dimensions. In discrete-time systems, such an attractor can appear in first-order maps; however, this map must be noninvertible,[16] as the logistic map. In two dimensions, invertible maps can display chaotic behavior.

A system can be considered chaotic if there is at least one positive Lyapunov exponent[17]; thus, nearby trajectories or orbits exponentially diverge with time. Since a strange chaotic attractor occupies a bounded region of the phase space, such an exponential divergence cannot occur in all directions; thus, a folding process must also occur. In a three-dimensional continuous-time system, the signs of the three Lyapunov exponents related to a strange chaotic attractor must be $L_1 > 0, L_2 = 0, L_3 < 0$; also $L_1 + L_2 + L_3 < 0$, because the system must be dissipative to present an attractor.

In accordance with Poincaré, Birkhoff, and Kolmogorov (and others), chaotic regions can appear in the phase space of conservative systems; however, these systems cannot exhibit (regular or strange) attractors. For instance, in a conservative three-dimensional continuous-time system with chaotic behavior, the Lyapunov exponents must satisfy the conditions $L_1 > 0, L_2 = 0, L_3 < 0$ and $L_1 + L_2 + L_3 = 0$.

A steady state is represented by a point in the phase space; hence, its dimension is zero. A periodic solution corresponds to a closed curve; therefore, its dimension is one. A toroidal surface containing a periodic or quasi-periodic motion has dimension equal to two. A strange attractor is an object with non-integer *Hausdorff dimension*. Felix Hausdorff (1868-1942) proposed in 1918 to determine the dimension of an object by using n-dimensional cubes of variable edge length to completely cover the object with the minimum number of such cubes. According to a proposal by Kolmogorov in 1958, this calculation is usually approximated by computing the *box-counting dimension* (or *capacity dimension*) $D_0 \equiv \lim_{\epsilon \to 0}(\log N(\epsilon))/(\log 1/\epsilon))$, in which $N(\epsilon)$ is the minimum number of identical n-dimensional cubes of size ϵ required to enclose the set. The value of D_0 is 0 for a point, 1 for a line, 2 for a surface, and 2.06 for the strange attractor appearing in the phase space of the *Lorenz system* written as $\dot{x}_1 = 10(x_2 - x_1)$, $\dot{x}_2 = 28x_1 - x_2 - x_1 x_3$, $\dot{x}_3 = x_1 x_2 - 8x_3/3$. The corresponding Lyapunov exponents are $L_1 \simeq 0.9$, $L_2 = 0$, $L_3 \simeq -14.6$; and, in this dissipative system with $\nabla \cdot \mathbf{f} = $ constant, then $L_1 + L_2 + L_3 = \nabla \cdot \mathbf{f} = -13.7 < 0$. This was the first chaotic attractor to be drawn.

Notice that the calculation of D_0 is not influenced by the amount of points pertaining to the attractor in each n-dimensional cube. Hence, to better characterize attractors with heterogenous structures, other dimensions have been used; for instance, *information dimension* and *correlation dimension*. In addition, from time series, these quantities can be numerically estimated and these estimates give hints on the dimension of the phase space of the system producing those time series, which can be useful to mathematically model (to describe in terms of equations) such a system.

---

[16]The map $\mathbf{f}$ is invertible if it is one-to-one; consequently, its inverse $\mathbf{f}^{-1}$ is a function.

[17]There is no general agreement concerning the definition of a chaotic system. Another possible (and very popular) definition was proposed by Robert L. Devaney (1948-) [8].

3.5 Structural stability and bifurcation

This section is primarily based on [4, 17, 25, 28]. Consider a vector field $\mathbf{f_p}(\mathbf{x})$ dependent on a set of parameters $(p_1, p_2, ..., p_k) = \mathbf{p} \in \mathbb{R}^k$, in which k is the dimension of the parameter space and $\mathbf{p} = \mathbf{p}'$ is a particular choice of parameter values. A dynamical system ruled by $\mathbf{f_p}(\mathbf{x})$ is *structurally stable* for $\mathbf{p} = \mathbf{p}'$ if there is $\epsilon > 0$ such that $\mathbf{f_p}(\mathbf{x})$ and $\mathbf{f_{p'}}(\mathbf{x})$ generate phase portraits that are topologically equivalent for every $\mathbf{p}$ satisfying the restriction $\sqrt{\sum_{i=1}^{k}(p_i - p_i')^2} < \epsilon$. Therefore, a dynamical system is structurally stable if a ϵ-perturbation in its vector field does not qualitatively alter its phase portrait; thus, there is a homeomorphism preserving of sense of time that relates the trajectories or orbits of $\mathbf{f_p}(\mathbf{x})$ and $\mathbf{f_{p'}}(\mathbf{x})$. By 1885, Poincaré introduced the term *bifurcation* to describe qualitative changes in the phase portrait of a dynamical system caused by variation of parameter value(s).

Important results in this matter are, for instance, the derivation of the conditions to a two-dimensional autonomous continuous-time system be structurally stable, by Andronov and Pontryagin in 1937, and then generalized by Maurício M. Peixoto (1921-) in 1962 [31]; and the *catastrophe theory*, originated from works by René Thom (1923-2002) on bifurcations in gradient continuous-time systems; that is, systems in which $\dot{x}_j = f_{j\,\mathbf{p}}(\mathbf{x}) = -\partial G_\mathbf{p}(\mathbf{x})/\partial x_j$, with $G_\mathbf{p}(\mathbf{x}) \in \mathbb{R}$ [42]. Notice that structural stability concerns the stability of vector fields $\mathbf{f_p}(\mathbf{x})$ by disturbing $\mathbf{p}$; Lyapunov stability concerns stability of solutions $\mathbf{x}(t)$ by disturbing $\mathbf{x_0}$. A system with an unstable equilibrium point can be structurally stable.

Bifurcations can cause the creation or destruction of equilibrium or fixed points, changes on the Lyapunov stability of steady states or periodic solutions, the appearing or disappearing of chaotic behavior, etc. In the phase space, local bifurcations can be studied by investigating $\mathbf{f_p}(\mathbf{x})$ in the neighborhood of a steady point or a closed path. Usually, global bifurcations cannot be understood from a local analysis; they can involve, for instance, the formation or breakdown of an entire trajectory joining distinct equilibrium points (known as *heteroclinic trajectory*) or an equilibrium point to itself (known as *homoclinic trajectory*). Invariant sets with new qualitative features can be created from global bifurcations, which can be caused, for instance, by applying ϵ-perturbation to a conservative system. To determine the existence of homoclinic and heteroclinc trajetories is important because, for instance, according to Andronov and Pontryagin, in a Cartesian plane, a two-dimensional autonomous continuous-time system is structurally stable if in its phase portrait there is no homoclinic or heteroclinic trajectory and the number of equilibrium points and closed trajectories is finite and every one is hyperbolic.[18] There is

---

[18]In a two-dimensional flow, a closed trajectory is hyperbolic if $\lambda \neq 1$, in which λ is eigenvalue of the fixed point of a Poincaré map used to describe the behavior of such a trajectory in a Poincaré section (a line, in this case).

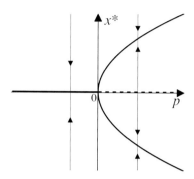

FIGURE 3.1
Bifurcation diagram of $\dot{x} = px - x^3$.

an additional condition, derived by Peixoto, for two-dimensional phase spaces that do not correspond to the Cartesian plane.[19] In three-dimensional flows, homoclinic and heteroclinic trajectories can be associated with chaos, as shown in 1965 by Leonid P. Shilnikov (1934-2011).

There is much to be done in bifurcation theory, mainly if $n, k \geq 3$. For simplicity, here $k = 1$; thus, there is a single *control parameter*.

For instance, the first-order differential equation $\dot{x} = f_p(x) = px - x^3$, with $x, p \in \mathbb{R}$, presents three equilibrium points: $x_a^* = 0$ and $x_{b,c}^* = \pm\sqrt{p}$. The linearization of $f_p(x)$ gives $df_p(x)/dx = p - 3x^2$. For $x_a^* = 0$, then $\lambda = df_p(x)/dx|_{x=0} = p$; therefore, $x_a^* = 0$ is locally asymptotically stable if $p < 0$ and unstable if $p > 0$. For $x_{b,c}^* = \pm\sqrt{p}$, then $\lambda = df_p(x)/dx|_{x=\pm\sqrt{p}} = -2p$; therefore, $x_{b,c}^* = \pm\sqrt{p}$ are locally asymptotically stable if $p > 0$ and they do not exist if $p < 0$ (because $x \in \mathbb{R}$).

In a *bifurcation diagram*, features of the invariant sets are plotted in function of system parameters. Figure 3.1 is the bifurcation diagram of $\dot{x} = px - x^3$. In this figure, the value and the stability of each equilibrium point x^* are plotted in function of p (there is no periodic or chaotic solution in this system). An asymptotically stable steady state $x^* = x^*(p)$ is represented by a solid line; an unstable steady state is represented by a dashed line. The vertical lines with arrows are the phase portraits, indicating the time evolution for $p < 0$ and $p > 0$. Thus, if $p < 0$, then $x(t)$ converges to $x_a^* = 0$ for any initial condition $x(0)$[20]; if $p > 0$ and $x(0) > 0$, then $x(t)$ converges to $x_b^* = \sqrt{p}$; if $p > 0$ and $x(0) < 0$, then $x(t)$ converges to $x_c^* = -\sqrt{p}$.

In Figure 3.1, the *bifurcation point* is given by $(x^*, \bar{p}) = (0,0)$. In such a point, $x^* = 0$ is non-hyperbolic (because the corresponding eigenvalue λ is

---

[19]For instance, toroidal or cylindrical surfaces can represent two-dimensional phase spaces.
[20]If $p = 0$, $x(t)$ also converges to $x_a^* = 0$.

null) and the system loses its structural stability in the critical value $p = \bar{p} = 0$. Notice that as the value of p increases, the stability of $x_a^* = 0$ alters in $p = 0$ and, above this critical number, two equilibrium points of equal stability are created (in this case, the pair $x_{b,c}^*$ is asymptotically stable). This is called *pitchfork bifurcation*. The vector fields $f_p(x) = px \pm x^3$ are the *normal forms* of this bifurcation, because these $f_p(x)$ are the simplest polynomial forms which are topologically equivalent to any system exhibiting such a bifurcation in a neighborhood of the bifurcation point. The idea of simplifying a dynamical system by changes of coordinates without qualitatively modifying the main features of a local solution (in the vicinity of a bifurcation point) was proposed in 1879 by Poincaré in his doctorate.

First-order autonomous continuous-time systems depending on a unique parameter p can experience three kinds of bifurcation: *saddle-node bifurcation* (two equilibrium points of opposite stability are created or destroyed by varying the value p; normal forms: $\dot{x} = p \pm x^2$), *transcritical bifurcation* (two equilibrium points of opposite stability switch their stability by varying p; normal forms: $\dot{x} = px \pm x^2$), and pitchfork bifurcation. In all cases, $\lambda = 0$ in the bifurcation point.

In two-dimensional flows, a *Hopf bifurcation* can occur: by varying p, a limit cycle enclosing an equilibrium point can appear such that the point and the cycle present opposite Lyapunov stabilities. In the bifurcation point, the eigenvalues $\lambda_{1,2}$ corresponding to the equilibrium point must be complex conjugate imaginary numbers; that is, if $\lambda_{1,2} = \alpha(p) + i\beta(p)$ ($i = \sqrt{-1}$) and $\bar{p}$ is the critical value of p, then $\alpha(\bar{p}) = 0$ and $\beta(\bar{p}) \neq 0$ (there is an additional condition given by $d\alpha(p)/dp|_{p=\bar{p}} \neq 0$). This bifurcation was studied by Poincaré in 1892, by Andronov in 1929, and by Eberhard F.F. Hopf (1902-1983) in 1942. In polar coordinates (x, θ), the normal forms related to the radial direction x are $\dot{x} = px \pm x^3$, with $x \in \mathbb{R}_+$. Consider $\dot{x} = px - x^3$, the equation analyzed in the penultimate paragraph. The diagram bifurcation concerning the radial coordinate is the one shown in Figure 3.1 by taking $x \geq 0$ (obviously, radius is a nonnegative number). According to this figure, there is a transition between an asymptotically stable equilibrium point (with coordinate $x^* = 0$) and an asymptotically stable closed trajectory (with radius $x^* = \sqrt{p}$) in $p = 0$. This transition between stationary state and period solution corresponds to a bifurcation, a qualitative alteration in the dynamical behavior of the system.

First-order autonomous discrete-time systems depending on p can present saddle-node bifurcation (normal forms: $x(t+1) = x(t) + p \pm x(t)^2$), transcritical bifurcation (normal forms: $x(t + 1) = x(t) + px(t) \pm x(t)^2$), and pitchfork bifurcation (normal forms: $x(t + 1) = x(t) + px(t) \pm x(t)^3$). In these three cases, $\lambda = +1$ in the bifurcation point; thus, there is a non-hyperbolic fixed point and the system is structurally unstable for $(x^*, \bar{p}) = (0, 0)$. Another kind of bifurcation, called *flip* or *period-doubling bifurcation*, is associated with $\lambda = -1$. In this case, by varying p, the period of an orbit is multiplied by two. For instance, take $x(t + 1) = x(t) + p - x(t)^2$, with $x, p \in \mathbb{R}$. For $p > 0$,

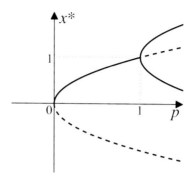

FIGURE 3.2
Bifurcation diagram of $x(t+1) = x(t) + p - x(t)^2$ for $0 \leq p < 3/2$.

this system has two fixed points: $x_a^* = -\sqrt{p}$ and $x_b^* = \sqrt{p}$. The eigenvalue of x_a^* is $\lambda = 1 + 2\sqrt{p} > 1$; therefore, x_a^* is unstable; the eigenvalue of x_b^* is $\lambda = 1 - 2\sqrt{p}$; hence, x_b^* is asymptotically stable if $0 < p < 1$ and unstable of $p > 1$. Notice that in $p = 0$ there occurs a saddle-node bifurcation (for $p < 0$, there is no fixed point; for $p > 0$, there are two fixed points of opposite stability), as shown in Figure 3.2. For $p = 1$, the eigenvalue associated with x_b^* is $\lambda = -1$, supporting the occurrence of a period-doubling bifurcation. For $p > 1$, the fixed point x_b^* loses its stability and an attracting orbit of period 2 appears. This orbit is represented by solid lines bifurcating in $p = 1$. For instance, for $p = 1.2$, the numbers 0.553 and 1.447 compose this orbit of period 2 and $x(t)$ tends to oscillate between these two values in permanent regime. Thus, in $(x^*, \bar{p}) = (1, 1)$, there is transition between an orbit of period $T = 1$ (the fixed point x_b^*) to an orbit of period $T = 2$. In $p \simeq 1.5$, there is another period-doubling bifurcation: in this critical value of p, the orbit of period 2 loses its stability and an asymptotically stable orbit of period 4 arises.

A *Neimark-Sacker bifurcation* (also known as *Hopf bifurcation for maps*) can occur in a two-dimensional map in which a fixed point is characterized by a pair of complex conjugate eigenvalues presenting unit modulus for a critical value of p. This name is due to the analyses by Yuri I. Neimark (1920-2011) in 1959 and Robert J. Sacker (1937-) in 1965. The birth of a periodic or quasiperiodic torus in the phase space of a continuous-time system can be associated to a Neimark-Sacker bifurcation in a corresponding Poincaré map.

Transition from regular behavior towards chaotic solution via bifurcations is usually called *route to chaos*. For instance, there are the route formed by a sequence of period-doubling bifurcations, also known as *Feigenbaum scenario*, due to the studies by Mitchell J. Feigenbaum (1944-) in the 1970s (this route can be found, for instance, in the logistic map and in the map corresponding to

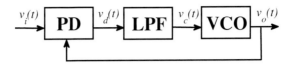

FIGURE 3.3
Block diagram of a basic PLL.

Figure 3.2); the route known as *Ruelle-Takens scenario* composed of a finite number of Hopf bifurcations leading to the system from steady to periodic to quasi-periodic to chaotic behavior (which could explain the emergence of turbulent motion in fluids by increasing its temperature), and the route called *Pomeau-Manneville scenario* associated with *intermittency*, an alternation in irregular time intervals between apparently periodic behavior and clearly aperiodic behavior (such a route is found in the logistic map for $p \lesssim 1 + \sqrt{8}$). By 1980, three types of intermittency were investigated by Yves Pomeau (1942-) and Paul Manneville (1946-) in first-order discrete-time systems.

Some concepts and results presented in the last sections are illustrated by investigating the dynamical behaviors of PLLs.

3.6   Examples: dynamics of PLLs

The *phase-locked loop* (PLL), introduced by Henri de Bellescize in 1932, is an electronic device designed to extract time signals from communication channels. It has been extensively used in applications requiring automatic control of frequency with the intent of obtaining synchronization, such as in computers, modems, motors, phones, radars, radio and television receivers, telecommunication networks, etc. (e.g., [13]). It is a closed loop made up of three elements: a phase detector (PD), a low-pass filter (LPF), and a voltage-controlled oscillator (VCO), as shown in Figure 3.3.

The aim of such a loop is to synchronize the VCO output signal $v_o(t)$ with the external input signal $v_i(t)$. For this purpose, the phase $\theta_o(t)$ of the output signal $v_o(t)$ is adjusted in function of the time-varying phase $\theta_i(t)$ of the input signal $v_i(t)$. Thus, as the time passes, the *phase error* $\phi(t) \equiv \theta_i(t) - \theta_o(t)$ tends to a fixed value or, equivalently, the *frequency error* $d\phi(t)/dt \equiv \omega(t) =$

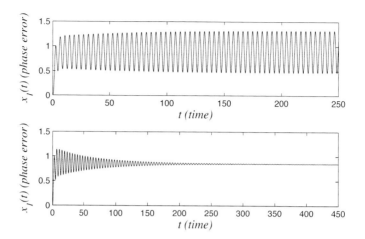

FIGURE 3.4
Time evolution of $x_1(t) = \phi(t)$ obtained from numerical simulations of the
third-order PLL for $p = 3/2$, $r = 2$, $q = 0.55$ (top) and $q = 0.60$ (bottom)
from the initial condition $(x_1(0), x_2(0), x_3(0)) = (0, 0, 0)$.

$d\theta_i(t)/dt - d\theta_o(t)/dt$ vanishes. This situation corresponds to a *synchronous
solution* (e.g., [13, 25, 40]).

In order to derive a mathematical model for PLL dynamics, assume that
$v_i(t) = V_i \sin[\omega_0 t + \theta_i(t)]$ and $v_o(t) = V_o \cos[\omega_0 t + \theta_o(t)]$, in which V_i and V_o
are amplitudes and ω_0 is the central frequency of both signals.

By considering the PD as a signal multiplier, then its output is given
by $v_d(t) = k_d v_i(t) v_o(t)$, in which k_d is the PD gain; therefore, $v_d(t) =
(k_d V_i V_o/2)[\sin(\theta_i(t) - \theta_o(t)) + \sin(2\omega_0 t + \theta_i(t) + \theta_o(t))]$. Usually, the second-
harmonic term of $v_d(t)$ is supposed to be fully cut out by the filter (for a
discussion, see [32]); thus, $v_d(t)$ is simplified to $v_d(t) = (k_d V_i V_o/2) \sin \phi(t)$.

The VCO is an integrator circuit, by which the phase $\theta_o(t)$ is adjusted by
the control signal $v_c(t)$ according to $\dot{\theta}_o(t) = k_o v_c(t)$. The constant k_o is the
VCO gain.

Consider the LPF with input-output relation ruled by [25–27, 33]:

$$\ddot{v}_c(t) + q\dot{v}_c(t) + v_c(t) = \dot{v}_d(t) + v_d(t),$$

in which q is a nonnegative number. By combining the expressions presented in
this section, the PLL dynamics can be described by the following third-order
nonlinear differential equation [25–27, 33]:

$$\dddot{\phi}(t) + q\ddot{\phi}(t) + (1 + r\cos\phi(t))\,\dot{\phi}(t) + r\sin\phi(t) = \dddot{\theta}_i(t) + q\ddot{\theta}_i(t) + \dot{\theta}_i(t) \equiv g(t),$$

in which $r \equiv (V_i V_o k_d k_o)/2$ is a nonnegative constant and $0 \leq \phi < 2\pi$.

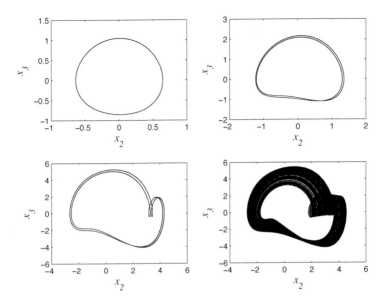

FIGURE 3.5
Attractors projected onto the plane $x_2 \times x_3$ for $p = 3/2$, $r = 2$ and $q = 0.550$ (top-left), $q = 0.355$ (top-right), $q = 0.285$ (bottom-left), and $q = 0.263$ (bottom-right). By decreasing the value of q, there is a transition from periodic solution to chaotic behavior.

Take $\theta_i(t) = pt + c$, with $p \geq 0$. Notice that when $\theta_i(t)$ varies as a ramp input ($p \neq 0$), then $g(t) = p = $ constant becomes a step input. Let $x_1(t) \equiv \phi(t)$, $x_2(t) \equiv \omega(t) = \dot{\phi}(t)$, and $x_3(t) \equiv \dot{\omega}(t) = \ddot{\phi}(t)$. This third-order differential equation can be rewritten in terms of these state variables as the following three first-order differential equations:

$$\begin{aligned}
\dot{x}_1(t) &= x_2(t) \\
\dot{x}_2(t) &= x_3(t) \\
\dot{x}_3(t) &= p - qx_3(t) - (1 + r\cos x_1(t))\, x_2(t) - r\sin x_1(t),
\end{aligned}$$

which are in the form $\dot{\mathbf{x}}(t) = \mathbf{f}(\mathbf{x}(t))$. Observe that $\nabla \cdot \mathbf{f} = -q$. Therefore, the system is dissipative for $q > 0$, and conservative for $q = 0$.

In PLL jargon, *capture range* is defined as the set of values of the velocity p for which the loop attains a synchronous state (e.g., [13, 25]). This state corresponds to a stationary solution $\mathbf{x}^*$ satisfying $\mathbf{f}(\mathbf{x}^*) = 0$; consequently, $x_1^* = \arcsin(p/r)$, $x_2^* = 0$, $x_3^* = 0$, which corresponds to equilibrium points in the phase space $x_1 \times x_2 \times x_3$. Notice that equilibrium points only exist if

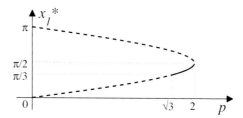

FIGURE 3.6
Bifurcation diagram for x_1^* in function of p for $r = 2$ and $q = 1/2$.

$0 \leq p \leq r$. In this system, there are two equilibrium points with coordinates $(x_{1a}^*, 0, 0)$ and $(x_{1b}^*, 0, 0)$, with $0 \leq x_{1a}^* < \pi/2$ and $\pi/2 < x_{1b}^* \leq \pi$.

The local stability of $(x_{1a}^*, 0, 0)$ and $(x_{1b}^*, 0, 0)$ depends on the eigenvalues $\lambda_{1,2,3}$ of the Jacobian matrix corresponding to the nonlinear system linearized around each point. These eigenvalues $\lambda_{1,2,3}$ are the roots of the polynomial $\lambda^3 + a_1 \lambda^2 + a_2 \lambda + a_3 = 0$, with $a_1 = q$, $a_2 = 1 + r \cos x_1^*$ and $a_3 = r \cos x_1^*$. By the *Routh-Hurwitz criterion*, all three eigenvalues have negative real parts (and the equilibrium point is locally asymptotically stable) if $a_1 > 0$, $a_2 > 0$, $a_3 > 0$, and $a_1 a_2 > a_3$ (e.g., [25, 29]).

Take $q > 0$ as the control parameter. If $0 \leq p < r$, the point with $x_1^* = x_{1b}^*$ is unstable ($\cos x_{1b}^* < 0$; therefore, $a_3 < 0$) and the point with $x_1^* = x_{1a}^*$ is an asymptotically stable synchronous state only if $q > r \cos(x_{1a}^*)/(1 + r \cos(x_{1a}^*))$ or $q > \sqrt{r^2 - p^2}/(1 + \sqrt{r^2 - p^2}) \equiv \bar{q}$. Hence, if $0 \leq q < \bar{q}$, x_{1a}^* is unstable. The PLL experiences a Hopf bifurcation for $q = \bar{q}$. For this critical value of q, $\lambda_{1,2} = \pm i\sigma$ ($\sigma > 0$) are purely imaginary numbers and λ_3 is a negative real number concerning the equilibrium point $(x_{1a}^*, 0, 0)$. For $q \lesssim \bar{q}$, an asymptotically stable limit cycle exists in the phase space $x_1 \times x_2 \times x_3$. Thus, the phase error $\phi(t) = x_1(t)$ periodically oscillates and the oscillation period T is given by

$T \simeq 2\pi/\sigma \simeq 2\pi/\sqrt{1 + \sqrt{r^2 - p^2}}$. For instance, for $p = 3/2$ and $r = 2$ (thus, $0 \leq p < r$), then $\bar{q} \simeq 0.57$. Figure 3.4 shows the time evolutions of $x_1(t) = \phi(t)$ numerically obtained by using the fourth-order Runge-Kutta integration method (e.g., [25, 40]) with time step of 0.01 for solving the three first-order differential equations for $q = 0.55$ and $q = 0.60$. For $q = 0.55 < \bar{q}$, the attractor is a limit cycle with period $T \simeq 4.1$; for $q = 0.60 > \bar{q}$, the attractor is a steady state with $x_{1a}* \simeq 0.85$. These observations are in agreement with the analytical results.

For q varying from $\bar{q}$ to zero, numerical simulations reveal that there occurs a cascade of bifurcations. Figure 3.5 shows the attractors projected onto the plane $x_2 \times x_3$ for $p = 3/2$, $r = 2$, and $q = 0.550$ (top-left), $q = 0.355$ (top-right), $q = 0.285$ (bottom-left), and $q = 0.263$ (bottom-right). For $q = 0.550$,

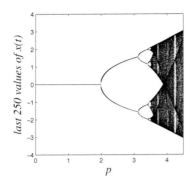 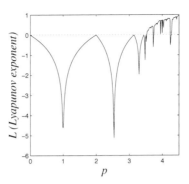

FIGURE 3.7

The "attractors" and the corresponding Lyapunov exponents for the PLL ruled by $x(t+1) = x(t) - p\sin x(t)$, with $0 < p \le 4.5$. The dotted line $L = 0$ was included just for reference.

$L_1 \simeq 0$, $L_2 \simeq -0.02$, $L_3 \simeq -0.53$; for $q = 0.355$, $L_1 \simeq 0$, $L_2 \simeq -0.007$, $L_3 \simeq -0.348$; for $q = 0.285$, $L_1 \simeq 0$, $L_2 \simeq -0.005$, $L_3 \simeq -0.280$; and for $q = 0.263$, $L_1 \simeq 0.063$, $L_2 \simeq 0$, $L_3 \simeq -0.326$. In all cases, the Lyapunov exponents L_j obey the relation $L_1 + L_2 + L_3 = \nabla \cdot \mathbf{f} = -q$ (for instance, for $q = 0.60$, the attractor is an equilibrium point and $L_1 \simeq -0.01$, $L_2 \simeq -0.01$, $L_3 \simeq -0.58$). These exponents were numerically calculated by using the algorithm proposed in the reference [47]. For $q = 0.550$, there is a limit cycle of period 1; for $q = 0.355$, a limit cycle of period 2 appears; for $q = 0.285$, the periodic attractor suffers another structural change; and for $0 < q \lesssim 0.263$, the attractor is chaotic. Simulations suggest that, for $q = 0$, there is chaos ($L_1 \simeq 0.03, L_2 \simeq 0, L_3 \simeq -0.03$ for $t = 1 \times 10^6$) but no attractor, because the system is conservative.

Now, consider p as the control parameter. There occurs a saddle-node bifurcation for $(x_1^*, \bar{p}) = (\pi/2, r)$, because if $p < r - \epsilon$ for $\epsilon \to 0_+$ there are two equilibrium points of opposite stability, and if $p > r + \epsilon$ there is no equilibrium point; hence, there is no synchronous state if $p > r$ [26, 27, 33]. The capture range for this PLL is given by $\sqrt{r^2 - (q/(q-1))^2} < p < r$ with $0 < q < 1$. For instance, for $r = 2$ and $q = 1/2$, the capture range is restricted to the interval $\sqrt{3} < p < 2$, as exhibited in Figure 3.6. In this bifurcation diagram, in which only the coordinate x_1^* of the steady states are shown, there is saddle-node bifurcation in $p = 2$ and Hopf bifurcation in $p = \sqrt{3}$ (concerning x_{1a}^*).

This third-order PLL can be employed for encoding binary messages in a chaos-based communication system [27].

In PLLs described by discrete-time equations, chaos can be found in first-order systems. For instance, consider the error phase $\phi(t)$ of a PLL ruled by $\phi(t+1) = \phi(t) - p\sin\phi(t) + \theta_i(t+1) - \theta_i(t)$ with $0 \le \phi < 2\pi$ and $p > 0$ [14,24]. Let $x(t) \equiv \phi(t)$. By taking a step input, then $\theta_i(t+1) = \theta_i(t)$; consequently

$x(t+1) = f(x(t)) = x(t) - p \sin x(t)$. There are two fixed points: $x_a^* = 0$ and $x_b^* = \pi$ (obtained from $f(x^*) = x^*$). Because $\lambda = df(x)/dx|_{x=x^*} = 1 - p \cos x^*$, then $x_a^* = 0$ is asymptotically stable for $0 < p < 2$ and unstable if $p > 2$. For $p = 2$, then $\lambda = -1$, supporting the occurrence of a period-doubling bifurcation. The fixed point $x_b^* = \pi$ is unstable for any value of $p > 0$.

Figure 3.7 shows the last 250 values of $x(t)$ after numerically solving the difference equation for 999 time steps from a random initial condition near zero $(x(0) \simeq 0)$, for $0 < p \leq 4.5$. Thus, the first 750 values of $x(t)$ were not taken into account in the plot in order to try to disregard the transient behavior. Figure 3.7 also shows the corresponding Lyapunov exponent obtained from (e.g., [4, 25, 40]) $L = (\sum_{t=t_1}^{t_2} \ln(|df(x)/dx|)|_{x=x(t)})/(t_2 - t_1 + 1)$, with $t_1 = 750$ and $t_2 = 999$. Chaos $(L > 0)$ exists for $p \gtrsim 3.532$.

Other chaotic PLLs described in terms of continuous or discrete-time systems can be found in the literature (e.g., [5, 12, 19]).

3.7 Concluding remarks

In this chapter, a superficial overview on Dynamical System Theory was presented. Some tools concerning this theory were applied for analyzing the dynamics of PLLs ruled by difference and differential equations. These analyses can guide investigations based on numerical simulations and practical applications on chaotic communication systems.

3.8 Acknowledgments

The author is partially supported by CNPq.

Bibliography

[1] L. A. Aguirre and C. Letellier. Modeling Nonlinear Dynamics and Chaos: A Review. *Mathematical Problems in Engineering*, 2009:238960, 2009.

[2] K. T. Alligood, T. Sauer, and J. A. Yorke. *Chaos: An Introduction to Dynamical Systems*. Textbooks in Mathematical Sciences. Springer, New York, 1996.

[3] A. A. Andronov, A. A. Vitt, and S. E. Khaikin. *Theory of Oscillators.* Dover, New York, 2011.

[4] J. Argyris, G. Faust, and M. Haase. *An Exploration of Chaos.* Elsevier Science, Amsterdam, 1994.

[5] T. Banerjee and B. C. Sarkar. Phase Error Dynamics of a Class of DPLLs in Presence of Cochannel Interference. *Signal Process.*, 85:1139–1147, 2005.

[6] L. Campbell and W. Garnett. *The Life of James Clerk Maxwell*, chapter 14. Cambridge University Press, Cambridge, 2010.

[7] L. O. Chua. The Genesis of Chua's Circuit. *AEU Int. J. Electron. Commun.*, 46:250–257, 1992.

[8] R. L. Devaney. *A First Course in Chaotic Dynamical Systems*, chapter 10. Perseus Books, Reading, MA, 1992.

[9] F. Diacu and P. Holmes. *Celestial Encounters: The Origins of Chaos and Stability.* Princeton University Press, Princeton, NJ, 1996.

[10] R. Dugas. *A History of Mechanics.* Dover, New York, 1988.

[11] J. P. Eckmann and D. Ruelle. Ergodic Theory of Chaos and Strange Attractors. *Rev. Mod. Phys.*, 57:617–656, 1985.

[12] T. Endo. A Review of Chaos and Nonlinear Dynamics in Phase-Locked Loops. *Journal of the Franklin Institute*, 331B:859–902, 1994.

[13] F. M. Gardner. *Phaselock Techniques.* John Wiley & Sons, New York, 2005.

[14] G. S. Gill and S. C. Gupta. First-Order Discrete Phase-Locked Loop with Applications to Demodulation of Angle-Modulated Carrier. *IEEE Transactions on Communications*, 20:454–462, 1972.

[15] K. Gleick. *Chaos: Making a New Science.* Penguin Books, London, 1987.

[16] C. Grebogi, S. M. Hammel, J. A. Yorke, and T. Sauer. Shadowing of Physical Trajectories in Chaotic Dynamics: Containment and Refinement. *Phys. Rev. Lett.*, 65:1527–1530, 1990.

[17] J. Guckenheimer and P. Holmes. *Nonlinear Oscillations, Dynamical Systems, and Bifurcations of Vector Fields.* Springer, New York, 1983.

[18] D. A. Hsieh. Chaos and Nonlinear Dynamics: Application to Financial Market. *The Journal of Finance*, 46:1839–1877, 1991.

[19] M. Y. Kucinski and L. H. A. Monteiro. Periodic Solutions of the Pendulum. *J. Phys. A: Math. Gen.*, 33:8489–8505, 2000.

[20] A. Lewis, M. D. Lipsitz, L. Ary, and M. D. Goldberger. Loss of 'Complexity' and Aging: Potential Applications of Fractals and Chaos Theory to Senescence. *JAMA*, 267:1806–1809, 1992.

[21] T. Y. Li and J. A. Yorke. Period Three Implies Chaos. *Amer. Math. Monthly*, 82:985–992, 1975.

[22] E. N. Lorenz. Deterministic Nonperiodic Flow. *J. Atmosf. Sci.*, 20:130–141, 1963.

[23] R. M. May. Simple Mathematical Models with Very Complicated Dynamics. *Nature*, 261:459–467, 1976.

[24] W. May and J. J. Leader. The Zero-Crossing Phase-Lock Loop: Results from Discrete Dynamical Theory. *Appl. Math. Lett.*, 14:495–498, 2001.

[25] L. H. A. Monteiro. *Sistemas Dinâmicos* (in portuguese), 3rd edition. Editora Livraria da Física, São Paulo, SP, 2011.

[26] L. H. A. Monteiro, D. N. Favaretto Filho, and J. R. C. Piqueira. Bifurcation Analysis for Third-Order Phase-Locked Loops. *IEEE Signal Processing Letters*, 11:494–496, 2004.

[27] L.H.A. Monteiro, A.C. Lisboa, and M. Eisencraft. Route to chaos in a third-order phase-locked loop network. *Signal Processing*, 89(8):1678–1682, 2009.

[28] A. H. Nayfeh and B. Balachandran. *Applied Nonlinear Dynamics*. John Wiley & Sons, New York, 1995.

[29] K. Ogata. *Modern Control Engineering*. Prentice-Hall, Englewood Cliffs, NJ, 2001.

[30] N. H. Packard, J. P. Crutchfield, J. D. Farmer, and R. S. Shaw. Geometry from a Time Series. *Phys. Rev. Lett.*, 45:712–716, 1980.

[31] M. M. Peixoto. Structural Stability on Two-Dimensional Manifolds. *Topology*, 1:101–120, 1962.

[32] J. R. C. Piqueira and L. H. A. Monteiro. Considering Second-Harmonic Terms in the Operation of the Phase Detector for Second-Order Phase-Locked Loop. *IEEE Transactions on Circuits and Systems I*, 50:805–809, 2003.

[33] J. R. C. Piqueira and L. H. A. Monteiro. All-Pole Phase-Locked Loops: Calculating Lock-in Range by Using Evan's Root-Locus. *International Journal of Control*, 79:822–829, 2006.

[34] F. J. Romeiras and E. Ott. Strange Nonchaotic Attractors of the Damped Pendulum with Quasiperiodic Forcing. *Phys. Rev. A*, 35:4404–4413, 1987.

[35] D. Ruelle. *Chance and Chaos.* Princeton University Press, Princeton, NJ, 1991.

[36] D. Ruelle and F. Takens. On the Nature of Turbulence. *Amer. Math. Monthly*, 20:167–192, 1971.

[37] M. Schroeder. *Fractals, Chaos, Power Laws.* Dover, New York, 2009.

[38] S. Smale. Differentiable Dynamical Systems I. Diffeomorphisms. *Bull. Am. Math. Soc.*, 73:747–817, 1967.

[39] S. Spielberg (Director). *Jurassic Park.* Universal Pictures, 2003.

[40] S. H. Strogatz. *Nonlinear Dynamics and Chaos.* Addison-Wesley, Reading, MA, 1994.

[41] F. Takens. Detecting Strange Attractors in Turbulence. In *Lecture Notes in Mathematics*, volume 898, pages 366–381. 1981.

[42] R. Thom. *Structural Stability and Morphogenesis.* Westview Press, Boulder, CO, 1994.

[43] W. Tucker. The Lorenz Attractor Exists. *C. R. Acad. Sci. Paris*, 328 (series I):1197–1202, 1999.

[44] S. M. Ulam and J. von Neumann. On the Combination of Stochastic and Deterministic Processes. *Bull. Am. Math. Soc.*, 53:1120, 1947.

[45] B. van der Pol and J. van der Mark. Frequency Demultiplication. *Nature*, 120:363–364, 1927.

[46] D. Wells. *The Penguin Dictionary of Curious and Interesting Numbers.* Penguin Books, London, 1995.

[47] A. Wolf, J. B. Swift, H. L. Swinney, and J. A. Vastano. Determining Lyapunov Exponents from a Time Series. *Physica D: Nonlinear Phenomena*, 16:285–317, 1985.

4

Basics of communications using chaos

Géza Kolumbán

Pázmány Péter Catholic University
The Faculty of Information Technology

Tamás Krébesz

Budapest University of Technology and Economics

Chi K. Tse and Francis C. M. Lau

The Hong Kong Polytechnic University

CONTENTS

In digital communications, either fixed, chaotic, or random, analog waveforms of finite duration are used to carry the information. To optimize a waveform communications system or determine its noise performance in analytic form, mathematical models of modulation and detection have to be developed.

This chapter extends the conventional waveform communications concept to chaotic communications. A common property of chaotic modulation schemes, namely, the estimation problem that, if not prevented, corrupts the noise performance of chaotic communications systems, is discussed. Then, introducing the Fourier analyzer concept, a received signal space is defined in which every received signal either deterministic, chaotic, or random can be represented. Finally a step-by-step approach is given for the derivation of different detection algorithms. As examples, the derivations of coherent, averaged optimum noncoherent, and autocorrelation detection algorithms and detector configurations are shown.

To show the feasibility of chaotic communications, an FM-DCSK data communications system operating in the 2.4-GHz ISM band is implemented. The FM-DCSK transceiver is implemented in software and a universal hardware device is used to convert the signals between the RF domain and baseband.

4.1   Introduction

The radio channel transmitting the information bearing signal from the transmitter to the receiver is analog, consequently, the digital information to be transmitted has to be mapped into analog waveforms of finite duration. This approach is referred to as *waveform communications* and the modulation is the process that maps the digital information into analog waveform referred to as carrier.

Chaotic signals are wideband signals that can be generated by simple circuits in any frequency band and at arbitrary power level. If the digital information to be transmitted is mapped directly into an inherently wideband chaotic carrier then a digital chaotic communications system, considered in this chapter, is implemented. The potential application areas of chaotic communications are those where the inherently wideband characteristic of chaotic carriers offers a special benefit.

There are two emerging applications where wideband signals have to be sent via the radio channel:

- *indoor radio communications*,
  where the radio coverage is limited by the multipath propagation and

- *frequency reuse*,
  where the radio communications is implemented in a frequency band already occupied by conventional narrowband radio links. The interference caused by

the new, referred to as Ultra-WideBand (UWB) radio system, is limited by keeping the Power Spectral Density (psd) of transmitted UWB signal below a threshold specified by the Federal Communications Commission (FCC) in the United States [6].

The digital information to be transmitted is mapped into sinusoidal waveforms in conventional digital communications. The required bandwidth is determined by the data rate and modulation scheme; the main design goal is to minimize the bandwidth required. In this sense the conventional telecommunications systems are narrowband. When the bandwidth of transmitted signal has to be increased then an additional signal, such as a pseudo-noise sequence, is used to get a Spread Spectrum (SS) system [8].

An alternative solution is the Impulse Radio (IR) approach where the digital information to be transmitted is mapped into Radio Frequency (RF) impulses [22]. The RF impulse used as carrier in UWB IR has an extremely short duration, consequently, the modulated UWB signal is an ultra-wideband signal with a very low psd in the frequency domain.

The last solution used in conventional communications is the Orthogonal Frequency-Division Multiplexing (OFDM) [23], where (i) a large number of subchannels are used, (ii) the data stream is split into parallel data substreams, and (iii) each substream is sent via an independent dedicated subchannel.

Let the solutions outlined above be referred to as conventional or non-chaotic approaches. Note, the common property of non-chaotic solutions is that fixed deterministic waveforms are used as carries. Consequently, if the same symbol is transmitted repeatedly then the same waveform is radiated.

In chaotic communications the digital information to be transmitted is mapped into a continuously varying chaotic carrier. In contrast to the conventional waveform communications, the transmitted waveform is continuously varying, even if the same symbol is transmitted repeatedly. The most significant properties of chaotic communications are as follows:

- chaotic carriers have no amplitude, frequency, or phase; consequently, brand *new modulation schemes* had to be elaborated;

- transmitted waveforms are continuously varying even if the same symbol is transmitted repeatedly; consequently, the *estimation problem* appears;

- robust solution to synchronization of chaotic signals has not yet been found; consequently, the chaotic carrier cannot be recovered from the received noisy waveform. Hence, only *non-coherent detection* schemes are feasible.

Because of the special properties listed above, the well-established theory of conventional digital telecommunications cannot be applied directly to chaotic communications. Starting from the well-known conventional theory and keeping its terminology and notation, this chapter extends the theory of conventional waveform communications to the chaotic carriers. The results derived can be applied to any kind of waveform communications systems.

Section 4.2 extends the theory of communications with fixed waveforms to the chaotic carriers by introducing the chaotic basis functions. It shows that in chaotic communications the basis functions are orthonormal only in mean. An undesirable consequence of this property is the estimation problem that corrupts the noise performance. A solution to the estimation problem is provided.

To derive the detection algorithms in analytical form a signal model has to be defined in which every transmitted waveform carrying a symbol defines a subspace. Two common basic properties possessed by each detector are exploited to construct the signal model, namely, (i) every detector observes the received signal corrupted by noise and interference in the radio channel only in the signaling time period and (ii) the bandwidth of observed signal is limited by a bandpass channel selection filter. Based on the outcome of this observation the detector makes an estimation for the transmitted symbol.

Section 4.3 introduces the Fourier analyzer concept where a finite-dimensional Hilbert space is defined. Its dimension is determined by the detector parameters, namely, by the product of observation time period and channel bandwidth. Any kind of received signal, either deterministic, chaotic, or random, can be fully *represented* in this finite-dimensional Hilbert space, referred to as *received signal space*, without any distortion.

Each element of symbol set is fully characterized by its Fourier coefficients in the received signal space. Section 4.4 starts from this *a priori* information and gives an analytical method for the derivation of detection algorithms and, from them, the detector configurations.

Recall, chaotic carriers have no amplitude, frequency, or phase; consequently, the integrated circuits developed to build conventional radio transceivers cannot be used to implement chaotic communications systems. Instead, a new approach is required. In Software Defined Electronics (SDE) [15], every RF bandpass signal processing algorithm is implemented in the Base-Band (BB) and a universal hardware device referred to as USRP unit is used to extract the BB information at the receiver or to reconstruct the RF bandpass signal from the BB waveform at the transmitter.

Section 4.5 uses the SDE approach to implement an FM-DCSK system in baseband. Both the transmitter and receiver are implemented and all relevant signals are measured. The implemented 2.4-GHz FM-DCSK radio link demonstrates the feasibility of chaotic communications systems.

4.2 Generalization of waveform communications

Since only analog waveforms can be transmitted over a physical telecommunications channel and the data rate is determined by the system specification, the modulator of a digital telecommunications system maps the symbols to be transmitted into analog waveforms of finite duration. Note, the waveform communications approach used here has nothing in common with the analog modulation techniques; here distinct waveforms, the elements of a *signal set*, are assigned to carry the different symbols.

To simplify the mathematical treatment of signal set let its elements be expressed as a weighted linear combination of N basis functions [8]. Consider a digital modulator where M symbols are transmitted. Each symbol m is mapped into a signal vector $\mathbf{s}_m = [s_{mn}]$ and the transmitted waveforms, i.e., the elements of signal set, are defined by

$$s_m(t) = \sum_{n=1}^{N} s_{mn} g_n(t), \quad \left\{ \begin{array}{l} 0 \leq t < T \\ m = 1, 2, \ldots, M \\ n = 1, 2, \ldots, N, \end{array} \right. \tag{4.1}$$

where $g_n(t)$, $n = 1, 2, \ldots, N$ are real-valued basis functions and $N \leq M$.

Each symbol is characterized by a distinct signal vector $\mathbf{s}_m = [s_{mn}]$, $m = 1, 2, \ldots, M$ and, according to Equation (4.1), by a distinct waveform. The signaling time period T, i.e., the duration of basis functions, is determined by the data rate. To avoid InterSymbol Interference (ISI), the value of basis functions is zero outside the signaling time interval.

During the design of a digital modulation scheme, either the elements $s_m(t)$ of signal set or the basis functions $g_n(t)$ can be chosen first. In the former approach, the basis functions are derived by the Gram-Schmidt orthogonalization procedure [8].

If fixed carriers are used then the basis functions are orthonormal:

$$\int_0^T g_i(t) g_n(t) dt = \left\{ \begin{array}{ll} 1, & \text{if } i = n \\ 0, & \text{otherwise} \end{array} \right. \tag{4.2}$$

which means that each basis function carries unit energy and each pair of distinct basis functions are orthogonal to each other in the signaling time period $[0, T]$. Recall, the $g_n(t) = 0$ outside the signaling time period.

Signal vector $\mathbf{s}_m$ and basis functions $g_n(t)$ are *a priori* known at the receiver. This knowledge is exploited to suppress channel noise and interference at the detector. The coherent and noncoherent receivers exploit fully and partly, respectively, the *a priori* information available. The more amount of *a priori* information exploited, the better the system performance.

The type of basis functions gives an upper bound on the *a priori* information. Based on the type of basis functions, three classes of waveform communications systems are distinguished, namely, communications with

- fixed waveforms, the conventional approach [8, 21];

- chaotic waveforms, considered here [7];

- random waveforms [3].

4.2.1   Types of basis functions

In *fixed waveform communications*, the basis functions are fixed. Consequently, every time when the same symbol is sent then the same waveform is transmitted. The basis functions and the elements of signal set are exactly known. In the built receivers the basis functions are recovered from the received signal (see the correlator receiver including a carrier recovery circuit) or stored at the receiver (see the matched filter approach) [21].

Note, the type of generator used to produce the fixed basis functions is irrelevant. The only important issue is that the basis functions are *fixed* waveforms. Even a windowed part of chaotic signal may be used as basis function in fixed waveform communications if it is stored in a memory.

In *varying waveform communications* (see chaotic communications as an example) each basis function is the actual output of a chaotic signal generator. Recall, chaotic signals are predictable only in short run because the chaotic systems have an extremely high sensitivity to the initial conditions and to the parameters of chaotic attractor [19]. The shapes of chaotic basis functions are continuously varying and different waveforms are radiated even if the same symbol is transmitted repeatedly.

Consider a chaotic waveform communications system where the basis functions are chaotic sample functions of *finite duration*. An important feature of chaotic communications is that the transmitted signal is never periodic. On the other hand, the energy carried by chaotic sample functions is not constant and the cross-correlation of two chaotic sample functions differs from zero. Even more, the energy and cross-correlation vary from sample function to sample function. This property referred to as estimation problem is discussed in the next subsection. If unsolved, the estimation problem corrupts the noise performance of every varying waveform communications system.

In the remaining part of this chapter, the continuously varying property of chaotic basis functions is reflected by the upper index q:

$$g_n^q(t), \quad q = 1, 2, \ldots, \tag{4.3}$$

where q identifies the basis function transmitting the qth element of a symbol stream.

4.2.2   Estimation problem

At the receiver, the detector observes the received signal in the signaling time period, also referred to as symbol duration T, and produces the observation variable. Due to the channel imperfections (noise, multipath propagation, interference, etc.) the observation variable is a random variable characterized by its probability density function. Considering the Bit Error Rate (BER) performance, the most important parameters are the mean and the standard deviation of this distribution. The higher the variance of observation variable, the worse the BER [8]. Any effect that increases the variance of observation variable corrupts the BER performance.

Chaotic signals are not deterministic signals, consequently, they can be characterized only by their statistical parameters. To get these statistical parameters, the stochastic signal model has to be adopted to the chaotic waveforms.

The chaotic stochastic signal model has been elaborated in [10] where the ensemble of sample functions is defined as the output of a chaotic attractor starting from each possible initial condition.

Although chaotic waveforms with *infinite duration* and generated by different attractors[1] or by the same attractor but started from different initial conditions are uncorrelated and their corresponding average powers are constant, the chaotic basis functions of *finite duration* are orthonormal only in mean,

$$E\left[\int_0^T g_i^q(t)g_n^q(t)dt\right] = \left\{ \begin{array}{ll} 1, & \text{if } i = n \\ 0, & \text{otherwise} \end{array} \right. \tag{4.4}$$

where $E[\cdot]$ denotes the expectation operator.

The integral

$$\int_0^T g_i^q(t)g_n^q(t)dt \tag{4.5}$$

gives estimates of auto- $(i = n)$ and cross- $(i \neq n)$ correlations of basis functions [4]. These estimates calculated from the sample functions of finite duration are random variables. Their variance increases the variance of observation variable and, consequently, corrupts the BER performance. To get the best BER performance, the variances of auto- and cross-correlation estimation have to be kept zero.

For $i = n$, Integral (4.5) estimates the energy of ith basis function. Equation (4.5) shows that if the chaotic signals are directly used as basis functions then the energy used to transmit one symbol is not constant but varies from symbol to symbol. This property is called autocorrelation estimation problem. The mean and standard deviation of estimation of $\int_0^T [g_n^q(t)]^2 dt$ versus the estimation time, i.e., the symbol duration, are plotted in Figure 4.1(a), where the chaotic basis function was generated by a third-order analog phase locked loop (APLL) [17].

The effect of cross-correlation estimation problem is shown in Figure 4.1(b), where the basis functions were generated by the same chaotic APLL but started from different initial conditions. Due to the cross-correlation estimation problem an interference among the different basis functions that corrupts the BER performance appears.

The standard deviations of both estimates are inversely proportional to the product of estimation time and the equivalent statistical bandwidth of chaotic basis functions [10]. The former and latter are equal to the symbol duration and RF bandwidth of chaotic basis functions, respectively.

---

[1]In this respect, attractors described by the same mathematical model but having different parameters are considered as different attractors.

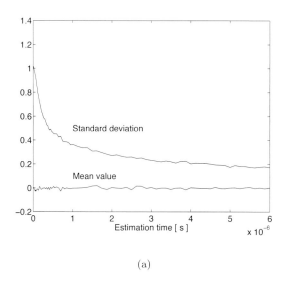

(a)

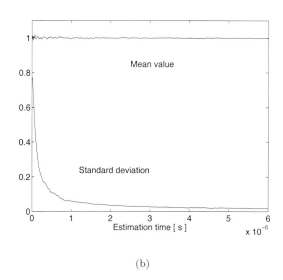

(b)

FIGURE 4.1
The mean and standard deviation of estimation versus the estimation time:
(a) autocorrelation and (b) cross-correlation estimation problems.

The estimation problem is present in every modulation scheme where the
basis functions are *orthonormal only in mean*. It appears in both the chaotic

and random modulation techniques. To solve the estimation problem there is no need to fix the shape of basis function; it is enough to satisfy the Condition (4.2) of orthonormality for each sample of basis functions.

4.2.3 Chaotic modulation schemes

Digital modulation using chaotic basis functions and coherent receiver was first introduced in 1992 [20] and referred to as Chaos Shift Keying (CSK) [5]. A coherent receiver needs a carrier recovery circuit being robust against the channel imperfections such as noise, interference, multipath propagation, etc. A lot of research effort has been done to develop a synchronization technique for chaotic carrier recovery but these efforts failed. Chaotic synchronization is not feasible because of its high sensitivity to channel imperfections.

The first robust noncoherent modulation technique referred to as Differential Chaos Shift Keying (DCSK) [18] was introduced in 1996, and later optimized as Frequency-Modulated DCSK (FM-DCSK) [14].

4.2.3.1 Antipodal binary CSK: modulation with one basis function

Let E_b denote the energy used to transmit one bit of information. In antipodal Binary CSK (BCSK), (i) only one chaotic basis function $g_1^q(t)$ is used, and (ii) bits "1" and "0" are represented by $s_{11} = \sqrt{E_b}$ and $s_{21} = -\sqrt{E_b}$, respectively. According to Equation (4.1), the modulated waveforms, i.e., the elements of signal set, are

$$s_m^q(t) = \pm\sqrt{E_b}\, g_1^q(t), \quad m = 1, 2. \tag{4.6}$$

To show the effect of estimation problem, the noise performance of a BCSK system implemented with varying and constant E_b has been evaluated [12]. Because only one chaotic sample function is used in BCSK, only the autocorrelation estimation problem is present. Figure 4.2 shows the manifestation of estimation problem; dash-dotted and solid curves give the noise performances versus the energy per bit-to-noise spectral density ratio E_b/N_0 for the cases when the energy of basis function is varying or is kept constant, respectively. As expected, the estimation problem increases the variance of observation variable and, consequently, corrupts the noise performance.

4.2.3.2 FM-DCSK: modulation with two basis functions

In binary FM-DCSK, the two elements of signal set representing bits "1" and "0," respectively, are defined by

$$s_1^q(t) = \sqrt{E_b}\, g_1^q(t) \quad \text{and} \quad s_2^q(t) = \sqrt{E_b}\, g_2^q(t), \tag{4.7}$$

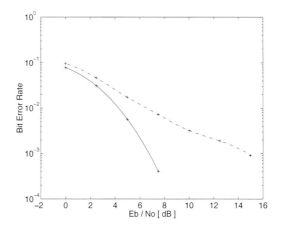

FIGURE 4.2
The noise performance of BCSK for varying (dash-dotted curve) and constant (solid curve) E_b.

where the two basis functions $g_1^q(t)$ and $g_2^q(t)$ are constructed from a frequency-modulated chaotic waveform $c^q(t)$ and the first two Walsh functions [13]

$$g_1^q(t) = \left\{ \begin{array}{ll} c^q(t), & 0 \leq t < T/2 \\ c^q(t - T/2), & T/2 \leq t < T, \end{array} \right. \tag{4.8}$$

$$g_2^q(t) = \left\{ \begin{array}{ll} c^q(t), & 0 \leq t < T/2 \\ -c^q(t - T/2), & T/2 \leq t < T. \end{array} \right.$$

Equations (4.7) and (4.8) show that FM-DCSK maps every bit to be transmitted into two consecutive waveforms of duration $T/2$. The first one serves as a reference while the second one carries the information. For bit "1," the information bearing waveform is identical to the reference one while for bit "0," it is an inverted copy of the reference waveform. Note, FM-DCSK belongs to the class of transmitted reference systems. FM-DCSK can be demodulated without carrier recovery.

To avoid the autocorrelation estimation problem, the frequency-modulated chaotic waveform $c^q(t)$ in Equation (4.8) is generated by an FM modulator. The chaotic signal is fed into the FM modulator input, $c^q(t)$ appears at the modulator output, and the bandwidth of $c^q(t)$ is set by the parameters of chaotic attractors and FM modulator [9]. Value of $c^q(t)$ is zero outside the time interval $[0, T/2]$. Note, the frequency-modulated chaotic waveform $c^q(t)$ is continuously varying, but because of the FM modulator, it is a constant envelope signal. Consequently, its energy is constant and $\int_0^{T/2} [c^q(t)]^2 dt = 1/2$.

In FM-DCSK both the autocorrelation and cross-correlation estimation problems are avoided. The constant energy and orthogonality of basis functions are assured by FM and the first two Walsh functions, respectively. The

FM-DCSK basis functions are always orthonormal regardless of the shape of continuously varying chaotic signals.

4.3 Signal model for detection

In digital communications, the elements of signal set carrying the symbols pass through a telecommunications channel in which they are corrupted by noise and interference, and may suffer from distortion and multipath effect. Observing the corrupted and distorted received analog waveform in the time interval of symbol duration, the detector must decide which symbol has been most likely transmitted.

According to Equation (4.1), the elements of signal set are constructed from the basis functions and signal vector. This *a priori* information is exploited to perform the detection and suppress channel noise and interference at the receiver.

This section defines a *finite-dimensional discrete* Hilbert space referred to as *received signal space* in which each signal passing through the channel selection filter, either deterministic or random, is represented by their Fourier coefficients. The *a priori* information available is also quantified by the Fourier coefficients. These coefficients will be used in Section 4.4 to derive the different detection algorithms.

4.3.1 General block diagram of a receiver

Consider an Additive White Gaussian Noise (AWGN) channel [8] and let N_0 denote the psd of channel noise. Symbol m is transmitted by sending the analog waveform $s_m^q(t)$ to the receiver via the analog radio channel where it is corrupted by Gaussian white noise or channel interference $n(t)$ as shown in Figure 4.3. The received signal $r_m^q(t) = s_m^q(t) + n(t)$ is bandlimited by the bandpass channel (selection) filter having an RF bandwidth of $2B$. The noisy filtered signal $\tilde{s}_m^q(t) + \tilde{n}(t)$ is observed by the detector in the symbol duration T to generate the *observation variable* $\mathbf{z}_m$, a random quantity.

Note, the (i) decision time instants, (ii) bit duration T, and (iii) RF bandwidth $2B$ of transmitted signal $s_m^q(t)$ are always known at the receiver.

4.3.2 Fourier analyzer concept

Consider the noise-free case first and assume that the channel filter passes the transmitted signal without any distortion, i.e., $\tilde{r}_m^q(t) = \tilde{s}_m^q(t) = s_m^q(t)$. To get a *discrete* Hilbert space in the frequency domain, a periodic signal has to be constructed in the time domain from the received signal.

Because the detector observes the received signal only in the time interval

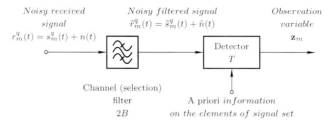

FIGURE 4.3
General block diagram of a digital waveform communications receiver.

$[0, T)$, $s_m^q(t)$ can be substituted by a periodic signal

$$s_{T,m}^q(t) = \begin{cases} s_m^q(t), & \text{for } 0 \leq t < T \\ s_m^q(t - CT), & \text{otherwise} \end{cases}, \qquad (4.9)$$

where C is an arbitrary nonzero integer. The introduction of the periodic signal in Equation (4.9) does not cause any distortion since the two signals, the received one and its periodic equivalent, *coincide* with each other in the observation time period, i.e., the symbol duration.

In the Fourier analyzer concept [16], the received signal space is a Hilbert space spanned by the harmonically related sinusoidal basis functions

$$\cos\left(k\frac{2\pi}{T}t\right) \quad \text{and} \quad \sin\left(k\frac{2\pi}{T}t\right),$$

where the frequencies k/T, $k = 1, 2, \cdots$ of this Fourier base are determined by the observation time period.

The detector projects the filtered received waveform $\tilde{r}_m^q(t) = s_m^q(t)$ in $0 \leq t < T$ into the Hilbert space and returns its Fourier coefficients

$$a_{mk}^q = \frac{2}{T}\int_0^T s_m^q(t)\cos\left(k\frac{2\pi}{T}t\right)dt,$$

$$b_{mk}^q = \frac{2}{T}\int_0^T s_m^q(t)\sin\left(k\frac{2\pi}{T}t\right)dt. \qquad (4.10)$$

When the channel noise and interference are also considered in the receiver model depicted in Figure 4.3 then the detector observes $\tilde{r}_m^q(t) = \tilde{s}_m^q(t) + \tilde{n}(t)$ and Equation (4.10) gives only *estimates*, denoted by hats, of the Fourier coefficients

$$\hat{a}_{mk}^q = \frac{2}{T}\int_0^T \tilde{r}_m^q(t)\cos\left(k\frac{2\pi}{T}t\right)dt,$$

$$\hat{b}_{mk}^q = \frac{2}{T}\int_0^T \tilde{r}_m^q(t)\sin\left(k\frac{2\pi}{T}t\right)dt. \qquad (4.11)$$

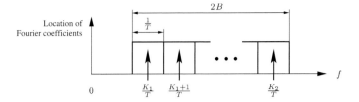

FIGURE 4.4
Location of Fourier series coefficients of $\tilde{r}_{T,m}^{q}(t)$ in the frequency domain.

If the Signal-to-Noise Ratio (SNR) is high enough then we obtain

$$\tilde{r}_{T,m}^{q}(t) = \sum_{k=K_1}^{K_2} \left[\hat{a}_{mk}^{q} \cos\left(k\frac{2\pi}{T}t \right) + \hat{b}_{mk}^{q} \sin\left(k\frac{2\pi}{T}t \right) \right] \approx s_{T,m}^{q}(t), \quad (4.12)$$

where, as will be shown below, the constants K_1 and K_2 are determined by the parameters of channel filter.

The periodicity introduced in Equation (4.9) gives a *discrete* received signal space. Now the dimension of this space has to be found.

The channel filter, an ideal bandpass filter, limits the bandwidth of the signal observed by the detector in $2B$. Figure 4.4 shows the location of Fourier series coefficients of $\tilde{r}_{T,m}^{q}(t)$ by arrows in the frequency domain. The channel filter suppresses each spectral component lying outside the frequency range

$$(2K_1 - 1)/2T \leq f \leq (2K_2 + 1)/2T,$$

where K_1 and K_2 are determined by the center frequency f_0 and bandwidth $2B$ of channel filter

$$K_1 = \frac{2(f_0 - B)T + 1}{2} \quad \text{and} \quad K_2 = \frac{2(f_0 + B)T - 1}{2}. \quad (4.13)$$

By definition, the signal dimension gives the number of harmonically related sinusoidal basis functions along which the receiver collects information on the received signal,

$$S_D = 2(K_2 - K_1 + 1) = 4BT. \quad (4.14)$$

In other words, the signal dimension S_D gives the dimension of Hilbert space spanned by the Fourier base which is required to represent *any signal* appearing at the detector input *in the observation time interval*. Note, the dimension of received signal space does not depend on the center frequency of telecommunications channel; it is given by the *always known* receiver parameters, namely, by the product of observation time period T and receiver bandwidth $2B$.

4.3.3 Quantifying *"a priori"* information

After channel filtering, the detector projects the received waveform into the received signal space and returns its Fourier coefficients. These Fourier coefficients are compared against the *a priori* information to get the observation variable.

To make the comparison possible, the *a priori* information also has to be expressed in the form of Fourier coefficients. They are obtained by projecting the basis functions $g_n^q(t)$, $n = 1, 2, \cdots, N$ into the received signal space:

$$
\begin{aligned}
\alpha_{nk}^q &= \frac{2}{T} \int_0^T g_n^q(t) \cos\left(k\frac{2\pi}{T}t\right) dt \\
\beta_{nk}^q &= \frac{2}{T} \int_0^T g_n^q(t) \sin\left(k\frac{2\pi}{T}t\right) dt.
\end{aligned}
\tag{4.15}
$$

These Fourier coefficients quantify the *a priori* information.

4.4 Derivation of detection algorithms

The Fourier coefficients of received signal are determined and compared against those of the *a priori* known basis functions. The tool of comparison is the inner product. The result, referred to as *observation variable*, is used to perform the decision.

This section focuses on the derivation of detection algorithms; consequently, only the transmission of a single isolated symbol is considered. Interference has to be prevented by forming a Nyquist channel as done in conventional telecommunications systems [8].

The type of detector depends on the extent to which the *a priori* information carried by α_{nk}^q and β_{nk}^q in Equation (4.15) is exploited during the derivation of detection algorithm. As the most important examples, three detection algorithms are derived here:

- *coherent detection*, where all *a priori* information carried by α_{nk} and β_{nk} are exploited;

- *optimum noncoherent detection*, where only the harmonic amplitudes $\gamma_{nk}^q = \sqrt{\left(\alpha_{nk}^q\right)^2 + \left(\beta_{nk}^q\right)^2}$ of basis functions are considered;

- *autocorrelation detection*, where only the separation of spectra of signals carrying bits "1" and "0" are exploited. Here the minimum amount of *a priori* information is exploited; consequently, this approach gives the worst BER performance.

4.4.1 Coherent detection algorithm

Coherent detector fully exploits the *a priori* information available; consequently, it offers the best BER performance.

Consider an antipodal BCSK modulation where the elements of signal set are given by Equation (4.6). The detector projects the noisy filtered received waveform $\tilde{r}_m^q(t)$ into the received signal space and returns a vector of estimates of its Fourier coefficients

$$(\hat{\mathbf{r_m}}^q)' = \left(\hat{a}_{mK_1}^q \ \hat{b}_{mK_1}^q \ \cdots \ \hat{a}_{mk}^q \ \hat{b}_{mk}^q \ \cdots \ \hat{a}_{mK_2}^q \ \hat{b}_{mK_2}^q \right), \tag{4.16}$$

where $\hat{a}_{m,k}^q$ and $\hat{b}_{m,k}^q$ are given by Equation (4.11), $K_1 \leq k \leq K_2$ are obtained from Equation (4.13), and the prime character $'$ denotes the transpose of a vector.

From Equation (4.15), a vector of Fourier coefficients of *a priori* information is obtained as

$$(\mathbf{g_1}^q)' = \left(\alpha_{1K_1}^q \ \beta_{1K_1}^q \ \cdots \ \alpha_{1k}^q \ \beta_{1k}^q \ \cdots \ \alpha_{1K_2}^q \ \beta_{1K_2}^q \right). \tag{4.17}$$

The signal space is a Hilbert space; consequently, the closeness of the two Fourier coefficient vectors defined by Equation (4.16) and (4.17) is expressed by their inner product as

$$C_{\hat{\mathbf{r}}_m \mathbf{g_1}}^q = (\hat{\mathbf{r_m}}^q)' \cdot \mathbf{g_1}^q = \sum_{k=K_1}^{K_2} \left(\hat{a}_{mk}^q \alpha_{1k}^q + \hat{b}_{mk}^q \beta_{1k}^q \right). \tag{4.18}$$

The observation variable is obtained from the inner product. Substituting Equation (4.11) into (4.18) and exchanging the order of the sum and integration, the observation variable is obtained as

$$
\begin{aligned}
z_{m1}^q &= \frac{T}{2} C_{\hat{\mathbf{r}}_m \mathbf{g_1}}^{\prime q} \\
&= \int_0^T \tilde{r}_m^q(t) \sum_{k=K_1}^{K_2} \left[\alpha_{1k}^q \cos\left(k\frac{2\pi}{T}t \right) + \beta_{1k}^q \sin\left(k\frac{2\pi}{T}t \right) \right] dt.
\end{aligned}
$$

Recognizing that the sum on the right-hand side (RHS) is the Fourier series representation of the basis function $g_1(t)$ in the observation time period, the detection algorithm for the coherent receiver is obtained as

$$z_{m1}^q = \int_0^T \tilde{r}_m^q(t) g_1^q(t) dt. \tag{4.19}$$

The block diagram of coherent detector constructed from Equation (4.19) is depicted in Figure 4.5. The decision circuit is a comparator with zero threshold and it generates the estimate $\hat{b}_m$ of transmitted bit.

Fixed waveforms are used in conventional communications; consequently,

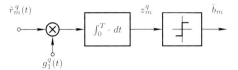

FIGURE 4.5
Block diagram of a coherent detector.

in that case q has to be suppressed in Equation (4.19) and in Figure 4.5. The basis function and decision time instants are recovered from the received signal by the carrier and clock recovery circuits, respectively. Note, the block diagram depicted in Figure 4.5 is identical with the coherent correlation receiver known from the literature [8].

Coherent chaos based communications systems are not feasible because a robust solution to the recovery of a chaotic basis function from the received signal has not yet been found.

4.4.2 Averaged optimum noncoherent detection algorithm

To get the averaged optimum noncoherent detector, the harmonic form of Fourier series representation has to be used in Equation (4.12),

$$\tilde{r}_{T,m}^q(t) = \sum_{k=K_1}^{K_2} A_k^q \cos\left(k\frac{2\pi}{T}t - \theta_k^q\right),$$

where each harmonic Fourier series component is estimated by its harmonic amplitude and phase angle

$$\hat{A}_{mk}^q = \sqrt{\left(\hat{a}_{mk}^q\right)^2 + \left(\hat{b}_{mk}^q\right)^2} \quad \text{and} \quad \hat{\theta}_{mk}^q = \tan^{-1}\left(\hat{b}_{mk}^q/\hat{a}_{mk}^q\right), \quad (4.20)$$

respectively. In the optimum noncoherent approach, the phase information is neglected and only the harmonic amplitudes are used to derive the detection algorithm.

Assume that the recovery of chaotic basis functions is not feasible; consequently, a noncoherent detection algorithm has to be developed. Because the basis functions are continuously varying, only the *averages* of harmonic amplitudes of basis functions are available:

$$\overline{\gamma_{nk}^q} = \mathrm{E}\left[\gamma_{nk}^q\right] = \mathrm{E}\left[\sqrt{(\alpha_{nk}^q)^2 + (\beta_{nk}^q)^2}\right], \quad (4.21)$$

where $\mathrm{E}[\cdot]$ denotes averaging over q and the averaged harmonic amplitudes $\overline{\gamma_{nk}^q}$ can be determined from the spectrum of chaotic basis function $g_n(t)$ [9].

Both the neglected phase information and averaging reduce the amount of exploited *a priori* information; consequently, the noise performance of averaged optimum noncoherent detector will be worse than that of the coherent one.

Projecting the received filtered waveform $\tilde{r}_m^q(t)$ into the received signal space, a vector of estimates of its harmonic amplitudes is obtained as

$$(\hat{\mathbf{r}}\mathbf{m}^q)' = \left(\hat{A}_{mK_1}^q \quad \cdots \quad \hat{A}_{mk}^q \quad \cdots \quad \hat{A}_{mK_2}^q \right). \tag{4.22}$$

The *exploited a priori* information is expressed by a vector of averaged harmonic amplitudes of basis functions

$$(\mathbf{g}\mathbf{n}^q)' = \left(\overline{\gamma_{nK_1}^q} \quad \cdots \quad \overline{\gamma_{nk}^q} \quad \cdots \quad \overline{\gamma_{nK_2}^q} \right). \tag{4.23}$$

The observation variable is the weighted inner product of Equations (4.22) and (4.23):

$$z_{mn}^q = \frac{T}{2} C_{\hat{\mathbf{r}}\mathbf{m}\mathbf{g}\mathbf{n}}^q = (\hat{\mathbf{r}}\mathbf{m}^q)' \cdot \mathbf{g}\mathbf{n}^q = \frac{T}{2} \sum_{k=K_1}^{K_2} \hat{A}_{mk}^q \overline{\gamma_{nk}^q}. \tag{4.24}$$

Substituting Equation (4.11) into (4.20), then substituting Equations (4.20) and (4.21) into (4.24), the observation variable is obtained as

$$z_{mn}^q = \sum_{k=K_1}^{K_2} \overline{\gamma_{nk}^q}$$
$$\times \sqrt{ \left[\int_0^T \tilde{r}_m^q(t) \cos\left(k\frac{2\pi}{T}t \right) dt \right]^2 + \left[\int_0^T \tilde{r}_m^q(t) \sin\left(k\frac{2\pi}{T}t \right) dt \right]^2 }. \tag{4.25}$$

Recall, the exploited *a priori* information is represented by $\overline{\gamma_{nk}^q}$, i.e., by the *averages of harmonic amplitudes* of chaotic basis functions.

Consider a binary modulation scheme where two basis functions are used and the two elements of signal set are defined by Equation (4.7). The observation variable in Equation (4.25) has to be determined for both basis functions and the decision is done in favor of bit "1" if

$$z_{11}^q > z_{22}^q \quad \text{or} \quad z_{11}^q - z_{22}^q > 0. \tag{4.26}$$

The block diagram of binary averaged optimum noncoherent detector constructed from Equations (4.25) and (4.26) is depicted in Figure 4.6.

The block diagram shown in Figure 4.6 is also valid for fixed waveform communications. Consider a binary frequency shift keying (FSK) [8] modulation with the signaling frequencies $f_1 = k_a/T$ and $f_2 = k_b/T$. In fixed waveform communications q has to be suppressed and $\overline{\gamma_{nk}^q} = \sqrt{2/T}$. The circuits included in dashed boxes in Figure 4.6 are the quadrature equivalents of *noncoherent matched filters* [8]. Each of them is matched to one of the signaling frequencies, the upper and lower ones to f_1 and f_2, respectively. Note, the block diagram depicted in Figure 4.6 is the generalization of the optimum noncoherent receiver concept to the varying waveform communications.

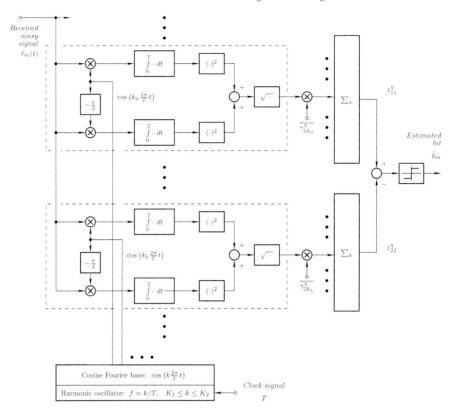

FIGURE 4.6
Block diagram of an averaged optimum noncoherent detector.

4.4.3   Autocorrelation detection algorithm

Binary DCSK/FM-DCSK uses two basis functions, the orthogonality of $g_1^q(t)$ and $g_2^q(t)$ is assured by the first two Walsh functions. The spectra of $g_1^q(t)$ and $g_2^q(t)$ are separated in the frequency domain; this separation is exploited to derive the autocorrelation detection algorithm. Because the least amount of *a priori* information is exploited in autocorrelation detector, it offers the worst noise performance. To find the separation of spectra in frequency domain, the Fourier coefficients of the two basis functions have to be determined.

Let $T_C = T/2$ denote the chip duration. The Fourier series coefficients of $g_{T,1}^q(t)$ are obtained from Equation (4.8) as

$$\alpha_{1k}^q = \frac{2}{T}\int_0^T g_{T,1}^q(t)\cos(k\frac{2\pi}{T}t)dt = \frac{1}{T_C}\int_0^{T_C} c^q(t)\cos(k\frac{2\pi}{T}t)dt$$
$$+ \frac{1}{T_C}\int_{T_C}^T c^q(t-T_C)\cos\left[k\frac{2\pi}{T}(t-T_C+T_C)\right]dt,$$

where $k\frac{2\pi}{T}(T_C - T_C)$ has been added to the argument of second cosine function on the RHS. By introducing a new variable $\hat{t}=t - T_C$ in the second term of the RHS and using the trigonometric identity $\cos(\alpha + \beta) = \cos(\alpha)\cos(\beta) - \sin(\alpha)\sin(\beta)$, we get

$$\alpha_{1k}^q = \frac{1}{T_C}\int_0^{T_C} c^q(t)\cos(k\frac{2\pi}{T}t)dt + \frac{\cos(k\pi)}{T_C}\int_0^{T_C} c^q(\hat{t})\cos(k\frac{2\pi}{T}\hat{t})d\hat{t}$$
$$-\frac{\sin(k\pi)}{T_C}\int_0^{T_C} c^q(\hat{t})\sin(k\frac{2\pi}{T}\hat{t})d\hat{t}.$$

Since $\sin(k\pi) = 0$ and $\cos(k\pi) = (-1)^k$, $k = 1, 2, \cdots$, we obtain

$$\alpha_{1k}^q = \begin{cases} \frac{2}{T_C}\int\limits_0^{T_C} c^q(t)\cos(k\frac{2\pi}{T}t)dt, & \text{for even } k \\ 0, & \text{for odd } k. \end{cases} \tag{4.27}$$

In a similar manner we get

$$\beta_{1k}^q = \begin{cases} \frac{2}{T_C}\int\limits_0^{T_C} c^q(t)\sin(k\frac{2\pi}{T}t)dt, & \text{for even } k \\ 0, & \text{for odd } k. \end{cases} \tag{4.28}$$

The following essential conclusions may be drawn from Equations (4.27) and (4.28): (i) the fundamental period of the Fourier series representation of $g_1(t)$ is equal to the observation time period T, (ii) the spectrum of the first basis function $g_1(t)$ has only *even* harmonic components.

Applying the same approach to the second basis function, the Fourier coefficients of $g_{T,2}(t)$ are obtained as

$$\alpha_{2k}^q = \begin{cases} 0, & \text{for even } k \\ \frac{2}{T_C}\int\limits_0^{T_C} c^q(t)\cos(k\frac{2\pi}{T}t)dt, & \text{for odd } k \end{cases} \tag{4.29}$$

and

$$\beta_{2k}^q = \begin{cases} 0, & \text{for even } k \\ \frac{2}{T_C}\int\limits_0^{T_C} c^q(t)\sin(k\frac{2\pi}{T}t)dt, & \text{for odd } k. \end{cases} \tag{4.30}$$

Note, the spectrum of second basis function $g_2(t)$ has only *odd* harmonic components.

To verify Equations (4.27)–(4.30), the spectrum of a modulated FM-DCSK signal with bit duration $T = 2$ μs and $2B = 17$ MHz was determined by computer simulation. The center frequency of modulated FM-DCSK signal was 2.4 GHz. Figures 4.7(a) and 4.7(b) show the spectrum when a pure bit "1" sequence and a pure bit "0" sequence, respectively, were generated. As predicted, the spectra of two FM-DCSK signals are fully separated in the frequency domain, although they overlap each other. The two spectra may be interpreted as the teeth of two combs fitted into one another.

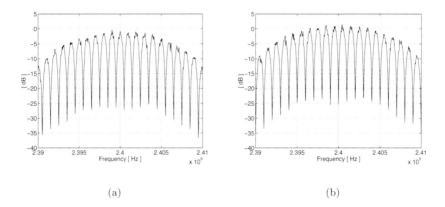

(a)   (b)

FIGURE 4.7
Spectrum of FM-DCSK signal with $1/T = 500$ kHz when (a) a pure bit "1" and (b) a pure bit "0" sequence are transmitted.

The autocorrelation detector is a *noncoherent* detector which exploits *only the separation* of spectra of two basis functions in the frequency domain. Let $\overline{\gamma^q_{nk}} = 1$ be set in Equation (4.25) for that subspace in which the received energy has to be determined. Then Equation (4.25) can be rearranged as

$$\frac{2}{T}\left(z^q_{mn}\right)^2 = \frac{T}{2}\sum_{k=K_1}^{K_2}\left(\left[\frac{2}{T}\int_0^T \tilde{r}^q_m(t)\cos\left(k\frac{2\pi}{T}t\right)dt\right]^2 \right. \\ \left. + \left[\frac{2}{T}\int_0^T \tilde{r}^q_m(t)\sin\left(k\frac{2\pi}{T}t\right)dt\right]^2\right). \tag{4.31}$$

Substituting Equation (4.11) into (4.31) and applying the Parseval theorem, the received energy measured in the mth subspace is obtained as

$$\frac{2}{T}\left(z^q_{mn}\right)^2 = \frac{T}{2}\sum_{k=K_1}^{K_2}\left[\left(\hat{a}^q_{mk}\right)^2 + \left(\hat{b}^q_{mk}\right)^2\right] \equiv E^q_m. \tag{4.32}$$

The FM-DCSK autocorrelation detector determines the received energy in the "*even*" and "*odd*" subspaces defined by Equations (4.27)–(4.28) and (4.29)–(4.30), respectively, and makes the decision in favor of subspace, and bit, that collects the larger amount of energy.

Let the energy measured in the even subspace be calculated first. The even subspace is defined by the first basis functions and is identified by $m = 1$. According to Equations (4.27)–(4.28), the energy components have to be collected along the even harmonics in the received signal space; consequently,

$$\overline{\gamma_{1k}^{q}} = \begin{cases} 1, & \text{for even } k \\ 0, & \text{for odd } k \end{cases}$$

has to be submitted into Equation (4.25).

Consider the FM-DCSK modulation scheme here. The energy per bit is kept constant by FM in FM-DCSK; consequently, the upper index q is suppressed in the remaining part of this section. The energy measured in the even subspace is obtained from Equations (4.31) and (4.32) as

$$E_1 = \frac{T_C}{4} \sum_{k=\hat{K}_1}^{\hat{K}_2} \left(\left[\frac{2}{T_C} \int_0^{T_C} \tilde{r}_m(t) \cos(\hat{k}\frac{2\pi}{T_C}t)dt + \right. \right.$$

$$+ \frac{2}{T_C} \int_{T_C}^{T} \tilde{r}_m(t) \cos(\hat{k}\frac{2\pi}{T_C}t)dt \right]^2 + \left[\frac{2}{T_C} \int_0^{T_C} \tilde{r}_m(t) \sin(\hat{k}\frac{2\pi}{T_C}t)dt + \right.$$

$$\left. \left. + \frac{2}{T_C} \int_{T_C}^{T} \tilde{r}_m(t) \sin(\hat{k}\frac{2\pi}{T_C}t)dt \right]^2 \right), \quad (4.33)$$

where $T_C = T/2, \hat{K} = K/2, \hat{K}_1 = K_1/2$ and $\hat{K}_2 = K_2/2$ have been substituted and each integral in Equation (4.31) has been decomposed into two parts: first the reference, then the information bearing parts of received signal are integrated. Since k is even for bit "1," $\hat{k}$ is a positive integer.

To have only one observation time period, i.e., the simplest detector configuration, each integral in Equation (4.33) should be evaluated from T_C to T.

Let a new variable $\hat{t} = t + T_C$ be introduced. Since $\cos[\hat{k}\frac{2\pi}{T_C}(\hat{t} - T_C)] = \cos(\hat{k}\frac{2\pi}{T_C}\hat{t} - 2\pi\hat{k}) = \cos(\hat{k}\frac{2\pi}{T_C}\hat{t})$, the first term in Equation (4.33) becomes

$$\frac{2}{T_C} \int_0^{T_C} \tilde{r}_m(t) \cos(\hat{k}\frac{2\pi}{T_C}t)dt = \frac{2}{T_C} \int_{T_C}^{T} \tilde{r}_m(\hat{t} - T_C) \cos(\hat{k}\frac{2\pi}{T_C}\hat{t})d\hat{t}. \quad (4.34)$$

In a similar manner, the third term in (4.33) may be written as

$$\frac{2}{T_C} \int_0^{T_C} \tilde{r}_m(t) \sin(\hat{k}\frac{2\pi}{T_C}t)dt = \frac{2}{T_C} \int_{T_C}^{T} \tilde{r}_m(\hat{t} - T_C) \sin(\hat{k}\frac{2\pi}{T_C}\hat{t})d\hat{t}. \quad (4.35)$$

Substituting Equations (4.34) and (4.35) into (4.33) and introducing $t = \hat{t}$

as a new variable we get

$$
E_1 = \frac{T_C}{4} \sum_{\hat{k}=\hat{K}_1}^{\hat{K}_2} \left(\left[\frac{2}{T_C} \int_{T_C}^{T} \tilde{r}_m(t - T_C) \cos(\hat{k}\frac{2\pi}{T_C}t)dt + \right. \right.
$$
$$
\left. + \frac{2}{T_C} \int_{T_C}^{T} \tilde{r}_m(t) \cos(\hat{k}\frac{2\pi}{T_C}t)dt \right]^2 + \left[\frac{2}{T_C} \int_{T_C}^{T} \tilde{r}_m(t - T_C) \sin(\hat{k}\frac{2\pi}{T_C}t)dt + \right.
$$
$$
\left. \left. + \frac{2}{T_C} \int_{T_C}^{T} \tilde{r}_m(t) \sin(\hat{k}\frac{2\pi}{T_C}t)dt \right]^2 \right). \quad (4.36)
$$

Recall, the received signal space is a discrete Hilbert space where every signal is represented by its periodic equivalent. The FM-DCSK signal is constructed from two chips; the first and second ones are referred to as reference and information bearing chips, respectively. These two chips are identified by "**R**" and "**I**" in Figure 4.8(a).

The expression in the bracket on the RHS of Equation (4.36) can be interpreted as follows: the second and fourth terms give the Fourier series coefficients of the information bearing chip, while the first and third terms define those of the *delayed* reference chip. These two chips are depicted in Figure 4.8(b).

Figure 4.8(b) shows the delayed reference and information bearing chips represented in the even subspace, a discrete Hilbert space. Similarly to Equation (4.9), the two chips are substituted by periodic ones where the two chips and their periodic equivalents are identical in the time period of $T_C \leq t < T$, in the time interval in which the integrals in Equation (4.36) are evaluated.

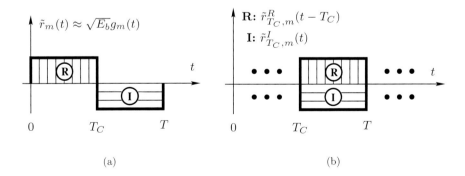

(a)   (b)

FIGURE 4.8
Structure of received FM-DCSK signal: (a) in its original form and (b) its equivalent represented in the even discrete subspace after time shifting.

Note, the period time is equal to T_C and not T in the even subspace. To avoid confusion, the reference and information bearing chips are plotted only in the time period of $T_C \leq t < T$ in Figure 4.8(b) and the periodicity of signals is marked by dots.

Let $\hat{a}_{1\hat{k}}^{R,del}$ and $\hat{b}_{1\hat{k}}^{R,del}$ denote the estimated Fourier series coefficients of the delayed reference $\tilde{r}_{T_C,m}^R(t - T_C)$ and let $\hat{a}_{1\hat{k}}^I$ and $\hat{b}_{1\hat{k}}^I$ define those of the information bearing $\tilde{r}_{T_C,m}^I(t)$ chips. Then from Equation (4.36) we obtain

$$E_1 = \frac{T_C}{4} \sum_{\hat{k}=\hat{K}_1}^{\hat{K}_2} \left[\left(\hat{a}_{1\hat{k}}^{R,del} + \hat{a}_{1\hat{k}}^I \right)^2 + \left(\hat{b}_{1\hat{k}}^{R,del} + \hat{b}_{1\hat{k}}^I \right)^2 \right]. \tag{4.37}$$

The signals $\tilde{r}_{T_C,m}^R(t - T_C)$ and $\tilde{r}_{T_C,m}^I(t)$ are periodic signals with a period time of T_C. The Parseval's identity expresses the relationship between the average power of a periodic signal and its Fourier series coefficients. The received signal is an RF bandpass signal; consequently, it has a zero DC component. Exploiting the Parseval's identity, Equation (4.37) can be interpreted as follows:

- RHS of Equation (4.37) shows that the sum of two RF bandpass periodic signals with the same periodicity T_C has to be considered in the time domain;

- the two bandpass signals are the delayed reference and the information bearing chips of received signal as depicted in Figure 4.8(b).

Based on these observations, Equation (4.37) can be reformulated as

$$E_1 = \frac{1}{2} \int_{T_C}^T \left[\tilde{r}_{T_C,m}^R(t - T_C) + \tilde{r}_{T_C,m}^I(t) \right]^2 dt.$$

The limits of integration show that the detector observes the received signal only in the time period of $T_C \leq t < T$. Note, the delayed reference and the information bearing chips are equal to $\tilde{r}_{T_C,m}^R(t - T_C)$ and $\tilde{r}_{T_C,m}^I(t)$, respectively, in this time interval. Hence, the energy received in the even subspace is obtained as

$$E_1 = \frac{1}{2} \int_{T_C}^T \left[\tilde{r}_m(t - T_C) + \tilde{r}_m(t) \right]^2 dt. \tag{4.38}$$

When bit "0" is transmitted then, according to Equations (4.29) and (4.30), the transmitted energy occupies only the odd subspace. Applying the steps discussed above, the energy received in the odd subspace is obtained as

$$E_2 = \frac{1}{2} \int_{T_C}^T \left[-\tilde{r}_m(t - T_C) + \tilde{r}_m(t) \right]^2 dt. \tag{4.39}$$

The decision is done in favor of bit "1" if

$$E_1 > E_2 \quad \text{or} \quad E_1 - E_2 > 0. \tag{4.40}$$

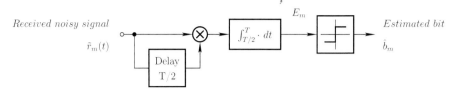

FIGURE 4.9
Block diagram of autocorrelation detector.

In FM-DCSK $T_C = T/2$. Substituting Equations (4.38) and (4.39) into (4.40), the autocorrelation detection algorithm is obtained. The decision is done in favor of bit "1" if

$$\int_{T/2}^{T} \tilde{r}_m(t - T/2)\tilde{r}_m(t)dt > 0 . \qquad (4.41)$$

The block diagram of autocorrelation detector constructed from Equation (4.41) is shown in Figure 4.9. Note, the autocorrelation detector determines the sign of correlation measured between the reference and information bearing chips, and the decision is done according to the sign of correlation. The only *a priori* information exploited by the detector is the sign of correlation. Because of the minimum amount of *a priori* information exploited, the autocorrelation detector has the worst noise performance [11].

4.5   Implementation of a microwave FM-DCSK system

Although a lot of materials have been published on chaotic communications systems, each of them uses only computer simulations to verify the results derived and the feasibility of chaos-based communications. The reason is very simple: the modulation schemes and demodulation algorithms required in chaotic communications are completely different from those used in conventional communications, hence, the integrated circuits developed for conventional radio communications cannot be used to implement a chaotic transceiver.

The advent of software defined electronics (SDE) has made the implementation of chaotic communication systems easy and possible. In the SDE approach (i) the entire transceiver is implemented in baseband in software (SW) and (ii) a universal RF hardware device is used to reconstruct the analog microwave/RF signal from its complex envelope and to extract the complex envelope from the received microwave/RF signal at the transmitter and receiver, respectively.

Bandpass microwave/RF signals are used in radio communications where, typically, the center frequency of modulated signal is much greater than its bandwidth. These signals can be *fully represented* without any distortion in BB by their complex envelopes [8]. The complex envelope is a complex-valued signal; its real and imaginary parts are referred to as the I/Q components of BB signals. The I/Q components are *lowpass signals* and their bandwidth is equal to the half of the bandwidth of RF bandpass signal. Hence, the sampling frequency required to process the BB signal is determined by the *bandwidth* of bandpass RF signal. The center frequency of RF bandpass signal has no influence on the required sampling rate, hence, the complex envelope concept assures the minimum sampling rate required.

The lowpass complex envelopes, i.e., the BB equivalent signals, are digitized and processed by a host computer or field programmable gate array (FPGA) and the entire transceiver is implemented in SW.

This technology is used here to demonstrate the feasibility of an FM-DCSK microwave radio transceiver. The details of SDE approach and the derivation of BB equivalents are not discussed here; for details refer to [15].

4.5.1 HW platform of implementation: The USRP device

The Universal Software Radio Peripheral (USRP) device [2] is a computer-hosted hardware platform developed for the implementation of software defined radio or low-accuracy virtual instrumentation. Each USRP device includes a receive and a transmit block. These blocks are used to generate the BB signals from the received RF one in the receive block and to reconstruct the RF bandpass signal from its BB equivalent in the transmit block. The USRP device performs all the waveform-specific signal processing tasks (such as modulation and demodulation) in BB on a host-computer while all the general purpose operations requiring high-speed data processing (such as interpolation and decimation) are carried out on an FPGA available on the main board of a USRP device [15].

4.5.2 Software platform for accessing the USRP device

LabVIEW [1] is a data-flow type graphical programming software environment used to get access to the USRP device. It offers a graphical user interface and an icon-based graphical code development environment referred to as "Front Panel" and "Block Diagram," respectively. LabVIEW platform provides icons to perform any kinds of signal processing tasks, data acquisition, and even remote control of stand-alone equipment.

The USRP device implements the physical layer in the OSI basic reference model; consequently, it has to offer two kinds of services for the host-computer: (i) USRP management service to set the USRP parameters such as receiver gain, carrier frequency, transmitted power, etc., and (ii) USRP data service to transfer the BB signal components. The former can be done via the configura-

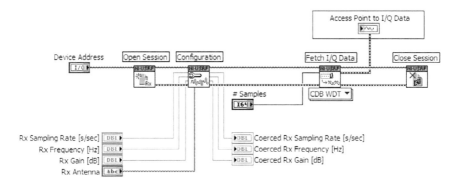

FIGURE 4.10
Complete control chain of the USRP device in receive mode.

tion icons, while the latter is performed by the "Fetch I/Q Data" icon for the receive operation and "Write I/Q Data" icon for the transmit operation. The software implementation of the receive control chain is depicted in Figure 4.10 where each icon is identified by its function shown above the icon.

The USRP devices are connected to the host computer via a Gigabit Ethernet interface and are identified by their IP addresses entered via the "Device Address" icon. This icon also opens the session with the USRP device. Then the following USRP parameters are entered via the "Configuration" icon:

- Rx Sampling Rate [s/sec]: sample rate applied to the BB waveforms.
- Rx Frequency [Hz]: assigns the RF center frequency.
- Rx Gain [dB]: specifies the gain of RF amplifiers.
- Rx Antenna: selects the transceiver configuration (duplex or simplex).

The USRP device keeps these parameters under control. If a desired combination of parameters is not allowed then the USRP device selects the closest parameter set available. These actual, referred to as "Coerced," parameters are returned by the "Configuration" icon. The "Fetch I/Q Data" icon uploads the digitized BB signal from the USRP device to the host-computer. It is also used to select the number of BB samples to be uploaded. Note, using the OSI terminology, this icon provides the access point to the I/Q components of BB equivalent of received RF bandpass signal. Finally, the "Close Session" icon terminates the session with the USRP device.

4.5.3   Measured spectra of FM-DCSK signals

Recall, the spectra of an FM-DCSK signal carrying pure bit "1" and "0" sequences are completely separated in the frequency domain. The theoretical results concluded in Equations (4.27)–(4.30) have been verified in Section 4.4.3 by computer simulations. The results of simulations are shown in Figure 4.7.

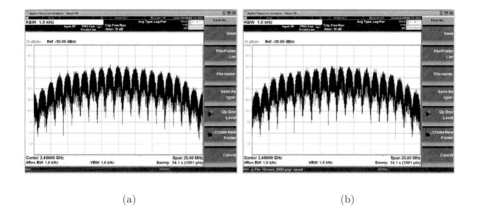

(a)   (b)

FIGURE 4.11
Measured spectra of a 2.4-GHz FM-DCSK signal when (a) a pure bit "1" and
(b) a pure bit "0" sequence are transmitted. The data rate is 500 kHz.

To verify the results of theoretical investigations and the feasibility of
chaotic radio communications by *measurements*, an FM-DCSK transceiver
with an autocorrelation detector has been implemented using the USRP HW
and LabVIEW SW platforms. The center frequency and data rate of implemented FM-DCSK radio system are 2.4 GHz and 500 kHz, respectively.

Figures 4.11(a) and 4.11(b) show the spectra measured with a stand-alone
microwave spectrum analyzer. The analyzer settings are shown in the figures.
As expected from the theory, the spectra belonging to the pure bit "1" and
"0" sequences are fully separated.

4.5.4   2.4-GHz FM-DCSK transceiver

The Front Panel of the implemented 2.4-GHz FM-DCSK transceiver is shown
in Figure 4.12. The configuration parameters of the USRP device are entered
and the coerced parameter values are returned on the left side of the Front
Panel. The bit duration T is also entered on the left subpanel.

The top left and right figures show I/Q components of BB signal in the time
domain and the spectrum of BB signal when a random bit stream is received.
The left figure in the bottom shows the observation signal generated by the
autocorrelation detector. This signal, denoted by E_m in Figure 4.9, is used by
the timing recovery and decision algorithms to perform the demodulation.

The RF bandwidth of FM-DCSK signal, set to 17 MHz, is determined by
the chaotic signal generator implemented by a Bernoulli map and by the FM
modulator [9].

The bottom right plot shows the value of observation signal at the decision

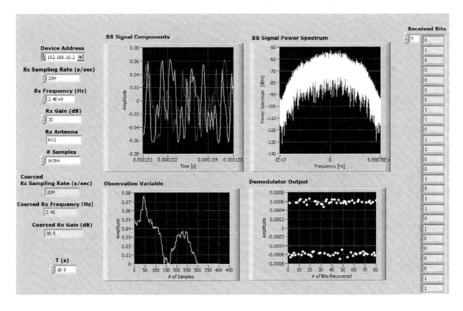

FIGURE 4.12

Front panel of the implemented FM-DCSK receiver.

time instants in Figure 4.9. The received bit stream is visualized on the right side of the Front Panel.

The autocorrelation detector of Figure 4.9 has been derived for the reception of a single isolated bit. To avoid InterSymbol Interference (ISI) the 2.4-GHz FM-DCSK receiver has been designed in such a way that the Nyquist ISI criterion is satisfied [8].

4.6   Conclusions

The waveform communications concept establishes a common framework for the description of any digital modulation scheme using either sinusoidal, chaotic, or random carriers. In waveform communications every symbol to be transmitted is mapped into an analog waveform of finite duration.

The most important feature of chaotic communications is that the transmitted signal is never periodic; instead, it is continuously varying even if the same symbol is transmitted repeatedly.

Transmitted waveforms are constructed from basis functions. Unfortunately, chaotic basis functions are orthonormal only in mean. This leads to the auto- and cross-correlation estimation problems which, if not prevented, corrupt the BER performance of the chaotic communications system.

Chaotic signals have no amplitude, frequency, or phase; consequently, the conventional modulation schemes cannot be used. The most popular chaotic modulation schemes are CSK, DCSK, and FM-DCSK. Among them the FM-DCSK offers the best BER performance.

The mathematical framework that makes the derivation of detection algorithms possible is the received signal space which is a finite dimensional Hilbert space. Both the *a priori* information and the received signal are projected into the received signal space and the decision is done in favor of symbol that is the closest to the received signal.

To show the applicability of the theory developed, the derivation of coherent, averaged optimum noncoherent and autocorrelation detection algorithms and detector configurations have been shown.

Note, going from coherent detection algorithm to the autocorrelation one there is a continuous loss in *a priori* information exploited. The ability of a detector to suppress noise and interference depends on the amount of *a priori* information available and *exploited* by the detection algorithm. Hence, if only the noise performance and interference rejection capability are considered then the coherent and autocorrelation detectors offer the best and worst, respectively, system performance.

To verify the feasibility of chaotic communications, an FM-DCSK radio link operating in the 2.4-GHz ISM band has been implemented using the SDE approach. In the SDE approach each RF signal processing block including the modulator and demodulator is implemented in software and in the baseband, and a universal hardware device referred to as USRP is used to convert the signals between the RF domain and baseband.

4.7   Acknowledgments

The NI USRP devices and NI LabVIEW software used to implement the FM-DCSK radio link have been donated by the National Instruments. This work has been sponsored by the Hungarian Scientific Research Found (OTKA) under Grant number T-084045.

Bibliography

[1] LabVIEW System Design Software. *National Instruments.* Online: <http://www.ni.com/labview/>.

[2] NI Universal Software Radio Peripheral (USRP). *National Instruments*. Online: <http://sine.ni.com/nips/cds/view/p/lang/en/nid/209947>.

[3] B. L. Basore. *Noise-Like Signals and Their Detection by Correlation*. PhD thesis, Massachusetts Institute of Technology, Cambridge, 1952.

[4] J. S. Bendat and A. G. Piersol. *Measurement and Analysis of Random Data*. John Wiley & Sons, New York, 1966.

[5] H. Dedieu, M.P. Kennedy, and M. Hasler. Chaos shift keying: modulation and demodulation of a chaotic carrier using self-synchronizing chua's circuits. *IEEE Transactions on Circuits and Systems II*, 40:634–642, October 1993.

[6] Federal Communications Commission. *Part 15 of the Commission Rs Rules Regarding Ultra-Wideband Transmission Systems; Subpart F*. FCC–USA, Online: <http://sujan.hallikainen.org/FCC/FccRules/2009/15/>.

[7] M. Hasler, G. Mazzini, M. Ogorzalek, R. Rovatti, and G. Setti (Eds.), Special Issue on Applications of Nonlinear Dynamics to Electronic and Information Engineering. *Proceedings of the IEEE*, 90(5), May 2002.

[8] S. Haykin. *Communication Systems*. John Wiley & Sons, New York, 3rd edition, 1994.

[9] M. P. Kennedy, G. Kolumbán, G. Kis, and Z. Jákó. Performance evaluation of FM-DCSK modulation in multipath environments. *IEEE Transactions on Circuits and Systems I*, 47(12):1702–1711, December 2000.

[10] G. Kis and G. Kolumbán. Constraints on chaotic oscillators intended for communications applications. In *Proc. NOLTA'98*, pages 883–886, Crans-Montana, Switzerland, September 14–17, 1998.

[11] G. Kolumbán. Theoretical noise performance of correlator-based chaotic communications schemes. *IEEE Transactions on Circuits and Systems I*, 47(12):1692–1701, December 2000.

[12] G. Kolumbán and M. P. Kennedy. The role of synchronization in digital communication using chaos—Part III: Performance bounds. *IEEE Transactions on Circuits and Systems I*, 47(12):1673–1683, December 2000.

[13] G. Kolumbán, M. P. Kennedy, Z. Jákó, and G. Kis. Chaotic communications with correlator receiver: Theory and performance limit. *Proceedings of the IEEE*, 90(5):711–732, May 2002.

[14] G. Kolumbán, G. Kis, Z. Jákó, and M. P. Kennedy. FM–DCSK: A robust modulation scheme for chaotic communications. *IEICE Transactions on Fundamentals of Electronics, Communications and Computer Sciences*, E81-A(9):1798–1802, October 1998.

[15] G. Kolumbán, T. Krébesz, and F. C. M. Lau. Theory and application of software defined electronics: Design concepts for the next generation of telecommunications and measurement systems. *IEEE Circuits and Systems Magazine*, 12(2):8–34, Second Quarter 2012.

[16] G. Kolumbán, F. C. M. Lau, and C. K. Tse. Generalization of waveform communications: The Fourier analyzer approach. *Circuits, Systems and Signal Processing*, 24(5):451–474, September/October 2005.

[17] G. Kolumbán and B. Vizvári. Nonlinear dynamics and chaotic behaviour of the analog phase-locked loop. In *Proc. NDES'95*, pages 99–102, Dublin, Ireland, July 1995.

[18] G. Kolumbán, B. Vizvári, W. Schwarz, and A. Abel. Differential chaos shift keying: A robust coding for chaos communication. In *Proc. NDES'96*, pages 87–92, Seville, Spain, June 27–28, 1996.

[19] T. S. Parker and L. O. Chua. *Practical Numerical Algorithms for Chaotic Systems*. Springer Verlag, New York, 1989.

[20] U. Parlitz, Leon O. Chua, L. Kocarev, K. S. Halle, and A. Shang. Transmission of digital signals by chaotic synchronization. Technical Report UCB/ERL M92/129, EECS Department, University of California, Berkeley, 1992.

[21] J. G. Proakis. *Digital Communications*. McGraw-Hill, Singapore, 1995.

[22] K. Siwiak and D. McKeown. *Ultra-Wideband Radio Technology*. Wiley, Chichester, UK, 2004.

[23] R. van Nee and R. Prasad. *OFDM for Wireless Multilmedia Communications*. Artech House Publishers, Boston, 2000.

5

Chaos in optical communications

Apostolos Argyris and Dimitris Syvridis
Department of Informatics and Telecommunications
National and Kapodistrian University of Athens

CONTENTS

5.1 Introduction

Security in communications is currently a field of particular importance. As computational power increases significantly each year, the software safety measures against eavesdroppers is consistently evaluated and upgraded. Various cryptographic methods are consistently used to protect all types of data transferred through intranet and internet links. Such cryptographic techniques are based on number theory or algebraic models. Secret key cryptography uses a secret key, such as the data encryption standard (DES) and the advanced encryption standard (AES) algorithms, in which both sender and receiver use the same key to encrypt and decrypt. Despite the speed of this method, getting the secret key to the recipient in the first place is a problem that is often handled by a public-key procedure. On the other hand, public-key cryptography is used to protect sensitive data during transmission over various channel types that support personalized communication [31, 92] and includes tasks such as message encryption, key exchange, digital signatures, and digital certificates [76]. The above algorithmic types of cryptography secure the upper layers of any type of communications, including the optical communication systems, regardless of the transmission medium. In the last decade strong research activity has been recorded in securing data transmission at the hardware level, by taking into account the properties of the transmission medium. Despite that fiber optic networks are considered more secure than standard wired or wireless networks, fiber cabling is just as vulnerable to eavesdropping as in other transmission media. Tapping into fiber optic cables with relatively inexpensive and appropriate equipment, an experienced eavesdropper can perform a successful attack. Although most of this cabling is difficult to access - it's underground, undersea, encased in concrete, etc. - there are plenty of accessible points for eavesdroppers. The potential of unauthorized access on the fiber infrastructure has motivated the development of systems that provide transmission security by hardware encryption. Two main categories of this type of security have been identified and investigated so far: *quantum cryptography* exploits the quantum nature of light while *chaos encryption* exploits the potential of optical emitters to operate under chaotic conditions.

Quantum cryptography is a technique for two parties to form a key on an open optical network [21,41,117]. Such keys can be used for the encryption of data exchanged between the two parties. The advantage of quantum cryptography is that fundamental laws of quantum mechanics guarantee its security. It allows the detection of an unauthorized eavesdropper, while it guarantees security when there is no eavesdropper present. This is not possible using any other form of key distribution, which relies either upon the difficulty of factorizing large numbers, or the assumed privacy of the network. In optical quantum cryptography the bits used to form the key are carried by single photons traveling either along an optical fiber or in an optical free space link. Information can be encoded on the photons in a variety of ways, such as by their polarization or phase. Since the information is carried by a single photon, it is not possible for a hacker to tap in and remove part of the signal. If a hacker collects a photon in the tapped fiber, the photon will not be received at the other end, alerting the intended recipient of the presence of the hacker. Furthermore, the technique is also secure from a slightly more sophisticated type of eavesdropping where the hacker first measures the photons and then retransmits them. This is because the laws of quantum mechanics tells us quantum bits (or qubits) of information, such as encoded single photons, have the peculiar characteristic that they are disturbed by measurement. This fact allows the legitimate receiver of the message to test whether it has been intercepted or altered by a hacker on the channel. Quantum cryptography belongs to the class of hardware-key cryptography and thus can be used only to exchange a secret key and is not suitable for real time data encryption, at least up to now [111]. The reason is related to the low bit-rate (in the order of tens of kHz) and the incompatibility with some key components of the optical communication systems - the optical amplifiers - that are needed for long distance transmission links.

A different approach to strengthen the security of optical high-speed data transmitted in fiber networks has been investigated extensively within the last fifteen years. Within this approach, data are encoded at the physical layer utilizing chaotic carriers generated by lasers that operate in the nonlinear regime. The objective of chaos hardware encryption is to encode the information signal within a chaotic carrier generated by components whose physical, structural, and operating parameters form the secret key. Once the information encoding is carried out, the chaotic carrier is sent for transmission to the authorized receiver. Decoding is achieved directly in real-time through the *chaos synchronization* process: the optical chaotic carrier is reproduced at the receiver - excluding any external interference, such as the encrypted message - and is canceled. The concept of the chaos-based optical communications systems is shown in Figure 5.1.

In conventional optical communications a semiconductor laser generates a coherent optical carrier on which the information is encoded using various modulation schemes. In the proposed approach of the chaos-based optical communications, the semiconductor laser is driven to operate in the chaotic regime by applying a feedback loop that leads it to coherence collapse. With this method, the generated optical carrier can exhibit an extremely broadband spectrum up to tens of GHz. The information is encoded on this chaotic carrier using different techniques (e.g., a simple yet efficient method is to use an external optical modulator electrically driven by the information bit stream while at its input is coupled the optical chaotic carrier). The amplitude of the encrypted message in any case should be kept small in respect to the amplitude fluctuations of the chaotic carrier, so that it would be practically impossible to extract

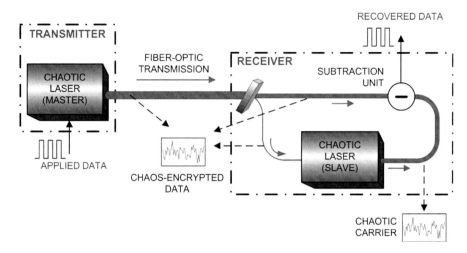

FIGURE 5.1

The concept of an optical communication system secured by chaotic data encryption.

this encoded information using conventional techniques like linear filtering, frequency domain analysis, or phase-space reconstruction. Especially the latter assumes a high complexity of the chaotic carrier and is directly dependent on the method that the chaos dynamics are generated. At the receiver side of the system a second semiconductor is used, identical to that of the transmitter. This identity refers to the semiconductor laser structural, emission (emitting wavelength, slope efficiency, current threshold, etc.), and intrinsic (linewidth enhancement factor, non-linear gain, photon lifetime, etc.) parameters, as well as to the feedback loop characteristics that contribute to chaos generation (cavity length, cavity losses, possible non-linearity, etc.) and the operating parameters (bias currents, feedback strength, etc.). The above set of hardware-related parameters constitutes the key of the encryption procedure.

The message extraction procedure is a result of the so-called *synchronization process*. Synchronization expresses that the irregular time evolution of the chaotic emitter's output in the optical power can be reproduced by the receiver, provided that both transmitter and receiver chaotic oscillators are identical in terms of the above set of parameters. Even minor discrepancies between the two oscillators can result in degraded synchronization, which means deviation from a perfect reproduction of the emitter's chaotic carrier. The key process for efficient message decoding resides in the fact that the receiver synchronizes to the chaotic oscillations of the emitter's carrier without being affected by the encoded message, referred to in the literature as *chaos filtering effect*. In the implementation of the receiver, an appropriate part of the incoming signal with the encoded information is injected into the receiver. Assuming all those conditions that lead to a sufficiently good synchronization quality, the receiver generates at its output a chaotic carrier almost identical to the injected, without the encoded information. Therefore, by subtracting the chaotic carrier from the incoming chaotic signal with the encoded information, the useful information is accessible.

Chaos-based secure communications systems provide some major advantages. They support real-time high-bit rate message encoding since the data encoding process does not introduce any additional delay relative to that of the conventional optical communication systems. The same holds for the receiver at least for bit rates up to 10 Gb/s since the synchronization process relies on the ultrafast dynamics of semiconductor lasers, the time response of the detection photoreceivers, etc. This is a significant advancement relative to the conventional software based approaches, where real time encoding of the bit stream - exploiting fast processors and sufficiently long bit series key - would result in much lower effective bit rate, and increased complexity and cost of the system. Moreover, chaos-encrypted optical communications can act complementarily to software encryption, providing a higher level of security. Compared to the quantum cryptography, the chaos-based approach provides the apparent advantage of significantly faster data transmission speed. In terms of security, a potential eavesdropper can exploit two methods to extract the chaos encoded information. The first one is to reconstruct the chaotic attractor in the phase space using strongly correlated points densely sampled in time. In this case, the number of needed samples increases exponentially with the chaos dimension. Taking into account the attractor dimension of the generated chaotic optical carriers, in which, its Lyapunov dimension may be of the order of a few hundred, and considering the potential of today's electronics and computational power this solution seems to be impossible. The second one is to identify the key for reconstruction of the chaotic time series. In the case of the chaotic encryption, the key is the hardware used and the full set of operating parameters. This means that if a semiconductor laser coupled to an external cavity is, e.g., the chaotic oscillator in the emitter, the eavesdropper must have an identical laser diode, with identical external resonator providing the same amount of feedback, and know the complete set of operating parameters. Finally, chaos optical encryption allows compatibility with the installed network infrastructure since there is no fundamental reason to preclude its application on installed optical network infrastructure. With proper compensation of fiber transmission impairments the chaotic signal that arrives at the receiver triggers the synchronization process successfully. All feasibility experiments showed that the use of erbium-doped fiber amplifiers (EDFAs) might induce some power penalty in the decoded data due to noise addition, but in any case does not prevent from synchronizing and extracting the encrypted information.

The concept of chaos synchronization was first proposed theoretically by Pecora and Carroll in 1990 [90]. This pioneering work triggered a burst of activities covering the early 1990s, introducing mainly electronic chaotic oscillators [28]. The first theoretical work and preliminary reports and possibility of synchronization between optical chaotic systems came out in the 1990s [9, 27, 77]. Since then the activities in the area of optical chaotic oscillators increased exponentially. Numerous research groups worldwide reported a large amount of theoretical and experimental work, covering mainly fundamental aspects related to synchronization of optical non-linear dynamical systems [14, 69, 85, 110, 113]. Special focus was given to semiconductor laser-based systems [47, 97], but there was also work on fiber-ring laser systems [2] and optoelectronic schemes [3, 43]. The applicability of the concept in optical communication systems was initially proved by encoding and recovery of single frequency tones, starting from frequencies of a few kHz [63] up to several GHz [88]. However, it's worth mentioning that single frequency encoding is much less demanding in terms of chaos complexity than pseudorandom bit sequences used in conventional communication systems. In 2002, a 2.5-Gb/s non-return to zero (NRZ) pseudorandom bit sequence was referred to be masked in a chaotic carrier, produced by a 1.3-μm distributed feedback (DFB) diode laser subjected to optoelectronic feedback, and recovered in a back-to-back configuration without including any fiber transmission [67]. The achieved bit-error rate values of that system were of the order of 10^{-4}. The above performance was improved in 2005 with a successful encryption at 3-Gb/s pseudorandom message into a chaotic carrier and system's decoding efficiency with bit error rate (BER) values at 10^{-9} [64]. A transmission system based on the above configuration has been implemented in laboratory conditions [15, 16], as well as in an installed optical fiber network with length over 100 km [17]. These works provided so far chaos-based methods appropriate for high bit-rate data encryption but not as an integrated, compact solution. The possibility of a realistic implementation of networks with advanced security and privacy properties based on chaotic encryption depends strongly on the availability of either hybrid optoelectronic or photonic integrated components. Such systems have been recently demonstrated, employing efficient photonic integrated circuits (PICs) [13], operating at 2.5 Gb/s and including transmission links over 100 km [11, 12]. The latter could be further integrated in communication transceivers compatible with conventional computer and communication systems. Finally, advanced optoelectronic configurations that employed, instead of amplitude, phase modulation techniques operated efficiently up to 10 Gb/s, including fiber transmission links [65].

5.2   Chaos generation from optoelectronic devices

The technology of semiconductor materials and crystals - such as InGaAs lasers and LiNbO3 modulators - has provided with devices that are used explicitly in optical communications. Such devices have inherent nonlinearities that are in general suppressed for conventional communication applications. However under appropriate configurations these devices could deviate from this behavior, revealing operating conditions that lead to complex behavior. Biasing the standalone devices one may not succeed much; however, an external contribution from perturbation signals could provide the desirable result. Application of optical or optoelectronic feedback is a common method for complex signal generation, as analyzed in the next paragraphs.

In semiconductor edge emitting lasers with side-mode suppression, the temporal evolution of the electric field's amplitude of the solitary longitudinal mode is described by a time-delayed rate equation model. This field equation is complemented by specifying the evolution of the total carrier population $N(t)$. The evolution of the field and carrier variables is governed by the following equations:

$$\frac{dE(t)}{dt} = \frac{1 - ia}{2} \cdot \left[g(t) - t_p^{-1}\right] \cdot E(t) + F_E(t), \qquad (5.1)$$

$$\frac{dN(t)}{dt} = \frac{I}{e} + \frac{N(t)}{t_n} - G(t) \cdot |E(t)|^2 + F_N(t), \qquad (5.2)$$

$$G(t) = \frac{g \cdot (N(t) - N_0)}{1 + s \cdot |E(t)|^2}, \qquad (5.3)$$

where $E(t)$ is the complex slowly varying amplitude of the electric field at the oscillation frequency ω_0, $N(t)$ is the carrier number within the cavity, and t_p is the photon lifetime of the laser. The detailed derivation of these equations can be found in [5,94]. In Equation (1.2) I/e is the number of injected electron-hole pairs by current biasing the laser, tn is the rate of spontaneous recombination (as also known as carrier lifetime), and $G(t)|E(t)|^2$ describes the processes of the stimulated recombination. The above set of equations takes into account gain suppression effects through the non-linear gain coefficient s, and also Langevin noise sources $F_E(t)$, $F_N(t)$. These spontaneous emission processes are described by white Gaussian random numbers [107] with zero mean value:

$$\langle F_E(t) \rangle = 0 \qquad (5.4)$$

and delta-correlation in time:

$$\langle F_E(t) \cdot F_E^*(t') \rangle = 4 \cdot t_n^{-1} \cdot \beta_{sp} \cdot N \cdot \delta(t - t'). \qquad (5.5)$$

The spontaneous emission factor β_{sp} represents the number of spontaneous emission events that couples with the lasing mode. The noise term in the carrier equation $F_N(t)$, coming from spontaneous emission as well as shot noise contribution, is generally small and thus usually neglected.

Semiconductor lasers are in general sensitive to external optical light; small external reflections may cause unstable operating behavior in the CW emitted output [84,103]. This is the reason why all types of commercial semiconductor lasers are provided with an optical isolator at the output, suppressing any reflections from the external environment. In applications, however, where instabilities are desirable - such as in chaos secured communications - the

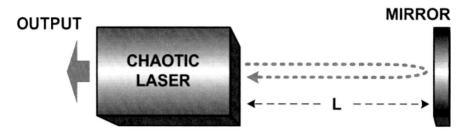

FIGURE 5.2
Block diagram of a laser subjected to optical feedback.

isolation stage is omitted and the semiconductor lasers are driven intentionally to unstable operation.

Semiconductor lasers that receive coherent optical feedback from their own output are very exciting configurations, not only from the viewpoint of non-linear dynamics but also for their potential to support novel applications. Optical feedback is practically the process in which a part of the laser's output field reflected by a mirror in distance L is re-injected into the laser's active region (Figure 5.2). The optical feedback system is a phase-sensitive delayed-feedback autonomous system for which all three known routes, namely, period-doubling, quasi-periodicity, and route to chaos through intermittency, can be found. Many types of lasers (edge-emitting semiconductor lasers such as Fabry-Perot, MQW, and DFB) exhibit such dynamics. The measure of the feedback strength is usually expressed in the literature by the C parameter [4]:

$$C = \frac{k_f T}{t_{in}} \sqrt{1 + a^2} \tag{5.6}$$

that includes the feedback fraction (k_f), the round-trip time for light in the external cavity ($T = 2L/c_g$) where c_g is the speed of light within the medium of the external cavity and L is the distance between the laser facet and the external mirror, the linewidth-enhancement factor (a), and the round-trip time of light in the internal laser cavity ($t_i n$). A semiconductor laser with various levels of optical feedback may exhibit various dynamic behaviors [105]. For insignificant feedback fractions of the laser's field amplitude (up to 0.01%) the laser maintains its continuous wave operation. By increasing the feedback fraction to values up to 0.1%, and at the same time C parameter is smaller than 1, generation of external cavity modes gives rise to mode hopping among internal and external modes. For a narrow region around 0.1% feedback (depending on laser dynamics) the mode-hopping noise becomes suppressed and the laser may oscillate with a narrowed linewidth. Increasing to moderate or strong feedback values (around 1% to 10%) the relaxation oscillation becomes undamped and the laser linewidth is greatly broadened to GHz bandwidth. It is then when the laser shows chaotic behavior and evolves into unstable oscillations in the so-called *"coherence collapse"* regime. Finally, in extremely strong feedback regimes, usually defined for a feedback ratio significantly higher than 10%, the laser oscillates in a single mode with a greatly suppressed linewidth.

For the applications investigated in this chapter, coherence collapse regime is of great significance since it is where the non-linear properties of the laser generate chaotic dynamics. A semiconductor laser with optical feedback for this regime is modeled by the Lang-Kobayashi equations [4,62,68] that include the optical feedback effects in the laser rate equations model. The stability and instability of the laser oscillations have been theoretically studied in numerous works by the linear stability analysis around the stationary solutions for the laser variables [82, 108]. The dynamics of semiconductor lasers with optical feedback depend on the system parameters; the ones that define parameter C (Equation 5.6), as well as the biasing injection current I of the device. For variation of the external mirror reflectivity, the laser exhibits a typical chaotic bifurcation very similar to a Hopf bifurcation [84]. Another type of instability is sudden power dropouts and gradual power recovery in the laser output power, the so-called *low frequency fluctuations* (LFFs) [35, 58, 79, 86, 93]. LFF signals are observed at low biasing injection currents, just above the threshold current of the laser, and consist of frequency components up to 1 GHz. On the other hand, the spectral distribution of the chaotic carrier depends on the relaxation oscillation frequency of a semiconductor laser, which is directly determined by the biasing current of the laser. By increasing the optical feedback the chaotic carrier expands beyond the relaxation frequency of the laser, eventuating in a broadband fully developed chaotic carrier that may expand up to several tens of GHz. It has also been proved so far that the laser oscillates stably for a higher bias injection current. Thus, larger optical feedback strength is required to destabilize the laser at a higher bias injection current. The external cavity length also plays an important role in the chaotic dynamics of semiconductor lasers. There are several important scales for the length and change of the external mirror in the dynamics. Chaotic dynamics may be observed even for a small change of the external mirror position comparable to the optical wavelength λ [51]. For a small change, the laser output shows periodic undulations (with a period of $\lambda/2$) and exhibits a chaotic bifurcation within this period. When the external reflector is a phase-conjugate mirror, the phase is locked to a fixed value and the laser appears to be insensitive to small changes in the external cavity length and its dynamics are only defined by the absolute position of the external mirror [82]. This is observed for every external mirror position as far as the coupling between the external and internal optical field is coherent. When the external mirror is positioned within the distance corresponding to the relaxation oscillation frequency (on the order of several centimeters) and the mirror moves within a range of millimeters, the coupling between the internal and external fields is strong and the laser shows a stable oscillation. When the external mirror is positioned over a distance equivalent to the relaxation oscillation frequency of a laser but it is within the coherence length of the laser (on the order of centimeter to several meters), the laser is greatly affected by the external optical feedback. In this region, the number of modes related to the C parameter is large and the laser shows various dynamical behaviors even at moderate feedback rate [51]. Finally, when the external mirror is positioned at a distance beyond the coherence length of the semiconductor laser, it still exhibits chaotic oscillations, but the effects have a partially coherent or incoherent origin [101]. Instabilities and chaos generation are also induced by this type of incoherent feedback, which can originate not only from the laser itself but also from optical injection from another laser source.

In the cases of relatively weak optical feedback, the rate Equation (5.1) that describes the semiconductor laser electric field is altered accordingly in order to include also the effect of the external cavity mirror. The evolution of the emitted field is governed now by the following equation:

$$\frac{dE(t)}{dt} = \frac{1 - ia}{2} \cdot \left[G(t) - t_p^{-1} \right] \cdot E(t) + k_f \cdot E(t - T) \cdot e^{i\omega_0 T} + F_E(t). \quad (5.7)$$

The carrier Equation (5.2) does not need any modification with respect to the free-running case. This basic equation that includes the applied optical feedback was introduced by Lang and Kobayashi [62]. The understanding of delayed feedback systems has been boosted during the last years using semiconductor lasers. Fundamental nonlinear dynamical phenomena, such as period doubling and quasi-periodic route to chaos, have been characterized in these systems. Also high-dimensional chaotic attractors have been identified. Furthermore the analogy between delay differential equations and one-dimensional spatial extended systems has been established [40] and exploited for the characterization of the chaotic regimes [74].

Nevertheless, there are alternative methods to cause instabilities and chaos dynamics in semiconductor lasers. In optical injection systems (Figure 5.3), the optical output of an independent driving laser is fed into the laser of importance in order to destabilize it and under specific conditions force it to oscillate in the chaotic regime [68, 96]. Crucial parameters that determine the latter operation are the optical injection strength of the optical field - with values that are adequate to achieve injection locking condition - and the frequency detuning between the two lasers - which is usually below the

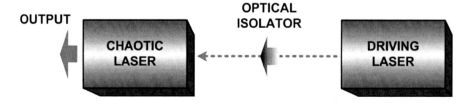

FIGURE 5.3
Block diagram of a laser subjected to optical injection by a second driving laser.

region of ± 10 GHz. Compared to the rate equations for the solitary laser, an additional term representing the injection field from the driving laser is added to the field equation. In this case of a weak optical injection condition the evolution of the field is modified accordingly:

$$\frac{dE(t)}{dt} = \frac{1 - ia}{2} \cdot \left[G(t) - t_p^{-1} \right] \cdot E(t) + k_{dr} \cdot E_{ext}(t) + F_E(t), \qquad (5.8)$$

where k_{dr} is the coupling coefficient of the driving laser to the master laser and E_{ext} is the injected electrical field of the driving laser. In contradiction to the optical feedback case, in which the time-delayed differential equations provide infinite degrees of freedom, optical injection provides low-complexity attractors.

A semiconductor laser with an applied delayed optoelectronic feedback loop is also an efficient technique of broadband chaos generation [102]. In such a configuration, a combination of photodetector and broadband electrical amplifier is used to convert the optical output of the laser into an electrical signal that is fed back through an electrical loop to the laser by adding it to the injection current (Figure 5.4). Since the photodetector responds only to the intensity of the laser output, the feedback signal contains the information on the variations of the laser intensity disregarding any phase information. Therefore, the phase of the laser field is not part of the feedback loop dynamics and consequently the dynamics of this system. The fact that part of the feedback loop is an electric path, the bandwidth response of this path may provide a filtered feedback due to the limited bandwidth of the photoreceiver, the electric amplifier, as well as the electric cables.

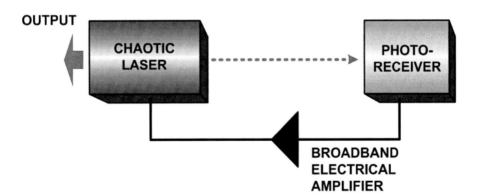

FIGURE 5.4
Block diagram of a laser subjected to optoelectronic laser feedback.

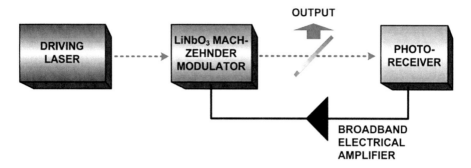

FIGURE 5.5
Block diagram of an electro-optic LiNbO$_3$ Mach-Zehnder modulator subjected to electro-optic feedback.

5.2.1 Non-linear dynamics from electro-optic crystals

Instead of biasing the laser with the delayed electrically converted output, a different approach can be used, exploiting the sinusoidal transfer function of a LiNbO$_3$ crystal included in a Mach-Zehnder amplitude modulator device (Figure 5.5). In this case, the non-linear element is the modulator, with the driving semiconductor laser just providing a continuous wave (CW) input to the close loop system. In order for the system to provide a chaotic output, the photodetection conversion efficiency and the electrical amplification unit that participates in the closed optoelectronic loop should fulfill some conditions. Typically the signal's voltage that is biasing the modulator should be several times the full-period voltage value (V_π) of the modulator. Analogue configurations have been also demonstrated using a phase modulator as the non-linear element, in cases that the chaotic waveform is desired to vary the phase of the emitted signal.

5.2.2 Non-linear dynamics from photonic integrated circuits

Based on the above chaos generation techniques various configurations of transmitters have been proposed and implemented based on standard optical components, providing high-dimensional chaotic carriers capable of message encryption. The miniaturization of the above configurations through photonic integration appears very attractive considering the efficiency of specifically designed PICs to generate non-linear dynamics. In [37] monolithic colliding pulse mode-locked lasers exhibited nonlinear behavior, from CW operation to self-pulsations and mode-locking, for the full range of the control parameters. In [20] a semiconductor laser, followed by a phase section and an active feedback element, form a very short complex photonic circuit that provides several types of dynamics and bifurcations under optical feedback strength and phase control. However, only a multiple-mode beating operation may transit the

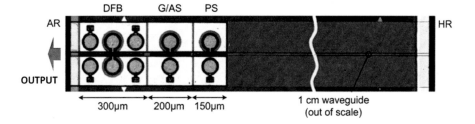

FIGURE 5.6 (SEE COLOR INSERT)
Photonic integrated circuit with straight waveguide for secure optical communications. DFB: Distributed feedback laser, G/AS: Gain - absorption section, PS: Phase section, AR: Anti-reflective coating, HR: Highly reflective coating.

dynamics beyond a quasi-periodic route to chaos with possible chaotic components. A simplified version of the aforementioned PIC, omitting though the active feedback element, was found to generate only distinct-frequency self-pulsations [112]. The implementation of an integrated colliding-pulse mode-locked semiconductor laser showed also nonlinear dynamics and low-frequency chaos, by controlling appropriately only the laser's injection current [121]. A photonic integrated circuit (Figure 5.6), capable to generate high-dimensional broadband chaos, has also been recently proposed, designed, and tested [13]. It consisted of four successive sections: a DFB InGaAsP semiconductor laser, a gain/absorption section, a phase section, and a 1-cm-long passive waveguide. The external cavity length is defined by the internal laser facet and the chip facet of the waveguide which is highly reflective coated. The effective feedback round-trip time allows the deployment of a fully chaotic behavior. The dynamics generated by this PIC could be easily controlled via phase conditions and feedback strength, establishing therefore this device as a compact integrated fully controllable chaos emitter. Chaotic signals emitted from this device can easily extend to a bandwidth over 10 GHz with a relatively flattened profile.

In another approach for chaos generating PICs, a compact chaos laser chip subjected to ring-type optical delayed feedback has been proposed and developed [100]. The structure of the PIC is shown in Figure 5.7; it is formed by a passive ring waveguide monolithically integrated with a single-frequency DFB laser, two semiconductor optical amplifiers (SOAs), and a fast-response photodiode (PD). The difference of the ring type configuration in respect to the PIC of Figure 5.6 is that precise cleaved facet mirrors and highly reflective (HR) coating are not needed for achieving strong optical feedback. Additionally, the photodetection stage is embedded in the device, providing only an electrical output. Also this device has been confirmed to generate broadband chaotic signals with a flat spectrum that extends over 10 GHz.

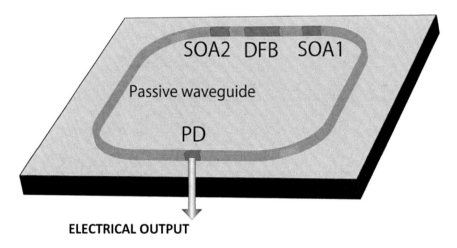

FIGURE 5.7 (SEE COLOR INSERT)
Photonic integrated circuit with ring waveguide and embedded photodetector
for secure optical communications applications. DFB: Distributed feedback
laser, SOA: Semiconductor optical amplifier, PD: Photodetector.

5.3   Broadband data encryption in chaos optical communications

5.3.1   Chaos synchronization

When exploiting chaotic carriers in communication systems for data encryption applications, these carriers should be finally canceled after transmission, in order to retrieve the useful information. Thus, the most crucial process in such systems is the carrier reproduction by the receiver and its elimination. This process is achieved through *chaotic carrier synchronization* and *cancellation.*

Synchronization of chaotic waveforms that are produced by two nonlinear systems should fulfill the following condition: the deviations of the corresponding parameters that characterize each system should be minor. Two types of configurations in all-optical systems have been studied for efficient synchronization (Figure 5.8) [85, 116]. The first one consists of a pair of external-cavity semiconductor lasers playing the role of the transmitter and the receiver, respectively (closed-loop scheme), while in the second approach an external-cavity laser transmitter produces the chaotic carrier and a single laser diode similar to the transmitter is used as the receiver (open-loop scheme) [26,69,85,116]. After arduous investigations, the closed-loop scheme

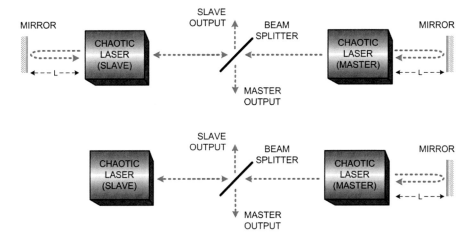

FIGURE 5.8
Closed-loop (top) and open-loop (bottom) synchronization configuration between a pair of semiconductor lasers. Chaos generation is based on external cavity feedback.

has been evaluated to be more robust in terms of synchronization. Prerequisite is the accurate matching of the external cavity lengths [26, 116]. On the contrary, the open-loop scheme is less robust, with simpler receiver architecture [26,85,116]. It requires a larger coupling strength between the transmitter and the receiver and demands loose constraints in terms of lasers' identity.

The rate equations that describe the coupled behavior between a transmitter and a receiver, based on the Lang-Kobayashi model, are:

$$\frac{dE_i(t)}{dt} = \frac{1 - ia}{2} \cdot \left[G_i(t) - t_{p,i}^{-1}\right] \cdot E_i(t) + k_{f,i} \cdot E_i(t - T) \cdot e^{i\omega_0 T} +$$
$$+ k_{inj} \cdot E_{ext}(t) + F_E(t), \quad (5.9)$$

$$\frac{dN_i(t)}{dt} = \frac{I}{e} + \frac{N_i(t)}{t_{n,i}} - G_i(t) \cdot |E_i(t)| + F_N(t), \quad (5.10)$$

$$G_i(t) = \frac{g \cdot (N_i(t) - N_{0,i})}{1 + s \cdot |E_i(t)|^2}, \quad (5.11)$$

where $i = t, r$ denotes the solution for the transmitter or the receiver. k_{inj} is the electrical field injection parameter applied to the receiver laser, and E_{ext} is the amplitude of the injected electric field. The term $k_{inj} \cdot E_{ext}(t)$ is applicable only in the rate equation of the receiver. For the case of open loop, no optical feedback is applied on the receiver, thus $k_{f,r} = 0$.

Two different types of synchronous responses of the receiver have been distinguished associated to the strength of the injection conditions [72,85,116].

The first one is the *complete chaos synchronization* in which the rate equations for the transmitter and the receiver are expressed through the same equations. Frequency detuning between the transmitter and receiver is considered to be zero while the rest crucial laser parameters should be identical [81]. This type of synchronization in semiconductor laser systems is realized when the optical injection fraction is small (typically less than a few percent of the chaotic intensity variations) [85, 116]. The mathematical solution emerges from the identity of the equations that describe the operation of the emitter and the receiver:

$$E_r(t) = E_t(t + T),\qquad(5.12)$$

$$N_r(t) = N_t(t + T),\qquad(5.13)$$

$$k_r = k_t - k_{inj}.\qquad(5.14)$$

These systems are considered secure for communication applications since the constraints on the devices' parameter mismatch are very severe. The time lag that exists in this type of synchronization is defined by the propagation time between the transmitter and the receiver, as well as the roundtrip time of the transmitter's external cavity. Specifically, the receiver laser anticipates the transmitter laser in terms of the emitted output, thus this synchronization type is also expressed as *anticipating chaos synchronization* [75, 97]. When much stronger injection is applied (kinj + kr ¿¿ kt), synchronization is achieved based on a driven response of the receiver to the transmitter's chaotic oscillations. This type of synchronization is called *isochronous* or *generalized* [70, 71, 85, 116] and is based on the optical injection locking. This phenomenon in semiconductor lasers depends on the detuning between the frequencies of the master and slave lasers. If the detuning is up to a few tens of GHz, a frequency pulling effect in the master-slave configuration appears [70]. The time lag of the synchronization process is now equal to the propagation time. This type of synchronization exhibits a large tolerance to lasers' parameter mismatch and consequently it can be more easily observed in experimental conditions. The relation between the electric fields of the pair of lasers in this type of synchronization is written as in [38]:

$$E_r(t) = A \cdot E_t(t).\qquad(5.15)$$

Most experimental results in laser systems including semiconductor lasers reported up to now were based on this type of chaos synchronization.

5.3.2 Data encoding techniques

The technique adopted for data encoding and encryption at the emitter side of chaos based optical communication systems is one of the crucial factors that determine the performance of the system. One major concern in all encoding schemes is not to disturb the inherent synchronization process, since the encrypted message is always an external perturbation in a fragile synchronized system. Various schemes have been used so far in the way the message is encoded within the chaotic carrier; the decoding process, however, is usually the same for all schemes, based on subtracting the output of the receiver's laser from the received signal. Some of the most trivial data encoding methods found in chaos-based encrypted communication systems are described below.

5.3.2.1 Additive chaos modulation (ACM)

In the ACM encoding method, data $m(t)$ are applied by externally modulating the electric field of the chaotic carrier generated by the emitter's oscillator:

$$E_{TR} = (1 + m(t)) \cdot E_M \cdot e^{i\phi_M}. \tag{5.16}$$

This is an analogue to the coherent amplitude modulation (AM) scheme [45, 87, 120]. The phase Φ_M of the chaotic carrier with the encrypted message is the same with the phase of the chaotic carrier in absence of any information. Consequently, the application of the message on the chaotic carrier is only a perturbation in amplitude and not in phase, which turns to be crucial for the efficient synchronization process of this type of phase-dependent system.

5.3.2.2 Multiplicative chaos modulation (MCM)

In the MCM encoding method, the message is also applied by external modulation. The difference with the ACM method is that message encoding is applied within the external cavity of the transmitter, providing a message-dependent chaotic carrier generation process.

5.3.2.3 Chaos masking (CMS)

In the CMS encoding method, the message is applied on an independent CW optical carrier which is coupled to the chaotic optical carrier [60]. Both carriers should correspond to exactly the same wavelength, the same polarization state, and the message carrier should be suppressed enough in respect to the chaotic carrier in order to ensure an efficient message encryption. Under this configuration, the message is a totally independent electric field which is added to the chaotic carrier according to the expression

$$E_M \cdot e^{i\phi_M} + E_{msg} \cdot e^{i\phi_{msg}}. \tag{5.17}$$

In this encryption method, the phase of the total electric field injected to the receiver consists of two independent components. The external phase of the message Φ_{msg} acts as an additional perturbation in the synchronization conditions. This phase mismatch is a significant perturbation in the synchronization process and for this reason this encoding method is not efficient.

5.3.2.4 Chaos shift keying (CSK)

In the CSK method, the bias current of the emitter's laser is digitally modulated between two levels, resulting in two different states of the chaotic attractor [78, 87]. The current of the emitter's laser is given by the equation:

$$I_M = I_B + m(t) \cdot I_{msg}, \qquad (5.18)$$

with $m(t) = \{-1, 1\}$ being the amplitude modulation index for the two bits $\{0, 1\}$. The condition $IB >> I_{msg}$ should be fulfilled in order to minimize synchronization error.

5.3.2.5   Phase shift keying (PSK)

The strong dependence of synchronization on the relative phase between the external cavities of emitter and receiver could be also used for message encoding [8]. Phase variations of the emitter's external cavity, that are small enough to be detectable by an eavesdropper, can substantially affect the correlation between the outputs of the system. Thus, if the emitter's phase is modulated by a message, the latter can be extracted by transferring the induced variation of the correlation coefficient into amplitude modulation.

5.3.2.6   Sub-carrier ACM modulation

All the above encoding techniques referred to baseband encryption within the chaotic carriers. However, the power spectral distribution of a generated chaotic carrier is not always suitable for base-band message encryption. For example, in cases where short external optical cavities are employed the most powerful spectral components of the carrier arise on the external cavity mode (ECM) frequencies. Using sub-carrier ACM message encryption in those frequencies where the chaotic carrier has powerful spectral components, a higher signal-to-noise ratio of the encrypted signal may be applied without any compromises in the quality of encryption, providing a better message recovery performance in comparison to the base-band techniques [10, 22].

5.3.3   Chaos-secured optical communication systems

Several optical communication systems have been so far deployed exploiting the properties of chaotic signals for data encryption. These systems utilized various types of chaotic oscillators as pairs of emitter/receiver, based on the architectures described in a previous paragraph. An insight into these systems is given below.

5.3.3.1   Erbium-doped fiber amplifier systems

Van Wiggeren and Roy developed a revolutionary chaotic laser system that generated and employed chaotic carriers with bandwidth of about 100 MHz [113]. In the proposed configuration the transmitter consisted of a fiber ring laser that included an EDFA (Figure 5.9). The optical signal generated by the EDFA was re-injected after circulating the fiber ring. Therefore the fiber laser is driven by its own delayed output leading finally to a chaotic behavior.

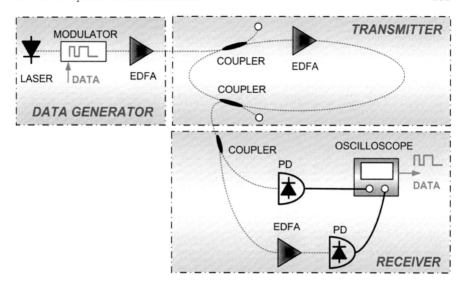

FIGURE 5.9 (SEE COLOR INSERT)
The optical chaos communication setup proposed by Van Wiggeren and Roy.

The encrypted message is coupled through another optical signal into the fiber ring of the transmitter, and was injected into the laser together with the time-delayed intrinsic signal. Thus, the data signal also participates in the generated dynamics of the transmitter. As the combined data/chaotic signal travels around the transmitter's ring, part of it is sent to the receiver. At the receiver the signal is split into two parts. One part is fed into an identical EDFA to the one used in the transmitter and the output is converted into an electrical signal by a photodiode. The other part is fed directly into another photodiode providing a duplicate of the chaos signal with the data. After canceling the time delay discrepancies, the chaotic signal is subtracted from the signal containing the information, revealing the data sequence. In this work, a 10-MHz square wave was encrypted and successfully decoded.

5.3.3.2 All-optical systems

Since then researchers have worked towards increasing the bandwidth of the chaotic carriers as well as the encrypted message bit rates, by using lasers or amplitude/phase modulators as the nonlinear element in time-delayed feedback loops. In [61] a 1.5-GHz sinusoidal tone transmission was performed based on synchronizing chaos in nonlinear systems that used semiconductor lasers with optical feedback. In [88], encoding, transmission, and decoding of a 3.5-GHz sinusoidal message in an external-cavity chaotic optical communication scheme operating at 1550 nm were demonstrated. Beyond testing sinusoidal carriers, contemporary optical chaotic systems employed pseudorandom bit sequences, also designating the feasibility of this encryption method to secure

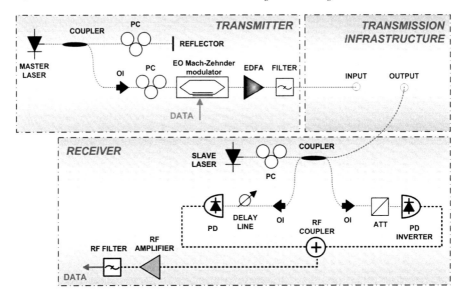

FIGURE 5.10 (SEE COLOR INSERT)
Experimental setup of an all-optical communication transmission system based on chaotic carriers. PC: polarization controller, OI: optical isolator, PD: photoreceiver, ATT: attenuator.

actual optical links. These systems are in principle more demanding especially in synchronization efficiency, since the chaotic carriers should be proficiently synchronized not only in a specific narrow frequency region but in a wider spectral region - the one that covers the encrypted message. Such systems are presented in the following paragraphs.

Initial all-optical chaotic communication systems were built on the concept of an open-loop receiver as is shown in Figure 5.10 and was described in the work presented in [17]. The emitter/receiver DFB lasers come from the same fabrication wafer, with almost identical characteristics and emitting at the same wavelength (1552.1 nm). Chaos generation is born within a 6-m-long fiber-optic external cavity formed between the emitter's (master) laser and a digital variable reflector that determined the amount of optical feedback sent back. Biasing of the lasers not over 1.5 times their current threshold ensured a powerful chaotic carrier at the first few GHz, appropriate for sufficient baseband data encryption. For such biasing currents and for the long length of the cavity, only a moderate optical feedback value (up to a few percent) is enough to raise broadband chaos dynamics. The data encrypted in this configuration were non-return-to-zero pseudorandom sequences applied by externally modulating the chaotic carrier using a Mach-Zehnder LiNbO$_3$ modulator (ACM encoding technique). Transmission impairments, when including links with a length around 100 km, are suppressed as long as the chaotic carrier bandwidth is below 10 GHz and appropriate dispersion management and amplification stages are adopted. The receiver's identical laser (slave) is synchronized to the chaotic dynamics of the master laser through the appropriate value of injection of the received complex signal. Low values of optical injection power are insufficient to force the receiver to synchronize, while higher values of injection power minimize the *chaos pass filtering* effect, leading to reproduction of both chaotic carrier and data. Photodetection and electrical subtraction of the received and locally generated signals, at the receiver, lead to data recovery. Appropriate compensation between two subtracted signals is needed in respect of optical power and time delay. The electrical subtraction product from such a system needs further amplification and filtering in order to compare it with the applied data sequence. The initial data amplitude for a given chaotic carrier determines the encryption quality. Large amplitude data could be recovered more efficiently, however, encryption would be only partial. For example, as studied in [17], for a high BER value of the encrypted data (0.06) - albeit not completely encrypted as a BER value of 0.5 would indicate - data recovery with a BER value 10^{-7} was measured around 1 Gb/s bit rate. As the bit rate is increased to a multi-Gb/s scale, BER values for data recovery are also increased monotonically. This has a threefold explanation: the chaos pass filtering effect has been confirmed to be larger for lower frequencies and decreases as message spectral components approach the relaxation oscillation frequency of the laser. Additionally, synchronization to a larger spectral region is less effective. Another important reason is that the decoding process is based on signal subtraction and not on signal division that should match the ACM encoding concept. The encrypted signal is of the form of $[1+m(t)] \cdot E_t(t)$, while the receiver reproduces the chaotic carrier $E_r(t) = E_t(t)$. Thus, the output is not the encrypted message m(t) but the product: $m(t) \cdot E_t(t)$.

The above configuration has been evaluated in real world conditions, by sending chaos-encrypted data in a commercially available fiber optic network. The utilized infrastructure included an installed optical network of single mode

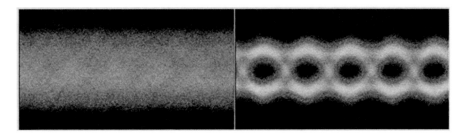

FIGURE 5.11 (SEE COLOR INSERT)
Eye diagrams for a $2^{31} - 1$, 1-Gb/s pseudorandom encrypted message with BER 0.06 (left) while the decrypted message is recovered with BER 10^{-7} (right).

fiber that covered the wider metropolitan area of Athens, Greece and had a total length of 120 km. The system's efficiency on the encryption and decryption performance was studied, by bit-error rate analysis of the encrypted/decoded message (Figure 5.11), and was compared with an in-situ system without transmission medium. The performance was evaluated to be almost the same, since such transmission lengths and signal bandwidths do not impose significant degradation in the system's performance.

The above concept of an all-optical chaos optical communication system has been further improved by considering monolithically integrated broadband chaotic oscillators. Such PICs were identified to be capable of exhibiting an excellent performance in closed-loop synchronization architectures providing extreme stability [10]. The philosophy of the presented closed-loop configuration is twofold. On one hand, the emitter increases its encryption efficiency since a completely hidden data stream is sent within a chaotic carrier along the transmission link (BER value = 0.5), and this is achieved by minimizing the message amplitude. On the other hand, the well-synchronized receiver detects data with a native bit-error rate as low as needed (10^{-3} to 10^{-4}) in order to acquire error rates below 10^{-12} using forward error correction (FEC) coding techniques. Reported results provided secure 2.5-Gb/s data exchange with bit-error rates below 10^{-12} [95, 118]. The FEC method used in that system posed a digital bit-error-rate threshold in its operation, discriminating decisively the data recovery efficiency between authorized and unauthorized users.

Besides single channel systems, wavelength division multiplexing (WDM) transmission systems have been also evaluated for chaos encryption. Preliminary studies have appeared to this direction [123] in order to numerically study such a potential. Experimental investigations have been deployed in chaos-encrypted installed systems with Gb/s encrypted data in [11]. This study evaluated three neighboring channels of a dense-WDM (DWDM) communication system: two chaotic optical channels with encrypted 1.25-Gb/s data stream, along with a conventional channel that carried unsecured 10-Gb/s digital data sequences. The inclusion of additional channels in the installed transmission medium changes the performance of the chaotic channel's synchronization and decoding efficiency due to inter-channel interference and cross-phase or cross-polarization effects. Specifically, at channel spacing of 0.8 nm, the relative input polarization state of the neighboring channels affected significantly the decoding performance. On the contrary, when the polarization of each channel was set orthogonal to its neighbors, the interference effects were minimized. In the latter case, the recovery performance was identified to be almost equivalent to the back-to-back performance, while, if the polarization was set to be parallel to its neighbors, the decoding of the chaos-encrypted channel was downgraded significantly.

5.3.3.3 Optoelectronic systems

In parallel with all-optical systems, optoelectronic systems have been implemented based on oscillators that exploited the sinusoidal properties of amplitude or phase modulators. In the system presented in [39] the emitter consisted of a laser diode whose output was modulated in a strongly nonlinear way by an electro-optic feedback loop through an integrated electro-optic Mach-Zehnder modulator (MZ) (Figure 5.12). This system is also known as a delayed dynamical system - inspired by the pioneering work of Ikeda [50] - due to the delay in the optical/electrical fiber loop, which is much larger than the response time of the MZ [43,63]. The digital data are coupled into the encryption system by a third laser that emits at the same wavelength. For efficient encryption the data signal must reside exactly at the optical frequency of the chaotic carrier, in order to be indistinguishable with spectral filtering. Moreover, to avoid coherent interaction between data signal and chaotic carrier, their polarization states should be orthogonal to each other. Thus, in order to prevent eavesdropping through polarization filtering, a fast polarization scrambler performing random polarization rotation should be used before transmitting the combined output. Half power of the data signal is coupled directly into the emitter's optoelectronic feedback loop, while the rest half is sent to the receiver. The receiver of this communication system consists of an identical optoelectronic feedback loop that has been split apart. A fraction of the incoming signal is photodetected. The rest of the signal is delayed by an amount identical to the delay produced by the transmitter's loop. The resulting signal is photodetected and used to drive an identical MZ optical modulator. The laser at the receiver provides the carrier for the modulated signal that is finally photodetected. The resulting subtraction signal reveals the initial data.

An efficient temporal chaos replication between the transmitter and the

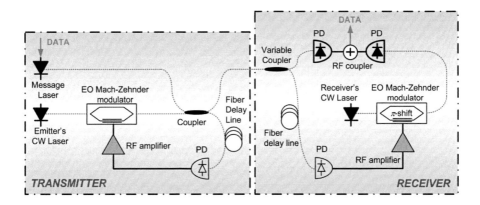

FIGURE 5.12
Experimental configuration of an optoelectronic communication system with chaos encryption.

receiver has been observed using this configuration, with an electrical cancellation of chaos equal to $c_{\Delta f}^{E} = 18$ dB for the first 5 GHz. Data sequence is a NRZ pseudorandom bit sequence (PRBS) of $2^7 - 1$ bit up to 3 Gbit/s. This encryption scheme allowed for a decoding BER performance equal to 7×10^{-9} at 3 Gb/s [39]. Phase modulation chaotic systems have been studied numerically in the recent past [23], while experimental investigations have been reported in [65, 66]. An implementation of a phase chaos communication system is shown in Figure 5.13 [65]. The concept is based on a standard differential phase shift keying (DPSK) modulation scheme. In such a system, any WDM channel can be selected through the use of the proper external laser source, independently of the subsequent chaos communication processing. A message phase modulator stage (MΦM) performs the binary DPSK phase modulation before seeding the emitter's chaotic oscillator. The chaotic masking at the phase chaos generator (ΦCG) consists of the superposition of the data phase modulation and the chaos masking phase modulation, the latter being partly determined by the data itself. At the receiver side, phase chaos cancellation (ΦCC) is processed from the input signal, while at a next stage a standard DPSK demodulation (MΦD) recovers the initial data. Note that if ΦCG and ΦCC were omitted in Figure 5.13, the communication system becomes practically a standard optical DPSK transmission system.

This approach provided the first experimental demonstration of 10 Gb/s chaos communication, and was tested also at field trials [65]. The first field experiments were performed on an all-optical fiber ring network installed in the city of Besançon, France. BER as low as 3×10^{-10} (2×10^{-7}) were obtained at 10 Gb/s for partially (fully) encrypted data using a transmission fiber link ring of 22 km. The second experiment was performed in an optical network in the metropolitan area of Athens, Greece. In this case the fiber link was around 120 km and involved intermediate amplification and dispersion compensation stages. The achieved BER value was of the order of 10^{-6} ($2 \cdot 10^{-4}$) at 10 Gb/s for partially (fully) encrypted data. By reducing the message bit rate to 3 Gb/s BER values less than 10^{-10} have been obtained for fully encrypted data. Dispersion management and polarization control in such configurations are critical, while long-term stable operation of the encoding and decoding requires the use of control systems for stabilizing the operating point of the DPSK demodulators.

5.4 Ultrafast random bit generation in chaos optical communications

Optical chaos generators can contribute to the implementation of secure communication systems not only via data encryption within complex optical analog signals, but also through the generation of ultra-fast physical random

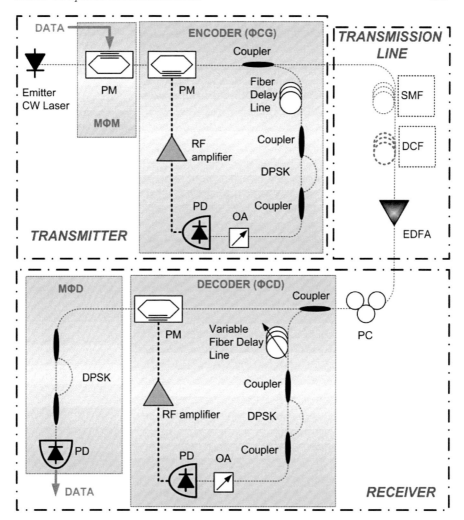

FIGURE 5.13
Communication system using optoelectronic phase chaos for data encryption. PM: phase modulator, DPSK: differential phase shift keying modulator, OA: optical attenuator, PD: photodiode, SMF: single mode fiber channel, EDFA: erbium doped fiber amplifier, DCM: dispersion compensation module, PC: polarization controller.

sequences. Random number generators play a fundamental role in most algorithms and systems used for cryptographic applications. Communications based on secret and public key cryptography, user authentication, as well as electronic lottery-based applications rely on the quality of the randomness and the generation speed provided by these generators. The most common approach to such applications is to deal with pseudorandom number generators (PRNGs) [1]. PRNGs are initiated by a relatively random short key (also known as the *seed*) and their output is expanded into a long sequence of random bits using computational deterministic algorithms [34]. However, in cases that the *entropy pool*, from which the seed is selected, is of low complexity - no matter how strong the encryption algorithm is - a malicious attacker could succeed in guessing correct. True random number generators (TRNGs), on the contrary, produce random bits exploiting random physical phenomena or noise sources [25, 30, 32, 42, 49, 106]. Such non-deterministic generators had limited efficiency in generation rates due to limitations of the mechanisms for extracting bits from the physical procedures. However, the latter drawback has been annulled due to the exploitation of broadband chaotic signals as seeds for ultra-fast true random bit generation. Such sequences could be utilized for securing not only optical, but in general communication systems with non-algorithmic cryptographic protocols.

5.4.1 Chaotic signal generation and prerequisites

Chaotic waveforms should fulfill critical prerequisites in order to be considered as potential seeds for ultra-fast random number generation processes. For example, their electrical bandwidth should be several GHz, in order to claim Gb/s digital sampling generation rate. High optical power that ensures powerful spectral components in the electrical domain is desirable, especially in multi-bit quantization systems that interpret several bits from each sample. Finally, randomness emerges from a uniform spectral distribution of the chaotic signal and minimization or elimination of residual periodicities. The statistical properties of the light emitted by semiconductor optical chaos generators - as the ones investigated in this chapter - are determined by the intrinsic characteristics of the semiconductor lasers (e.g., the relaxation frequency) and the characteristics of the external cavity (e.g., the strength of optical feedback or the external cavity delay). These parameters may result in the oscillator operating under completely diverse dynamic regimes, from stable operation and periodic solutions to complex non-linear dynamics and broadband chaos. Strong optical feedback drives such unit to behave chaotically, as also described by the deterministic time-delayed differential model (Equation 5.7). Such level of feedback, that was not manageable in achieving synchronization of communication systems, is favored for random number generation. Since the pre-described optical chaos generators exploit physical processes for emitting broadband optical signals, the characteristic frequencies of these processes may be evident in the output signal. Such drawbacks of the analog chaotic signal are dealt in the relevant works so far either by optical de-correlation techniques or by post-processing after sampling.

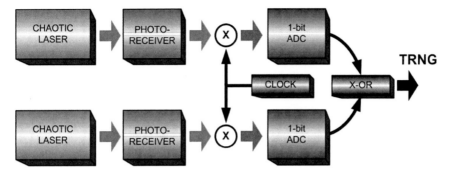

FIGURE 5.14
A schematic diagram of Uchida's chaotic lasers' configuration for TRNG.

5.4.2   Single-bit sampling systems

Uchida et al. [109] demonstrated for the first time the exploitation of broadband chaotic signals as the seed for high-rate generation of random bit sequences with verified randomness. In the specific work, a 1.7-Gb/s TRNG was presented based on the binary digitization of two independent chaotic semiconductor lasers (SLs), which were finally combined under a logic gate exclusive-OR (XOR) operation (Figure 5.14). The output intensity of each laser was converted to an AC electrical signal by photodetection, amplification, and at a next stage converted to a binary signal using a 1-bit analog-to-digital converter (ADC) driven by a fast clock. The ADC converted the input analog signal into a binary signal by comparing the sample detected value with a threshold voltage and then sampled the binary signal at the rising edge of the clock (Figure 5.15). The binary bit signals obtained from the two lasers were combined by a logical exclusive-OR (XOR) operation to generate a single random bit sequence. This stage is mandatory in order to provide the appropriate de-correlation that appears in each signal due to non-uniform spectral profile. Especially when using chaotic lasers based on external cavities, the periodicities of the ECMs do not allow standalone chaotic lasers to be adopted as a random source [109].

5.4.3   Multi-bit sampling systems

Following the previous work several configurations appeared targeting on increasing the bit-rate generation, as well as simplifying the used architectures (Figure 5.16). Kanter et al. [91] increased the bit generating rate of their proposed TRNG to 12.5 Gb/s. In this configuration, a multimode semiconductor laser operated in the coherence collapse regime due to feedback from an external mirror. Due to feedback the laser is chaotic, with a broadened lasing frequency spectrum and intensity fluctuating in time. Only one incommensurate ratio between the external cavity length and an external clock rate was

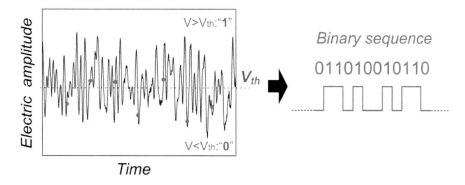

FIGURE 5.15
1-bit analog-to-digital conversion. A threshold voltage V_{th} should be carefully and strictly determined in order to eliminate bias between the "0"s and "1"s.

required. The electrically converted output was sampled by an 8-bit ADC and was used to generate a Boolean sequence in the following way: the difference between consecutive sampled 8-bit values was obtained and the m least significant bits (LSBs) of the difference value served as the next m random bits of the sequence. This method is insensitive to variations of parameters - such as the average laser power - while it does not require the use of a decision threshold.

Hirano et al. [48] adopted this multi-bit sampling methodology and by using bandwidth-enhanced chaos - through optical injection techniques - in semiconductor lasers, increased the generation ratio of the TRN implementations to 75 Gb/s. In this work, the chaotic signal of a semiconductor laser was injected into a second laser, enhancing thus its bandwidth up to 16 GHz. At the same time Kanter et al. have improved their TRNG performance at

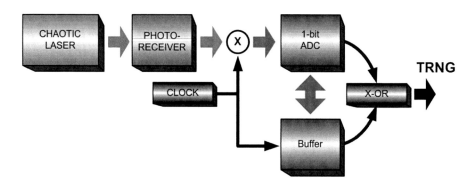

FIGURE 5.16
A schematic diagram of Kanter's et al. TRNG.

higher bit rates, up to 300 Gb/s, using a single chaotic semiconductor laser by retaining a number of the LSBs of the value of a high derivative of the digitized chaotic laser intensity [54]. However, the bandwidth of the laser chaos was only a few GHz in their scheme. This suggests that the reported bit rate exceeds the capacity of the laser chaos to generate non-deterministic random bits. Moreover, extracting more bits from high derivatives could be more susceptible to the effects of physical noise in the ADC - a potential additional source of randomness which is separate from the laser chaos. A strong motivation for using direct sampling of optical chaos for random number generation is to reduce the dependence on digital electronic operations, which may be difficult to implement at high frequencies, and which in principle cannot increase the rate for generation of nondeterministic bits. In an alternative of the single laser emitting sources, compact PICs that exploited broadband chaotic signals were used by Argyris et al. to generate true random bit sequences (TRBS) at 140 Gb/s [10]. The proposed generator was a simple configuration, consisting of the PIC, a photodetector and a 40-GSa/s oscilloscope, without including any optical de-correlation methods (Figure 5.17). Depending on the operating conditions of the PIC and by using most significant bit (MSB) elimination post-processing, real time bit sequences were extracted. The proposed configuration provided a significant advance in terms of simplicity, performance, and especially robustness. The oscilloscope's 8-bit A/D converter, along with the internal 16-bit digital-to-analog convertor (DAC) and the rest of the processing units, provided a noise-enhanced, 16-bit output binary sequence for each sample (Figure 5.18). An external down-sampling to 10 GSa/s was applied - only one out of four samples was considered - in order to eliminate any effect of interpolation samples of the oscilloscope. In such a way the initial bandwidth of the chaotic signal was preserved.

Lately, Akizawa et al. [6] proposed a scheme with bandwidth-enhanced chaotic semiconductor lasers. The chaotic laser's intensity output and its time-delayed signal are sampled at 50 GSa/s and converted into 8-bit values. The order of the 8-bit samples of the time delayed signal was reversed, and bitwise exclusive-or operation was executed between the bit-order-reversed samples and the original 8-bit samples. This method does not require any bit elimina-

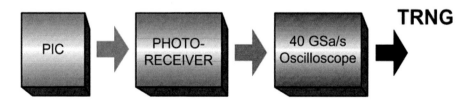

FIGURE 5.17
A TRNG based on photonic integrated circuit that emits chaotic optical signals.

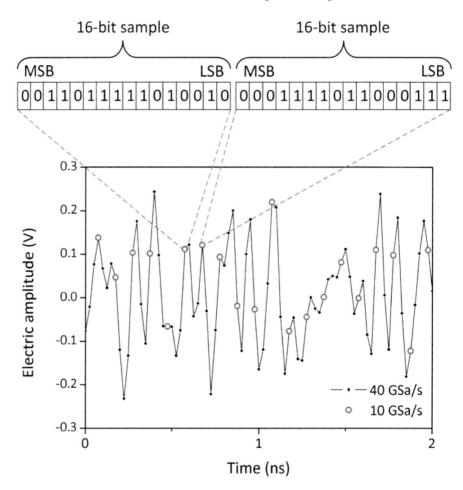

FIGURE 5.18

Multiple bit sampling of a chaotic analog waveform. Only k LSBs out of the 16-bit digitized representation of each sample are exploited and included in the final output sequence in order to preserve randomness. (Reproduced from A. Argyris, © OSA, 2010.)

tion methodology in order to obtain good-quality random bit sequences. The equivalent generation bit rate of 8×50 Gb/s has been achieved.

5.4.4 Randomness verification

In order to claim and prove the randomness of a bit sequence strict conditions must be fulfilled. The most representative statistical tests performed for this reason are included in the NIST SP800-22 (last revision in April 2010) test suite [1]. In this suite at least 1000 samples of 1-Mbit sequences - constructed by the appropriate number of the k LSBs adopted in each case - are evaluated. Pass criteria are determined by the sequence length and the significance level. A significance level $\alpha = 0.01$ is set for the p-values of each sequence test, with a desirable uniformity p-value larger than 0.0001. For the 1000 samples, the proportion of sequences that satisfies p-value $\gg \alpha$ is estimated to be 0.99 ± 0.0094. Another statistical test suite for random number generators that is commonly used is the so-called "Diehard" tests [73]. The Diehard test suite consists of 18 statistical tests and is performed using 74-Mbit sequences and significance level of $\alpha = 0.01$. An example of randomness mapping for the case of the PIC-based TRBG of [10] in terms of operating conditions is illustrated in Figure 5.19. In order to fulfill the randomness of the sequence all 15 NIST tests must be successful; however, such a performance is strictly determined by the operational conditions of the signal generator - in the studied case the laser's biasing current. By exploiting the processing potential - for example by considering different numbers of LSBs that participate in the evaluated sequence - operating criteria may be more flexible, not in favor though of the final bit error generation.

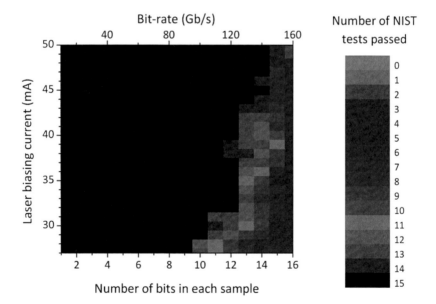

FIGURE 5.19 (SEE COLOR INSERT)
Mapping of the TRBG performance in terms of bit rate and number of NIST randomness tests passed, versus the laser current and the number of LSBs of each sample included in the output sequence. Black color grade regions designate operating conditions where all NIST randomness tests are successful. (Reproduced from A. Argyris, © OSA, 2010.)

TABLE 5.1
Typical results of NIST SP800-22 (rev.1a) statistical test suite for randomness verification.

STATISTICAL TEST	P-VALUE (min)	PROPORTION	RESULT
Frequency	0.375313	0.986	Passed
Block frequency	0.674543	0.990	Passed
Runs	0.773405	0.988	Passed
Longest run	0.087162	0.988	Passed
Rank	0.291091	0.991	Passed
Discrete Fourier transform	0.134355	0.984	Passed
Non-overlapping templates	0.002186	0.986	Passed
Overlapping templates	0.989425	0.990	Passed
Universal	0.705466	0.991	Passed
Linear complexity	0.883171	0.989	Passed
Serial	0.123038	0.995	Passed
Approximate entropy	0.607993	0.993	Passed
Cumulative sums	0.514124	0.991	Passed
Random excursions	0.023140	0.9886	Passed
Random excursions variant	0.015993	0.9869	Passed

A representative analysis of the NIST statistical tests results, for the worst value of the obtained p-values for each test, is presented in Table 5.1, as submitted in the work of [10].

An analogous analysis of the Diehart statistical tests results has been presented in the work by Hirano et al. [48], indicating the worst value of the obtained p-values for each test (Table 5.2).

5.4.5   Chaos-based TRNGs applied to cryptography

In the typical scenarios of secured channels the communicating parties have to hold a common key in the form of a bit string which is known only to the two parties. Physical mechanisms based on quantum mechanics have been suggested recently for a secure key-exchange protocol with the important and unique ability of the two communicating parties to detect the presence of any third party trying to gain knowledge of the key. The first layer of the quantum protocol is based on quantum ingredients such as entangled pairs of photons and results in correlated keys for both partners. The second classical layer consists of information reconciliation and privacy amplification (error correcting code and source coding). These result in identical keys for the communicating pair while leakage of information to an eavesdropper is eliminated; however, such procedures lower the rate at which random bits can be generated. By exploiting the previously reported ultra-fast random number generators based on optical chaos, Kanter et al. proposed a secure synchronization method of two high bandwidth TRNG over a public channel using a classical mechanism [55]. The focus of this work was to propose a secure synchronization zero lag synchronization (ZLS) of two mutually coupled chaotic lasers. The ZLS mechanism is not sufficiently secure in its simple form to act as a key-exchange protocol, and it serves only as an information carrier to gen-

TABLE 5.2
Typical results of the statistical test suite Diehard for randomness verification.

STATISTICAL TEST	P-VALUE (min)	RESULT
Birthday spacing	0.882291	Success (KS)
Overlapping 5-permutaion	0.483639	Success
binary rank for 31×31 matrices	0.658636	Success
binary rank for 32×32 matrices	0.391334	Success
binary rank for 8×8 matrices	0.367852	Success (KS)
Bit stream	0.056500	Success
Overlapping-Paris-Spares-Occupancy	0.000700	Success
Overlapping-Quadruples-Spares-Occupancy	0.015800	Success
DNA	0.015800	Success
Count-the-1's on a stream of bytes	0.049820	Success
Count-the-1's for specific bytes	0.353940	Success
Parking lot	0.260828	Success (KS)
Minimum distance	0.326556	Success (KS)
3D spheres	0.059882	Success (KS)
Speeze	0.458916	Success
Overlapping sums	0.965410	Success (KS)
Runs	0.181109	Success (KS)
Craps	0.812245	Success

Note: "KS" indicates that a single P-value is obtained by the Kolmogorov-Smirnov (KS) test [48]. For the tests which produce multiple P-values without the KS test, the worst case is shown.

erate correlated random bit sequences. Identical random bit sequences could be constructed from these correlated sequences via information reconciliation and privacy amplification. Furthermore, the proposed mechanism allows the secure generation of a synchronized random bit string amongst a small network of communicating parties. This application forecasts that ultra-fast random number generators are expected to play a significant role in the future security of high-speed communications. Compact chaos-on-chip devices, which will be integrated in computer cards or motherboards with direct access to Ethernet ports, will provide each user an unlimited source of random codes for every potential application.

5.5 Chaos synchrony in multi-laser optical networks

The preceding theoretical and experimental works presented in this chapter have demonstrated the capability of chaotic oscillator synchrony and their application in securing optical communication links. These works followed an emitter/receiver configuration, usually employing a pair of identical oscillators. Aside from these communication configurations, many works have focused lately on the joint behavior of mutually coupled SLs (semiconductor lasers) in various configurations, employing two, three, or many more oscillators with symmetric or asymmetric coupling [7, 19, 33, 38, 44, 53, 115, 124].

Lacking optical feedback, the complex nature of the generated signals rises from the optical injection process among the coupled oscillators. Following a network concept that multiple SLs are mutually coupled, several types of synchronization may occur, by solving the coupled model of semiconductor laser equations: isochronous or zero-lag synchronization, time-lag synchronization, group or cluster synchronization, and generalized synchronization. Special interest lies on cases where all nodes are completely synchronized at the same time or synchronized under classified clusters.

5.5.1 Synchrony of three-laser mutually coupled systems

Experiments and numerical simulations have demonstrated that the coupling induced dynamics of two semiconductor lasers interacting through the mutual injection of their coherent optical fields exhibit a symmetry breaking [46,52,80]. Instead of showing an identical behavior spontaneously, the two twin lasers develop a generalized synchronization between them. This type of sync is mainly characterized by peaks at $\pm\tau$, with τ being the coupling time, in the cross-correlation function between the laser intensities. From a different perspective, investigations of three instantaneously coupled semiconductor or solid-state lasers interacting through their overlapping optical fields were performed by Winful and Rahman [119] and Terry et al. [104]. These authors observed that when arranged in a linear array, an identical synchronization between the first and third lasers showed up, while the temporal traces of any of the extreme lasers and the central one appeared rather uncorrelated. These interesting results motivated in recent years a deeper study on how the number of lasers and the network of couplings modify the synchronization properties of SLs interacting with a finite time delay. Questions set by Vicente [114] need to be addressed. *"Can the isochronous solution between the first and third lasers be maintained even when the lasers need a time to communicate? How will the sync look if we add a fourth laser to the array? What is the role of the symmetry on selecting possible sync patterns?"* With a view to answer these questions, various configurations studied the synchronization properties of semiconductor lasers arranged in open-end (linear chain), mesh, and ring configurations. Other works adopted also analyses of graphs.

FIGURE 5.20
Mutually (or bi-directionally) coupled semiconductor lasers in linear configuration synchronized via a mediating element.

An optimal case in the investigation of a three-laser mutually coupled system (Figure 5.20) with a hub laser is to consider identical parameters and under a highly symmetric configuration, within a perfect free-running frequency tuning and a moderate and identical current pump (one to twice the threshold current). Once coupled under standard conditions the three semiconductor lasers are observed to enter into a fully chaotic regime known as coherence collapse (CC). Remarkably, after coupling increase the traces of the outer lasers (SL1 and SL3 in Figure 5.20) start to become more and more alike up to the point that they end up being perfectly synchronized at zero-lag. Symmetry under the exchange between SL1 and SL3 assures the existence of the synchronized solution but not its stability. Of course, in this zero-lag synchronization no superluminal effect is taking place, but this is another example of self-organizing dynamics where the units composing the system negotiate a complex and unexpected behavior. Moreover, the computation of the cross-correlation functions between the lasers reveals a high degree of similarity between any of the outer laser traces and the hub one at a specific time lag. This lag, for which the maximum of the cross-correlation function appears, corresponds to the coupling time. Surprisingly, the mediator element is lagging behind the synchronized outer units for which it is acting as a communicating bridge. This type of dynamics excludes the interpretation of the central element as a simple leader directly forcing the extreme lasers. A strict dependence on the synchronization scenarios lies on the strength of the interaction between lasers next. Some windows of periodic behavior are observed for small coupling rates. Nevertheless, for moderate or strong coupling cases the three lasers operate in the chaotic CC regime. Interestingly, regardless of the dynamics exhibited by the lasers, the perfect synchronization at zero-lag between the extreme units extends to all the coupling values explored. No coupling threshold for the transition from unsynchronized to synchronous motion could be identified. The central laser, on the other hand, varies its correlation coefficient with the outer units from a value of unity obtained for small couplings and periodic dynamics to a value that continuously lowers for large strengths where chaotic dynamics develop. The fluctuations in the lag value for small coupling strengths can be associated to the high symmetry of the cross-correlation functions that appear for regular dynamics stages [114].

Based on the zero lag synchronization property, Fischer et al. [36] showed experimentally the synchronization potential between two mutually coupled distant laser nodes, exploiting this mediating laser. The module consisted of three similar dynamical elements coupled bidirectionally along a line, in such a way that the central element acted as a relay of the dynamics between the outer elements. This type of network module could be expected to exist, for instance, as a unit within the complex functional architecture in the neural activity of the brain.

5.5.2   Synchrony of multi-laser mutually coupled systems

In most cases, the coupling surrounding leads to a synchronous behavior between the coupled elements. The dominant features that determine whether a coupling medium is able to elicit synchronization is the number of elements that are connected through it, their coupling strength, and of course the type and tolerances of the elements. The type of coupling among a number of laser oscillators that participate in such configurations could effectively resemble different types of interconnections of *"optical nodes"* within an optical network. At the extension of the three-laser concept, Munt et al. showed the first large-scale collective behavior on a set of up to 75 lasers that gave evidence that these SL *"nodes"* could operate in isochronous synchrony [122]. In this work, a symmetric coupling scenario was investigated in terms of the number of the lasers included. This setup is a generalization of the case investigated in the previous paragraph. The mediating (hub) laser when operating in a passive regime (below threshold) plays the role of a coupling passive medium, however with triggered dynamics from the incoming signals. As affirmed by Munt et al., *"the general properties of both the crowd synchrony and the quorum sensing transition are readily reproduced"* with the multi-laser configuration. Identity of laser elements could be claimed when carefully selecting two or three items. However, this is not a realistic case when building a network of tens or hundreds of devices. A small tolerance in the operating frequency and the internal characteristics of the devices should be admitted and taken into account. Munt et al. have included a tolerance in the operating frequency of the laser elements ω_j which is chosen from a Gaussian distribution with zero mean and standard deviation σ that corresponded to 14 pm. The conclusions of this work showed that for a small number of lasers N, the lasers oscillate independently. By increasing N, synchronized emission at near zero lag occurs for lasers with similar frequencies, forming clusters with similar dynamics. The number of synchronized lasers in those clusters grows as N increases, with an emission characterized by short pulses of irregular amplitudes with a repetition period around 2π. Those characteristics become more evident for large N, where almost all the lasers emit synchronously at zero lag, with emission pulses taking place simultaneously in most of the lasers, especially those with minimized frequency detuning [24, 122].

A multi-nodal star all-optical network topology based on mutual coupling has been also investigated recently in [24] in terms of synchronization, complexity, and robustness, by incorporating asymmetric coupling between the hub and the star lasers. The coupling defines the complexity of the generated within the network signal. For example, different coupling strengths for the star-to-hub coupling path can exhibit completely different dynamics, from noise-enhanced quasi-periodic states to bandwidth-enhanced chaotic signals (Figures 5.21 and 5.22). In this work analysis, two distinct regions of coupling strength values - one with low mutual interaction and low complexity signals and another with strong mutual interaction and high complexity signals - have

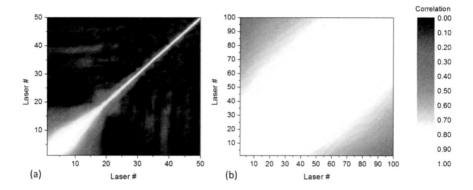

FIGURE 5.21 (SEE COLOR INSERT)
Zero-lag cross-correlation among the chaotic emitted optical signals from (a)
$N = 50$ and (b) $N = 100$ lasers that participate in the star coupling network
of Munt et al. configuration. Lasers had identical operational characteristics
and a frequency detuning within a linear spread of -40 GHz (laser $\sharp1$), 40 GHz
(laser $\sharp N$). The N lasers were biased just above threshold, hub laser was biased
below threshold, and all couplings were set to 30 GHz.

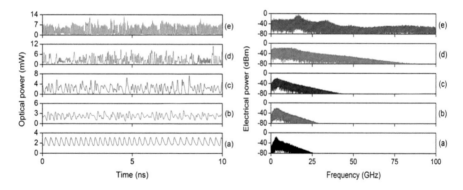

FIGURE 5.22
Timetraces and spectra of a randomly selected laser from the $N = 50$ star
configuration of [24], for an asymmetric network, operating with a hub-to-star
coupling equal to $k = 45$ GHz and a star-to-hub coupling equal to (a) $0.15k$,
(b) $0.2k$, (c) $0.4k$, (d) $0.8k$, and (e) k. (Reproduced from M. Bourmpos, ©
IEEE, 2012.)

been identified to advance high correlation among the nodes' output signals.
Moreover, specific operating conditions for which this star system exhibits robust and sensitive behavior have been identified, making this network suitable
for sensing authentication. In an initially established network operation, the
connection or disconnection of a node becomes, in general, transparent. Thus,

additional nodes that may be subsequently connected or disconnected from the network do not affect the overall behavior of the system. However, they can be detected from the network itself, through monitoring the changes in the overall correlation among signals, especially when appropriate optimization in their operating conditions includes them in the cluster with the synchronized nodes [24].

5.5.3   Cluster synchrony of multi-laser mutually coupled systems

In a network consisting of N identical nodes, we refer to cluster synchronization as the state where clusters of lasers (or nodes in a network concept) show isochronous synchronization internally, but synchronization between these clusters does not occur, or is based on non-isochronous solutions. Such solutions may induce a phase lag between the clusters. Group synchronization describes a similar state of synchrony, but the node dynamics differ from cluster to cluster. It is common to refer to these clusters as groups.

In general, more complicated synchronization patterns have been observed including cluster, group, and sub-lattice synchronization in several configurations based on graphs, motifs, and complex networks [29, 56, 57, 83, 98]. The characterization of stability of isochronous synchronization has been widely studied; the revolutionary work by Pecora and Carroll [89], which allows for a separation of network topology and local dynamics of the nodes, was recently also applied to networks with delays in the links. Such delay times can greatly change the synchronization properties and appear in many natural coupled systems. For group and cluster synchronization, attempts have been made to treat stability within a master stability approach. Sorrentino and Ott [99] considered initially two groups of nodes governed by different local dynamics. Cluster synchronization and its stability have been also investigated by Dahms et al. [29], through also the master stability approach, targeting on a generalization to a higher number of groups and investigating the restrictions set by the network topology that is adopted. These investigations also considered a large number M of SLs in a network topology, employing different coupling matrices.

The concept of cluster or sub-lattice synchronization — as defined in the work of Aviad et al. [18] — has been also experimentally appraised, in small scale networks of lasers. There has been verified evidence of two classes of zero-lag synchronization: ZLS of all units within the network, and sub-lattice synchronization, where the network units are split into two clusters, each in ZLS, with no synchronization between the clusters [18, 59]. Such classes of synchronization were recently experimentally observed for phase dynamics of laser networks as well as for neuronal circuits [83]. The most common approach of such investigations is to relegate complex or large-scale networks to network motifs as fundamental units. Thus, such motifs can play the role of major elements of larger networks of oscillators and the determination of their properties can provide an intermediate step to better understand the emergent behaviors in larger collections. A large number of works thus have been carried out lately investigating the behaviour of such motifs and, in the specific case of our interest, these motifs are associated with SL laser configurations that can be part of larger scale optical networks with coupled elements. Aviad et al. recently showed the concept of generalized ZLS in small networks employing SLs, for which the synchronized signals are as highly correlated with each other as in ZLS, despite allowing different feedback delay times for the individual oscillators [18]. In generalized ZLS, the correlation as a function of delay time has the same symmetry and features as in ZLS, but with the origin of the time axis shifted. In this type of synchronization, the constraining limitation is to use precise and specific ratios among delay times of the network in order for synchronization to occur. The results from these types of studies, either numerical or theoretical, suggest the applicability of these experimental results to other configurations and larger networks, opening a route to new multi-user communication schemes which can be of great significance to modern communication networks.

5.6   Future aspects on the field

Exploiting broadband chaotic signals for data encryption in optical communications proves to be a promising alternative of securing transmission lines at the physical layer. The plethora of implementations of communication systems deployed so far has proved the feasibility of the concept to strengthen security oriented aspects. Compact chaotic generators could be easily integrated in network components in order to provide a hardware level of security not only on single channel communications but also in multiple access networks. Chaos-based data encryption is not forecasted to substitute the cryptographic methods developed so far in an algorithmic level that shield any type of communications nowadays. However, it could provide an additional level of transmission security based on the hardware properties of the infrastructure. A deficiency, one could argue, in the so far deployed chaos configurations, is the analog nature of the chaotic signal that "intrudes" in a digital communications world. Digitization of chaotic carriers and appropriate methods of synchronization and data recovery could boost this concept to be readily adopted in current communication protocols or to apply in other transmission environments, e.g., wireless and mobile communications, with appropriate mitigation stages. Finally, the extension of the synchronization property to larger scales of optical networks, through mutual coupling among different clusters of the network, shows a large potential of creating a synchronized communication network through complex signals. ZLS synchronization may be used as an authentication protocol for communicating nodes or as a sensing media for monitoring the network operation.

Bibliography

[1] Security requirements for cryptographic modules (fips pub 140-2) – http://csrc.nist.gov/publications/fips/fips140-2/fips1402.pdf, 2001.

[2] H. D. I. Abarbanel, M. B. Kennel, M. Buhl, and C. T. Lewis. Chaotic dynamics in erbium-doped fiber ring lasers. *Phys. Rev. A*, 60:2360–2374, 1999.

[3] H. D. I. Abarbanel, M. B. Kennel, L. Illing, S. Tang, H. F. Chen, and J. M. Liu. Synchronization and communication using semiconductor lasers with optoelectronic feedback. *IEEE Journal of Quantum Electronics*, 37:1301–1311, 2001.

[4] G. P. Agrawal and N. K. Dutta. *Semiconductor lasers*. Van Nostrand Reinhold, New York, 1993.

[5] G. P. Agrawal and N. K. Dutta. *Semiconductor lasers*. Kluwer Academic Publishers, Boston, 2nd edition, 2000.

[6] Y. Akizawa, T. Yamazaki, A. Uchida, T. Harayama, S. Sunada, K. Arai, K. Yoshimura, and P. Davis. Fast random number generation with bandwidth-enhanced chaotic semiconductor lasers at $8, times$, 50 gb/s. *Photonics Technology Letters, IEEE*, 24(12):1042–1044, June 15, 2012.

[7] V. Annovazzi-Lodi, G. Aromataris, M. Benedetti, and S. Merlo. Private message transmission by common driving of two chaotic lasers. *IEEE Journal of Quantum Electronics*, 46(2):258–264, February 2010.

[8] V. Annovazzi-Lodi, M. Benedetti, S. Merlo, T. Perez, P. Colet, and C. R. Mirasso. Message encryption by phase modulation of a chaotic optical carrier. *Photonics Technology Letters, IEEE*, 19(2):76–78, January 15, 2007.

[9] V. Annovazzi-Lodi, S. Donati, and A. Scire. Synchronization of chaotic injected laser systems and its application to optical cryptography. *IEEE Journal of Quantum Electronics*, 32:953–959, 1996.

[10] A. Argyris, A. Bogris, M. Hamacher, and D. Syvridis. Experimental evaluation of subcarrier modulation in chaotic optical communication systems. *Optics Letters*, 35(2):199–201, January 2010.

[11] A. Argyris, E. Grivas, A. Bogris, and D. Syvridis. Transmission effects in wavelength division multiplexed chaotic optical communication systems. *Journal of Lightwave Technology*, 28(21):3107–3114, November 1, 2010.

[12] A. Argyris, E. Grivas, M. Hamacher, A. Bogris, and D. Syvridis. Chaos-on-a-chip secures data transmission in optical fiber links. *Optics Express*, 18:5188–5198, 2010.

[13] A. Argyris, M. Hamacher, K. E. Chlouverakis, A. Bogris, and D. Syvridis. Photonic integrated device for chaos applications in communications. *Phys. Rev. Lett.*, 100:194101, May 2008.

[14] A. Argyris, D. Kanakidis, and A. Bogris. Spectral synchronization in chaotic optical communication systems. *IEEE Journal of Quantum Electronics*, 41:892–897, 2005.

[15] A. Argyris, D. Kanakidis, A. Bogris, and D. Syvridis. Experimental evaluation of an open-loop all-optical chaotic communication system. *IEEE Journal of Selected Topics in Quantum Electronics*, 10:927–935, 2004.

[16] A. Argyris, D. Kanakidis, A. Bogris, and D. Syvridis. First experimental demonstration of an all-optical chaos encrypted transmission system. In *Proc. 30th European Conference of Optical Communications*, Stockholm, 2004.

[17] A. Argyris, D. Syvridis, L. Larger, V. Annovazzi-Lodi, P. Colet, I. Fischer, J. Garcia-Ojalvo, C. R. Mirasso, L. Pesquera, and K. A. Shore. Chaos-based communications at high bit rates using commercial fibre-optic links. *Nature*, 438(7066):343–346, November 2005.

[18] Y. Aviad, I. Reidler, M. Zigzag, M. Rosenbluh, and I. Kanter. Synchronization in small networks of time-delay coupled chaotic diode lasers. *Optics Express*, 20(4):4352–4359, February 2012.

[19] J. F. M. Ávila, R. Vicente, J. R. Rios Leite, and C. R. Mirasso. Synchronization properties of bidirectionally coupled semiconductor lasers under asymmetric operating conditions. *Phys. Rev. E*, 75:066202, June 2007.

[20] S. Bauer, O. Brox, J. Kreissl, B. Sartorius, M. Radziunas, J. Sieber, H. J. Wünsche, and F. Henneberger. Nonlinear dynamics of semiconductor lasers with active optical feedback. *Phys. Rev. E*, 69:016206, 2004.

[21] C. H. Bennett and G. Brassard. Quantum cryptography: public key distribution and coin tossing. In *Proc. International Conference Computer Systems & Signal Processing*, pages 175–179, Bangalore, India, 1984.

[22] A. Bogris, K. E. Chlouverakis, A. Argyris, and D. Syvridis. Subcarrier modulation in all-optical chaotic communication systems. *Optics Letters*, 32(15):2134–2136, August 2007.

[23] A. Bogris, P. Rizomiliotis, K. E. Chlouverakis, A. Argyris, and D. Syvridis. Feedback phase in optically generated chaos: A secret key for cryptographic applications. *IEEE Journal of Quantum Electronics*, 44(2):119–124, February 2008.

[24] M. Bourmpos, A. Argyris, and D. Syvridis. Sensitivity analysis of a star optical network based on mutually coupled semiconductor lasers. *Journal of Lightwave Technology*, 30(16):2618–2624, August 15, 2012.

[25] M. Bucci, L. Germani, R. Luzzi, A. Trifiletti, and M. Varanonuovo. A high-speed oscillator-based truly random number source for cryptographic applications on a smart card ic. *IEEE Transactions on Computers*, 52(4):403–409, April 2003.

[26] H. F. Chen and J. M. Liu. Open-loop chaotic synchronization of injection-locked semiconductor lasers with gigahertz range modulation. *IEEE Journal of Quantum Electronics*, 36(1):27–34, January 2000.

[27] P. Colet and R. Roy. Digital communication with synchronized chaotic lasers. *Optics Letters*, 19:2056–2058, 1994.

[28] K. M. Cuomo and A. V. Oppenheim. Circuit implementation of synchronized chaos with applications to communications. *Phys. Rev. Lett.*, 71(1):65–68, July 1993.

[29] T. Dahms. *Synchronization in Delay-Coupled Laser Networks*. PhD thesis, Technischen Universität Berlin, Berlin, 2011.

[30] J. L. Danger, S. Guilley, and P. Hoogvorst. High speed true random number generator based on open loop structures in fpgas. *Microelectronics Journal*, 40:1650–1656, 2009.

[31] W. Diffie and M. E. Hellman. New direction in cryptography. *IEEE Transactions on Information Theory*, 22:644–654, 1976.

[32] J. F. Dynes, Z. L. Yuan, A. W. Sharpe, and A. J. Shields. A high speed, post-processing free, quantum random number generator. *Applied Physics Letters*, 93:031109, 2008.

[33] A. Englert, W. Kinzel, Y. Aviad, M. Butkovski, I. Reidler, M. Zigzag, I. Kanter, and M. Rosenbluh. Zero lag synchronization of chaotic systems with time delayed couplings. *Phys. Rev. Lett.*, 104:114102, March 2010.

[34] N. Fergusson and B. Scheier. *Practical Cryptography*. John Wiley & Sons, New York, 2003.

[35] I. Fischer, G. H. M. van Tartwijk, A. M. Levine, W. Elsässer, E. Göbel, and D. Lenstra. Fast pulsing and chaotic itinerancy with a drift in the coherence collapse of semiconductor lasers. *Phys. Rev. Lett.*, 76(2):220–223, January 1996.

[36] I. Fischer, R. Vicente, J. M. Buldu, M. Peil, C. R. Mirasso, M. C. Torrent, and J. Garcia-Ojalvo. Zero-lag long-range synchronization via dynamical relaying. *Phys. Rev. Lett.*, 97:123902, 2006.

[37] T. Franck, S. D. Brorson, A. Moller-Larsen, J. M. Nielsen, and J. Mork. Synchronization phase diagrams of monolithic colliding pulse-modelocked lasers. *Photonics Technology Letters, IEEE*, 8(1):40–42, January 1996.

[38] H. Fujino and J. Ohtsubo. Synchronization of chaotic oscillations in mutually coupled semiconductor lasers. *Optics Review*, 8:351–357, 2001.

[39] N. Gastaud, S. Poinsot, L. Larger, J.-M. Merolla, M. Hanna, J.-P. Goedgebuer, and F. Malassenet. Electro-optical chaos for multi-10 gbit/s optical transmissions. *Electronics Letters*, 40(14):898–899, July 2004.

[40] G Giacomelli and A. Politi. Relationship between delayed and spatially extended dynamical systems. *Phys. Rev. Lett.*, 76:2686–2689, 1996.

[41] N. Gisin, G. Ribordy, W. Tittel, and H. Zbinden. Quantum cryptography. *Rev. Mod. Phys.*, 74:145–179, 2002.

[42] J. T. Gleeson. Truly random number generator based on turbulent electroconvection. *Applied Physics Letters*, 81:1949, 2002.

[43] J.-P. Goedgebuer, P. Levy, L. Larger, C.-C. Chen, and W. T. Rhodes. Optical communication with synchronized hyperchaos generated electooptically. *IEEE Journal of Quantum Electronics*, 38:1178–1183, 2002.

[44] C. M. González, M. C. Torrent, and J. García-Ojalvo. Controlling the leader-laggard dynamics in delay-synchronized lasers. *Chaos*, 17:033122, 2007.

[45] K. S. Halle, C. W. Wu, M. Itoh, and Leon O. Chua. Spread spectrum communication through modulation of chaos. Technical Report UCB/ERL M93/27, EECS Department, University of California, Berkeley, 1993.

[46] T. Heil, I. Fischer, W. Elsässer, J. Mulet, and C. R. Mirasso. Chaos synchronization and spontaneous symmetry-breaking in symmetrically delay-coupled semiconductor lasers. *Phys. Rev. Lett.*, 86:795–798, January 2001.

[47] T. Heil, J. Mulet, I. Fischer, C. R. Mirasso, M. Peil, P. Colet, and W. Elsasser. On/off phase shift-keying for chaos-encrypted communication using external-cavity semiconductor lasers. *IEEE Journal of Quantum Electronics*, 38:1162–1170, 2002.

[48] K. Hirano, T. Yamazaki, S. Morikatsu, H. Okumura, H. Aida, A. Uchida, S. Yoshimori, K. Yoshimura, T. Harayama, and P. Davis. Fast random bit generation with bandwidth-enhanced chaos in semiconductor lasers. *Optics Express*, 18(6):5512–5524, March 2010.

[49] W. T. Holman, J. A. Connelly, and A. B. Dowlatabadi. An integrated analog/digital random noise source. *IEEE Transactions on Circuit and Systems I*, 44(6):521–528, June 1997.

[50] K. Ikeda. Multiple-valued stationary state and its instability of the transmitted light by a ring cavity system. *Optics Communications*, 30(2):257–261, 1979.

[51] Y. Ikuma and J. Ohtsubo. Dynamics in a compound cavity semiconductor laser induced by small external-cavity-length change. *IEEE Journal of Quantum Electronics*, 34(7):1240–1246, July 1998.

[52] J. Javaloyes, P. Mandel, and D. Pieroux. Dynamical properties of lasers coupled face to face. *Phys. Rev. E*, 67:036201, March 2003.

[53] N. Jiang, W. Pan, L. Yan, B. Luo, S. Xiang, L. Yang, D. Zheng, and N. Li. Chaos synchronization and communication in multiple time-delayed coupling semiconductor lasers driven by a third laser. *Selected Topics in IEEE Journal of Quantum Electronics*, 17(5):1220–1227, September-October 2011.

[54] I. Kanter, Y. Aviad, I. Reidler, E. Cohen, and M. Rosenbluh. An optical ultrafast random bit generator. *Nature Photonics*, 4:58–61, 2010.

[55] I. Kanter, M. Butkovski, Y. Peleg, M. Zigzag, Y. Aviad, I. Reidler, Mi. Rosenbluh, and W. Kinzel. Synchronization of random bit generators based on coupled chaotic lasers and application to cryptography. *Optics Express*, 18(17):18292–18302, August 2010.

[56] I. Kanter, E. Kopelowitz, R. Vardi, M. Zigzag, W. Kinzel, M. Abeles, and D. Cohen. Nonlocal mechanism for cluster synchronization in neural circuits. *EPL (Europhysics Letters)*, 93(6):66001, 2011.

[57] I. Kanter, M. Zigzag, A. Englert, F. Geissler, and W. Kinzel. Synchronization of unidirectional time delay chaotic networks and the greatest common divisor. *EPL (Europhysics Letters)*, 93(6):60003, 2011.

[58] Y. H. Kao, N. M. Wang, and H. M. Chen. Mode description of routes to chaos in external-cavity coupled semiconductor lasers. *IEEE Journal of Quantum Electronics*, 30(8):1732–1739, August 1994.

[59] E. Klein, N. Gross, M. Rosenbluh, W. Kinzel, L. Khaykovich, and I. Kanter. Stable isochronal synchronization of mutually coupled chaotic lasers. *Phys. Rev. E*, 73:066214, June 2006.

[60] L. Kocarev, K. S. Halle, K. Eckert, L. O. Chua, and U. Parlitz. Experimental demonstration of secure communications via chaotic synchronization. *International Journal of Bifurcation and Chaos*, 2:709–713, 1992.

[61] K. Kusumoto and J. Ohtsubo. 1.5-ghz message transmission based on synchronization of chaos in semiconductor lasers. *Optics Letters*, 27(12):989–991, June 2002.

[62] R. Lang and K. Kobayashi. External optical feedback effects on semiconductor injection laser properties. *IEEE Journal of Quantum Electronics*, 16(3):347–355, March 1980.

[63] L. Larger, J.-P. Goedgebuer, and F. Delorme. Optical encryption system using hyperchaos generated by an optoelectronic wavelength oscillator. *Phys. Rev. E*, 57:6618–6624, 1998.

[64] L. Larger, J.-P. Goedgebuer, and V. Udaltsov. Ikeda-based nonlinear delayed dynamics for application to secure optical transmission systems using chaos. *Comptes Rendus Physique*, 5:669–681, 2004.

[65] R. Lavrov, M. Jacquot, and L. Larger. Nonlocal nonlinear electro-optic phase dynamics demonstrating 10 gb/s chaos communications. *IEEE Journal of Quantum Electronics*, 46(10):1430–1435, October 2010.

[66] R. Lavrov, M. Peil, M. Jacquot, L. Larger, V. Udaltsov, and J. Dudley. Electro-optic delay oscillator with nonlocal nonlinearity: Optical phase dynamics, chaos, and synchronization. *Phys. Rev. E*, 80:026207, August 2009.

[67] J.-M. Liu, H. F. Chen, and S. Tang. Synchronized chaotic optical communications at high bit-rates. *IEEE Journal of Quantum Electronics*, 38:1184–1196, 2002.

[68] J.-M. Liu and T. B. Simpson. Four-wave mixing and optical modulation in a semiconductor laser. *IEEE Journal of Quantum Electronics*, 30(4):957–965, April 1994.

[69] Y. Liu, H. F. Chen, J.-M. Liu, P. Davis, and T. Aida. Communication using synchronization of optical-feedback-induced chaos in semiconductor lasers. *IEEE Transactions on Circuit and Systems I*, 48(12):1484–1490, December 2001.

[70] Y. Liu, P. Davis, Y. Takiguchi, T. Aida, S. Saito, and J.-M. Liu. Injection locking and synchronization of periodic and chaotic signals in semiconductor lasers. *IEEE Journal of Quantum Electronics*, 39(2):269–278, February 2003.

[71] A. Locquet, C. Massoler, and C. R. Mirasso. Synchronization regimes of optical-feedback-induced chaos in unidirectionally coupled semiconductor lasers. *Phys. Rev. E*, 65:056205, 2002.

[72] A. Locquet, F. Rogister, M. Sciamanna, P. Mégret, and M. Bondel. Two types of synchronization in unidirectionally coupled chaotic external-cavity semiconductor lasers. *Phys. Rev. E*, 64:045203, 2001.

[73] G. Marsaglia. Diehard: A battery of tests of randomness – http://www.stat.fsu.edu/pub/diehard/, 1996.

[74] C. Masoller. Spatio-temporal dynamics in the coherence collapsed regime of semiconductor lasers with optical feedback. *Chaos*, 7:455–462, 1997.

[75] C. Masoller. Anticipation in the synchronization of chaotic semiconductor lasers with optical feedback. *Phys. Rev. Lett.*, 86:2782–2785, 2001.

[76] A. Menezes, P. Van Oorschot, and S. Vanstone, editors. *Handbook of Applied Cryp.* CRC Press, Boca Raton, FL, 1996.

[77] C. R. Mirasso, P. Colet, and P. García-Fernández. Synchronization of chaotic semiconductor lasers: Application to encoded communications. *IEEE Photonics Technology Letters*, 8:299–301, 1996.

[78] C. R. Mirasso, J. Mulet, and C. Masoller. Chaos shift-keying encryption in chaotic external-cavity semiconductor lasers using a single-receiver scheme. *Photonics Technology Letters, IEEE*, 14(4):456–458, April 2002.

[79] J. Mork, B. Tromborg, and P. L. Christiansen. Bistability and low-frequency fluctuations in semiconductor lasers with optical feedback: A theoretical analysis. *IEEE Journal of Quantum Electronics*, 24(2):123–133, February 1988.

[80] J. Mulet, C. Mirasso, T. Heil, and I. Fischer. Synchronization scenario of two distant mutually coupled semiconductor lasers. *Journal of Optics B: Quantum and Semiclassical Optics*, 6(1):97, 2004.

[81] A. Murakami and J. Ohtsubo. Synchronization of feedback-induced chaos in semiconductor lasers by optical injection. *Phys. Rev. A*, 65:033826, 2002.

[82] A. Murakami, J. Ohtsubo, and Y. Liu. Stability analysis of semiconductor laser with phase-conjugate feedback. *IEEE Journal of Quantum Electronics*, 33(10):1825–1831, October 1997.

[83] M. Nixon, M. Friedman, E. Ronen, A. A. Friesem, N. Davidson, and I. Kanter. Synchronized cluster formation in coupled laser networks. *Phys. Rev. Lett.*, 106:223901, June 2011.

[84] J. Ohtsubo. Feedback induced instability and chaos in semiconductor lasers and their applications. *Optics Reviews*, 6:1–15, 1999.

[85] J. Ohtsubo. Chaos synchronization and chaotic signal masking in semiconductor lasers with optical feedback. *IEEE Journal of Quantum Electronics*, 38:1141–1154, 2002.

[86] M.-W. Pan, B.-P. Shi, and G. R. Gray. Semiconductor laser dynamics subject to strong optical feedback. *Optics Letters*, 22(3):166–168, February 1997.

[87] U. Parlitz, L. O. Chua, L. Kocarev, K. S. Halle, and A. Shang. Transmission of digital signals by chaotic synchronization. Technical Report UCB/ERL M92/129, EECS Department, University of California, Berkeley, 1992.

[88] J. Paul, M. W. Lee, and K. A. Shore. 3.5-ghz signal transmission in an all-optical chaotic communication scheme using 1550-nm diode lasers. *Photonics Technology Letters, IEEE*, 17(4):920–922, April 2005.

[89] L. Pecora and T. Carroll. Master stability functions for synchronized coupled systems. *Phys. Rev. Lett.*, 80:2109–2112, 1998.

[90] L. M. Pecora and T. L. Carroll. Synchronization in chaotic systems. *Phys. Rev. Lett.*, 64:821–824, February 1990.

[91] I. Reidler, Y. Aviad, M. Rosenbluh, and I. Kanter. Ultrahigh-speed random number generation based on a chaotic semiconductor laser. *Phys. Rev. Lett.*, 103:024102, 2009.

[92] R. L. Rivest, A. Shamir, and L. M. Adleman. A method for obtaining digital signatures and public-key cryptosystems. *Communication of the ACM*, 21:120–126, 1978.

[93] T. Sano. Antimode dynamics and chaotic itinerancy in the coherent collapse of semiconductor-lasers with optical feedback. *Phys. Rev. A*, 50:2719–2726, 1994.

[94] M. Sargent, M. O. Scully, and J. E. Lamb. *Laser physics*. Addison-Wesley, Reading, MA, 1974.

[95] L. Shu and D. J. Costello Jr. *Error control coding: fundamentals and applications*. Prentice-Hall, Englewood Cliffs, NJ, 1983.

[96] T. B. Simpson and J.-M. Liu. Period-doubling cascades and chaos in a semiconductor laser with optical injection. *Phys. Rev. A*, 51:4185–4185, 1995.

[97] S. Sivaprakasam, E. M. Shahverdiev, P. S. Spencer, and K. A. Shore. Experimental demonstration of anticipating solution in chaotic semiconductor lasers with optical feedback. *Phys. Rev. Lett.*, 87:4101–4103, 2001.

[98] M. C. Soriano, G. Van der Sande, I. Fischer, and C. R. Mirasso. Synchronization in simple network motifs with negligible correlation and mutual information measures. *Phys. Rev. Lett.*, 108:134101, March 2012.

[99] F. Sorrentino and E. Ott. Network synchronization of groups. *Phys. Rev. E*, 76:056114, November 2007.

[100] S. Sunada, T. Harayama, K. Arai, K. Yoshimura, P. Davis, K. Tsuzuki, and A. Uchida. Chaos laser chips with delayed optical feedback using a passive ring waveguide. *Optics Express*, 19(7):5713–5724, March 2011.

[101] Y. Takiguchi, Y. Liu, and J. Ohtsubo. Low-frequency fluctuation and frequency-locking in semiconductor lasers with long external cavity feedback. *Optics Reviews*, 6:399–401, 1999.

[102] S. Tang and J.-M. Liu. Message encoding-decoding at 2.5 gbits/s through synchronization of chaotic pulsing semiconductor lasers. *Optics Letters*, 26:1843–1845, 2001.

[103] G. H. M. Van Tartwijk and G. P. Agrawal. Laser instabilities: A modern perspective. *Progress in Quantum Electronics*, 22:43–122, 1998.

[104] J. R. Terry, K. S. Thornburg, D. J. DeShazer, G. D. VanWiggeren, S. Zhu, and P. Ashwin. Synchronization of chaos in an array of three lasers. *Phys. Rev. E*, 59:4036–4043, 1999.

[105] R. W. Tkach and A. R. Chraplyvy. Regimes of feedback effects in 1.5 μm distributed feedback lasers. *Journal of Lightwave Technology*, LT-4:1655–1661, 1986.

[106] C. Tokunaga, D. Blaauw, and T. Mudge. True random number generator with a metastability-based quality control. *IEEE Journal of Solid-State Circuits*, 43(1):78–85, January 2008.

[107] R. Toral and A. Chakrabarti. Generation of gaussian distributed random numbers by using a numerical inversion method. *Computer Physics Communications*, 74:327–334, 1993.

[108] B. Tromborg, J. Osmundsen, and H. Olesen. Stability analysis for a semiconductor laser in an external cavity. *IEEE Journal of Quantum Electronics*, 20(9):1023–1032, September 1984.

[109] A. Uchida, K. Amano, M. Inoue, K. Hirano, S. Naito, H. Someya, I. Oowada, T. Kurashige, M. Shiki, S. Yoshimori, K. Yoshimura, and P. Davis. Fast physical random bit generation with chaotic semiconductor lasers. *Nature Photonics*, 2:728–732, 2008.

[110] A. Uchida, Y. Liu, and P. Davis. Characteristics of chaotic masking in synchronized semiconductor lasers. *IEEE Journal of Quantum Electronics*, 39:962–970, 2003.

[111] R. Ursin, F. Tiefenbacher, T. Schmitt-Manderbach, H. Weier, T. Scheidl, M. Lindenthal, B. Blauensteiner, T. Jennewein, J. Perdigues, P. Trojek, B. Ömer, M. Fürst, M. Meyenburg, J. Rarity, Z. Sodnik, C. Barbieri, H. Weinfurter, and A. Zeilinger. Entanglement-based quantum communication over 144 km. *Nature Physics*, 3:481–486, 2007.

[112] O. Ushakov, S. Bauer, O. Brox, H. J. Wünsche, and F. He. Self-organization in semiconductor lasers with ultrashort optical feedback. *Phys. Rev. Lett.*, 92:043902, 2004.

[113] G. D. Van Wiggeren and R. Roy. Communication with chaotic lasers. *Science*, 279:1198–1200, February 20, 1998.

[114] R. Vicente. *Nonlinear dynamics and synchronization of bidirectionally coupled semiconductor lasers*. PhD thesis, Universitat de les Illes Balears, Department de Fisica, Palma, 2006.

[115] R. Vicente, I. Fischer, and C. R. Mirasso. Synchronization properties of three delay-coupled semiconductor lasers. *Phys. Rev. E*, 78:066202, December 2008.

[116] R. Vicente, T. Perez, and C. R. Mirasso. Open-versus closed-loop performance of synchronized chaotic external-cavity semiconductor lasers. *IEEE Journal of Quantum Electronics*, 38(9):1197–1204, September 2002.

[117] S. Wiesner. Conjugate coding. *ACM SIGACT News*, 15(1):78–88, January 1983.

[118] S. G. Wilson. *Digital Modulation and Coding*. Prentice-Hall, Upper Saddle River, NJ, 1996.

[119] H. G. Winful and L. Rahman. Synchronized chaos and spatiotemporal chaos in arrays of coupled lasers. *Phys. Rev. Lett.*, 65:1575–1578, September 1990.

[120] C. W. Wu and L. O. Chua. A simple way to synchronize chaotic systems with applications to secure communication systems. *International Journal of Bifurcation and Chaos*, 3(6):1619–1627, December 1993.

[121] M. Yousefi, Y. Barbarin, S. Beri, E. A. J. M. Bente, M. K. Smit, R. Nötzel, and D. Lenstra. New role for nonlinear dynamics and chaos in integrated semiconductor laser technology. *Phys. Rev. Lett.*, 98:044101, 2007.

[122] J. Zamora-Munt, C. Masoller, J. Garcia-Ojalvo, and R. Roy. Crowd synchrony and quorum sensing in delay-coupled lasers. *Phys. Rev. Lett.*, 105:264101, December 2010.

[123] J.-Z. Zhang, A.-B. Wang, J.-F. Wang, and Y.-C. Wang. Wavelength division multiplexing of chaotic secure and fiber-optic communications. *Optics Express*, 17(8):6357–6367, April 2009.

[124] W.-L. Zhang, W. Pan, B. Luo, X.-H. Zou, and M.-Y. Wang. One-to-many and many-to-one optical chaos communications using semiconductor lasers. *Photonics Technology Letters, IEEE*, 20(9):712–714, May 1, 2008.

6

Chaotic convergence of unsupervised equalizers

Romis Attux, Everton Z. Nadalin, and João Marcos T. Romano

School of Electrical and Computer Engineering (FEEC)
University of Campinas (UNICAMP)

Diogo C. Soriano and Ricardo Suyama

Centro de Engenharia, Modelagem e Ciências Sociais Aplicadas (CECS)
Universidade Federal do ABC (UFABC)

CONTENTS

6.1 Introduction

The problem of data transmission can be defined, in simple terms, as that of allowing proper information exchange between a transmitter and a receiver interconnected by what is generally termed *communication channel*. In view of a number of physical factors, this channel is, in practice, responsible for distorting the messages it conveys, which establishes a concrete demand for efficient countermeasures. A popular strategy in this context is to make use of a filter — a systematically crafted information-processing device — to compensate for the channel effects. This filter is known as *equalizer*, and the aforementioned strategy is the description of the *raison d'être* of an *equalization* process.

Equalizer design is a task that can be approached from a number of standpoints that are insinuated by the character of the application at hand. For example, if the channel model is static and well known, the use of an inverse model is a straightforward possibility (which, nonetheless, must be considered *cum grano salis* whenever the existence of noise is a crucial aspect). On the other hand, the direct use of statistical information of the involved signals is natural when the channel in unknown or when its model has a prominent stochastic or time-variant character.

A classical statistical approach, which is feasible whenever a reference signal is available, is to build a stochastic criterion based on an error signal representing the difference between the desired response $d(n)$ and the actual filter output $y(n)$. The emblematic formulation of this class of supervised approaches is the Wiener criterion, the rationale of which is the minimization of the following cost function [16][1]:

$$J_W(\mathbf{w}) = E\left[e^2(n)\right] = E\left[(d(n) - y(n))^2\right],\tag{6.1}$$

$E[\cdot]$ being the statistical expectation operator and $\mathbf{w}$ the equalizer parameter vector. If a finite impulse response (FIR) filter with N taps is employed, $y(n) = \mathbf{w}^T\mathbf{x}(n)$, where $\mathbf{x}(n) = [x(n), x(n-1), \ldots x(n-N+1)]^T$ is the equalizer input vector and the gradient of the above cost function is [9]:

$$\nabla J_W(\mathbf{w}) = 2\mathbf{R}\mathbf{w} - 2\mathbf{p},\tag{6.2}$$

$\mathbf{R} = E\left[\mathbf{x}(n)\mathbf{x}^T(n)\right]$ being the input correlation matrix and $\mathbf{p} = E\left[\mathbf{x}(n)d(n)\right]$ the cross-correlation vector, which expresses the degree of correlation between the desired signal and all filter inputs. From Equation (6.2), and taking into account the fact that $\mathbf{R}$ is, as a rule, a positive definite matrix, it can be shown that the single minimum of $J_W(\mathbf{w})$ is located at

$$\mathbf{w}_{wiener} = \mathbf{R}^{-1}\mathbf{p}.\tag{6.3}$$

This vector, which is called *Wiener solution*, can be considered, under the adopted hypotheses, the optimal filter in the mean-squared error (MSE) sense.

In spite of the existence of a closed-form solution for this formulation, it can notwithstanding be more effectively employed in time-variant contexts and/or in situations where computational resources are relatively scarce with the aid of iterative optimization methods. A natural option is the *steepest descent method*, which makes use of the gradient vector to define a promising direction for taking a minimizing step:

$$\mathbf{w}(n+1) = \mathbf{w}(n) - \mu\nabla J_W\left[\mathbf{w}(n)\right],\tag{6.4}$$

where μ is the algorithm step-size. This iterative expression can be understood

---

[1]Throughout this work, we will always consider that all signals and systems are defined in terms of real values.

as a learning rule to automatically control the process of filter parameter selection, which evokes the important concept of *adaptive filtering* [9].

The update expression in Equation (6.4) is, in view of Equation (6.2), essentially a linear dynamical system. Hence, it is possible to analyze its behavior in accordance with two fundamental scenarios: convergence to a fixed point (the Wiener solution presented in Equation (6.3))/divergence towards infinity. In order to develop this point, let us rewrite Equation (6.4) using Equation (6.2) (the factor 2 in the gradient expression was incorporated, without loss of generality, to the step-size):

$$\mathbf{w}(n+1) = (\mathbf{I} - \mu\mathbf{R})\mathbf{w}(n) + \mu\mathbf{p}. \tag{6.5}$$

Asymptotic convergence towards the Wiener solution depends on the eigenstructure of the matrix $\mathbf{A} = \mathbf{I} - \mu\mathbf{R}$, where $\mathbf{I}$ is the $N \times N$ identity matrix. It is possible to relate the eigenvalues of $\mathbf{A}$ to those of $\mathbf{R}$ quite directly [9]:

$$\lambda_{\mathbf{A}} = 1 - \mu\lambda_{\mathbf{R}}. \tag{6.6}$$

If the Wiener solution is to be an attractor, it is necessary that all eigenvalues of $\mathbf{A}$ lie within the unit circle, i.e.,

$$-1 < \lambda_{\mathbf{A}} < 1. \tag{6.7}$$

In view of Equation (6.6), this means that

$$0 < \mu\lambda_{\mathbf{R}} < 2. \tag{6.8}$$

As both the eigenvalues and the step-size are expected to be positive, the first inequality is not relevant. The second inequality can be applied directly to the most stringent case, leading to

$$\mu < \frac{2}{\lambda_{\mathbf{R}_{max}}}. \tag{6.9}$$

This equation shows that the step-size μ is the key parameter in the convergence analysis of the steepest descent algorithm, and, moreover, that its choice must be regulated by $\lambda_{\mathbf{R}_{max}}$, the largest eigenvalue of the input correlation matrix. In view of the linear character of the algorithm, there are, as already mentioned, only two options: convergence to the Wiener solution or divergence.

The steepest descent framework, in spite of its soundness, is based on the availability of the correlation matrix and the cross-correlation vector, which, in practice, must be estimated both in time-invariant and time-variant environments. This necessity was an important motivation underlying the development of online adaptive methods, among which we highlight the *least mean square* (LMS) algorithm [5].

The LMS can be understood as a version of the steepest-descent algorithm in which the correlation matrix and the cross-correlation vector are replaced with instantaneous estimates of the form [9]:

$$\hat{\mathbf{R}}(n) = \mathbf{x}(n)\mathbf{x}^T(n), \tag{6.10}$$

$$\hat{\mathbf{p}}(n) = \mathbf{x}(n)d(n). \tag{6.11}$$

Using Equations (6.10) and (6.11), Equation (6.4) can be rewritten as

$$\mathbf{w}(n+1) = \mathbf{w}(n) + \mu \mathbf{e}(n)\mathbf{x}(n). \tag{6.12}$$

In contrast with the steepest descent algorithm, Equation (6.12) is, as a rule, a stochastic dynamical system, the study of which cannot be carried out along the previously employed lines. However, there are important points of contact between the stability condition obtained in Equation (6.9) and stochastic convergence results derived for the LMS. These points, however, will not be further discussed here, the reader being referred to references like [5] and [9], which address them in detail.

When it is not possible to make direct use of a reference (or desired) signal, it is necessary to resort to unsupervised (or blind) equalization methods. These methods typically aim at building the channel inverse model in accordance with information brought by higher-order statistics of the received signal. Among the various unsupervised approaches found in the literature, it is possible to highlight, in view of its historical and practical relevance, the class of Bussgang algorithms [10]. These algorithms can be seen as extensions of the emblematic LMS formulation in which the reference signal is estimated according to a memoryless nonlinear mapping operating on the equalizer output. Mathematically, this definition leads to an update expression of the following general form:

$$\mathbf{w}(n+1) = \mathbf{w}(n) + \mu \left[\Phi[y(n)] - y(n) \right] \mathbf{x}(n), \tag{6.13}$$

where $\Phi[\cdot]$ is the aforementioned mapping and μ is, once again, the step-size parameter.

In this chapter, we will concentrate our attention on two of the most widespread Bussgang methods, those defined by the decision-directed (DD) and constant modulus (CM) criteria [16]. The DD criterion arises from a choice regarding the nonlinear estimator $\Phi[\cdot]$ that is quite intuitive in view of the structure of a digital communication system - the decision device itself [13]. In other words, as a digital communication system will necessarily contain a decision device whose function is to ensure that the recovered information suit the features of the used digital alphabet, it should, at least under favorable conditions, provide reliable estimates of the transmitted symbols. Naturally, the same will not hold whenever the equalizer is not capable of properly mitigating the channel effects - this dilemma, as a matter of fact, holds the key to essential aspects regarding the study of the DD approach [12, 13].

The DD criterion can be defined in terms of the following cost function:

$$J_{DD}(\mathbf{w}) = E[dec[y(n)] - y(n)]^2,$$ (6.14)

which, if the equalization structure is a FIR filter and a gradient method using the kind of stochastic approximation described in Equations (6.10) and (6.11) is used, gives rise to the DD algorithm:

$$\mathbf{w}(n+1) = \mathbf{w}(n) + \mu [dec[y(n)] - y(n)] \mathbf{x}(n).$$ (6.15)

The DD formulation, in spite of its straightforwardness, is not devoid of aspects that, in some cases, may give rise to practical difficulties. The relationship between phase recovery and equalization, for instance, motivated a work by Godard [8] that, together with the contributions of Treichler and Agee [18], defined the CM equalization approach. The CM criterion is built from a cost function that quantifies the average equalizer output deviation from a constant value that incorporates information about the amplitude of the transmitted symbols:

$$J_{CM}(\mathbf{w}) = E\left[\left(R_2 - y^2(n)\right)^2\right],$$ (6.16)

where $R_2 = E\left[s^4(n)\right]/E\left[s^2(n)\right]$. The same line of reasoning that was followed to derive the DD algorithm leads to the constant modulus algorithm (CMA):

$$\mathbf{w}(n+1) = \mathbf{w}(n) + \mu \left[R_2 - y^2(n)\right] y(n)\mathbf{x}(n).$$ (6.17)

The study of the CMA was essential in the development of unsupervised equalization theory, both in view of the practical relevance of the algorithm itself [11] and of the connections it establishes with kurtosis-based equalization and source separation approaches [1], which have been consolidated with the pioneering efforts of Ding and Li [4] and the beautiful proof obtained by Regalia [15].

A direct inspection of the DD and CM criteria shown in Equations (6.14) and (6.16) reveal that the use of a gradient-based method for their optimization gives origin to a nonlinear update expression. In other words, the update expressions associated with both criteria are nonlinear dynamical systems that are *prima facie* deterministic and may, when instantaneous estimates are used for the gradient, become stochastic. However, despite their nonlinear character, the study of the dynamical behavior of these Bussgang methods has followed, as a rule, the dual perspective that characterizes the analysis of the linear Wiener-based methods: convergence to a fixed point/divergence. In the remainder of this chapter, we will show that this classical standpoint does not make justice to the plethora of dynamical possibilities — including chaotic behavior — that might occur whenever the DD and CM approaches are employed [2,7].

6.2 Analysis of the constant modulus and decision-directed algorithms

Before we proceed to the dynamical analyses of the constant modulus and decision-directed algorithms along the lines described in [2] and [17], it is important to introduce some hypotheses that will, throughout this work, be assumed to hold:

H1 – The equalizer is a FIR filter with N coefficients.

H2 – The transmitted signal obeys a binary and symmetrical digital modulation scheme.

H3 – The transmitted symbols are composed of i.i.d. samples drawn from a uniform distribution.

6.2.1 Analysis of the CMA

The study of the dynamical behavior of the CMA will begin with a very simple case, for which the equalizer is a single-tap filter, i.e., $N = 1$. In this case, the equalizer output can be simply written as

$$y(n) = w(n)x(n). \tag{6.18}$$

In view of hypotheses H2 and H3, if the absolute value of the symbols that compose $s(n)$ is assumed to be equal to a, the value of the constant R_2 is given by

$$R_2 = a^4/a^2 = a^2. \tag{6.19}$$

Using Equations (6.18) and (6.19), the update expression shown in Equation (6.17) becomes

$$w(n+1) = w(n) + \mu[a^2 - w^2(n)x^2(n)]w(n)x^2(n). \tag{6.20}$$

If the channel is also a single-tap filter, that is,

$$x(n) = ks(n), \tag{6.21}$$

it is possible to rewrite Equation (6.20) as

$$w(n+1) = w(n) + \mu k^2 a^4 w(n)[1 - k^2 w^2(n)]. \tag{6.22}$$

An interesting point is that this scenario engenders an update expression that is deterministic, in spite of the stochastic character of the transmitted signal. This is certainly a welcomed feature, as it becomes natural to perform the entire analysis along well-trodden paths of dynamical system theory [7,14].

The first step is to obtain the equilibrium points of the system. This is done by looking for solutions that make $w(n+1)$ identical to $w(n)$. From Equation (6.22),

$$\mu k^2 a^4 w(n)[1 - k^2 w^2(n)] = 0. \tag{6.23}$$

After manipulating Equation (6.23), it is possible to identify three solutions:

$$w_e = 0 \quad \text{and} \quad w_e = \pm 1/k. \tag{6.24}$$

The first solution, which is virtually ubiquitous in the study of unsupervised methods, can be considered trivial. On the other hand, the pair of symmetrical solutions is very relevant, as they allow perfect channel compensation - the sign ambiguity is unavoidable in view of structure of the CMA [16], but it should be kept in mind that it is information-preserving.

After having found the equilibrium points, the next step is to study their local stability. This is typically done with the aid of an analysis of the structure of the system Jacobian matrix, which, in this case, corresponds to a single scalar value. If Equation (6.22) is written as $w(n+1) = F[w(n)]$, this value is $F'[w(n)]$. If this function, at a given point w_e, is within the unit-radius circle (URC), that is, $|F'(w_e)| < 1$, the equilibrium point is asymptotically stable. The stability condition for the pair of ideal solutions is identical:

$$0 < \mu < 1/k^2 a^4. \tag{6.25}$$

It is interesting to notice that the upper limit value depends on two factors related to the scale of the received signal - the amplitude of the transmitted symbols and the channel single free parameter. This shows that, as in the case of the analysis of supervised algorithms [9], the step-size has an important role as a sort of stabilizing "gain control."

In view of the dual character of the stability analysis of supervised methods, we could be led to believe that, for $\mu > 1/k^2 a^4$, the algorithm would necessarily diverge towards infinity. However, if we recall that the CMA, in this scenario, is a nonlinear deterministic discrete-time dynamical system, it is licit to expect a broader range of possible behaviors. In order to investigate this issue, we present, in Figure 6.1, a bifurcation diagram generated by the evolution of the CMA from a single initial condition ($w(0) = 0.49$), under the parameters $a = 1$ and $k = 2$.

First, Figure 6.1 confirms the obtained stability range, as, for $0 < \mu < 1/k^2 a^4 = 0.25$, the algorithm converges towards the positive nontrivial equilibrium point $w_e = 0.5$. However, when the upper limit is passed, there is no immediate divergence - as a matter of fact, the system experiences a period-two limit cycle, which subsequently is expanded into a cascade of period-doubling bifurcations followed by aperiodic and chaotic behavior in a typical Feigenbaum scenario [7, 14]. To verify the soundness of these conclusions, we present, in Figure 6.2, the value of the Lyapunov exponent of the system for the same range of step-size values.

Notice that, when there is convergence towards either a fixed point or

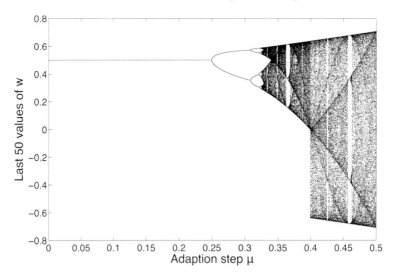

FIGURE 6.1
Bifurcation diagram — single-tap CM equalizer.

a limit cycle, the exponent assumes negative values. The aperiodic regions indicated in the diagram are associated with a positive exponent, which shows that they, indeed, correspond to chaotic behavior. After the chaotic zone, which is permeated by periodicity windows, there is actual divergence towards infinity.

Many conclusions can be drawn from this effort. Perhaps the foremost is that, as conjectured by Frater et al. [6], the CMA is indeed capable of showing chaotic behavior, thus revealing that the dual analytical framework employed in the study of supervised methods is far less representative in an unsupervised context.

It is also noteworthy that, for this simple case, it was possible to reach a closed-form expression for the maximum step-size. Its dependence on a fourth-order term is a consequence of the quartic structure of the CM cost function. To illustrate this point, let us consider what would be the step-size range for the supervised steepest descent algorithm [9]. To do so, let us calculate the correlation matrix, which will be, in this case, a scalar:

$$\mathbf{R} = k^2 a^2 \tag{6.26}$$

as well as the cross-correlation vector,

$$\mathbf{p} = k a^2. \tag{6.27}$$

The Wiener solution, as shown in Equation (6.3), is given by

$$w_{wiener} = k a^2 / k^2 a^2 = 1/k(28). \tag{6.28}$$

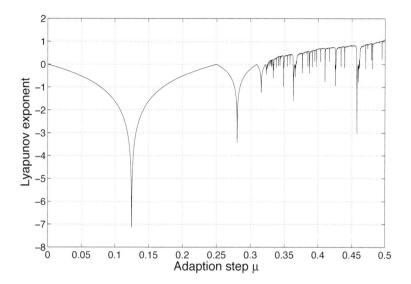

FIGURE 6.2
Lyapunov exponent — single-tap CM equalizer.

Notice that the Wiener solution is identical to one of the equilibrium points of the CMA - there is no sign ambiguity, as a reference signal is available. The maximum step-size is, in accordance with Equation (6.9),

$$\mu_{max} = 2/max = 2/k^2 a^2. \tag{6.29}$$

The inexistence of dependence with respect to a fourth-order term stresses the structural differences between both methods.

Let us now extend the analysis to the more complex case of a two-tap equalizer applied to the inversion of a two-tap channel. This case has some noteworthy features:

1– A two-tap channel effectively introduces intersymbol interference, which means that the scenario is more representative of a real-world equalization application.

2– It is not possible to attain an ideal solution (with complete intersymbol interference removal), which means that the equilibrium points will not be straightforwardly determinable [4].

3– The CMA will have a general stochastic update expression.

The first two aspects force us to resort to approximate values whenever dealing with the study of the equilibrium points. The third point poses an even more complex problem, one that, in principle, should necessarily force us to abandon the analytical line followed in the single-tap case and follow approaches like [3]. However, to prevent this, we will consider a steepest-descent/deterministic version of the CMA, which can be understood as a limit case of a batch version of this algorithm for a sufficiently elevated number of samples (more on this subject will be said in Section 6.2.3).

In view of the aforementioned difficulties, the analysis will be carried out in the context of a specific case that is, nonetheless, representative of the achievable dynamical behaviors [2]. This case is defined by a noiseless maximum-phase channel with a coefficient vector $\mathbf{h} = [1, 1.5]$ and transmitted samples with unit amplitude. For this case, an exhaustive analysis of the CM cost reveals the existence of two pairs of minima: $\mathbf{w}_{global} = \pm [0.246, -0.522]^T$ and $\mathbf{w}_{local} = [0.378, 0.065]^T$.

First, let us investigate the behavior of the deterministic CMA in the vicinity of one of the two global minima. In Figure 6.3, we present the bifurcation diagram associated with the first equalizer tap when the initial parameter is close to $\mathbf{w} = [0.246, -0.522]^T$.

It can be seen that the dynamical repertoire present in this two-tap case is similar to that verified for the single-tap case: the equilibrium point is stable for a range of step-size values and, when its stability is lost, a cascade of period-doubling bifurcations takes place, after which the equalizer becomes subject

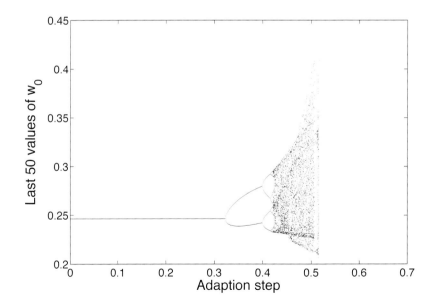

FIGURE 6.3
Bifurcation diagram — two-tap CM equalizer (global minimum).

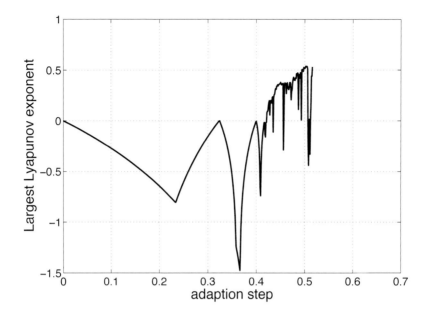

FIGURE 6.4
Largest Lyapunov exponent — two-tap CM equalizer (global minimum).

to chaotic behavior. This qualitative analysis is supported by the values of the largest Lyapunov exponent of the system, which is shown in Figure 6.4.

When the system is initialized at the vicinity of a local minimum, the general pattern is preserved, but some peculiar aspects must be pointed out. Figure 6.5 contains the necessary information.

It can be seen that, for small values of μ, the local minimum is locally stable, but, for a step-size value close to 0.22, the period-doubling cascade begins and leads to chaotic behavior. However, for μ close to 0.36, the strange attractor loses its stability and the system follows a pattern that corresponds to the evolution of the studied global minimum.

The general conclusions are that, for the more general case of a two-tap scenario, the dynamical possibilities of the CMA also transcend the dual framework that characterizes the supervised case in view of the existence of multiple limit cycles and also of strange attractors. In Section 6.2.3, we will try to investigate in a more systematic way points of contact with the stochastic case.

6.2.2 Analysis of the DD algorithm

The investigation concerning the decision-directed algorithm will follow the same pattern adopted for the CMA. First, we will consider the single-tap scenario and, subsequently, the deterministic two-tap case, being, as already

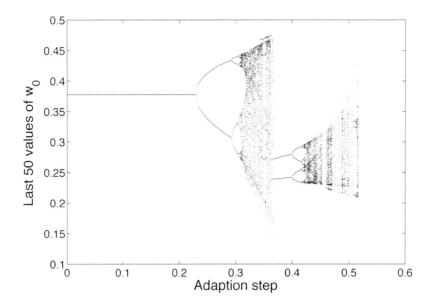

FIGURE 6.5
Bifurcation diagram — two-tap CM equalizer (local minimum).

mentioned, the discussion of points of contact with the stochastic version postponed until Section 6.2.3.

The single-tap scenario is similar to that used in the analysis of the CMA, with the channel corresponding to a simple gain k. To simplify the input-output expression associated with the decision device, we will consider, without loss of generality insofar as the adopted line of reasoning is concerned, that $a = 1$. The equalizer obeys the relationship described in (18), and the general update expression for the DD algorithm, shown in Equation (6.15), becomes:

$$w(n+1) = w(n) + \mu[sign[w(n)ks(n)] - w(n)ks(n)]ks(n). \qquad (6.30)$$

After some manipulations, we obtain:

$$w(n+1) = w(n) + \mu\left[sign[w(n)]|k| - w(n)k^2\right], \qquad (6.31)$$

which can be understood as a one-dimensional deterministic nonlinear map. This map can be analyzed with the aid of the same classical tools employed in the study of the CMA. The first step is, again, to obtain the equilibrium points, which arise from the canonical condition $w(n+1) = w(n)$. They must obey the following equality:

$$sign[w_e]|k| = w_e k^2. \qquad (6.32)$$

The trivial solution is valid, and the nontrivial solutions are:

$$w_e = \pm 1/|k|. \tag{6.33}$$

It is interesting to notice that these points are exactly the same obtained for the CMA: this is expected in view of the concrete possibility of perfect channel inversion and the simplicity of the scenario as a whole.

The stability of these points is determined by the first derivative of $F[w(n)]$ if Equation (6.31) is assumed to have the form $w(n+1) = F[w(n)]$. For the trivial solution, the range of step-size values for which $|F'(w_e)| < 1$ is, for the ideal solutions,

$$0 < \mu < 2/k^2. \tag{6.34}$$

Notice that this condition is the same obtained for the steepest descent algorithm in this scenario (see Equation (6.29)). This is due to the fact that the DD algorithm, in the vicinity of good Wiener solutions (i.e., solutions that are capable of promoting perfect signal reconstruction), generates a perfect estimate of the desired signal, becoming, in practice, equivalent to the LMS. For more details on this sort of equivalence, see, for instance, [12, 13].

To study the behavior of the DD algorithm beyond the obtained step-size limit, we will start, again, from a bifurcation diagram, the diagram shown in Figure 6.6 (we used again $k = 2$).

The diagram reveals a behavior that is different from that of the CMA. Instead of experiencing a cascade of period-doubling bifurcations, the algorithm starts, immediately after the equilibrium point ($w_e = 1/2$) stability limit is passed, to operate in a chaotic regime. Further increase in the step-size leads to modifications in the structure of the corresponding strange attractor. In

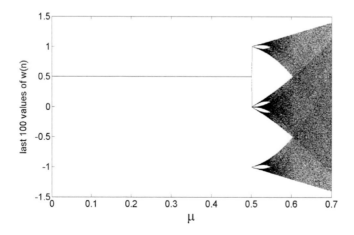

FIGURE 6.6
Bifurcation diagram — single-tap DD equalizer.

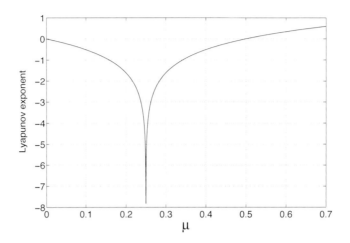

FIGURE 6.7
Lyapunov exponent — single-tap DD equalizer.

Figure 6.7, the Lyapunov exponent is presented for all values of μ, and its values confirm the presented analysis and also the range shown in Equation (6.34).

To gain further insight on the dynamical behavior of the DD algorithm, we will replace the hard decision-device considered so far with a soft decision-device, which will take the following form:

$$dec[y(n)] = tanh[py(n)]. \tag{6.35}$$

The free parameter p makes possible to control the abruptness of the transition between possible values (-1 and $+1$), being the $sign(\cdot)$ function obtained when $p \to \infty$.

For this nonlinearity, the equilibrium points are those who satisfy the following equation:

$$\begin{aligned}
w_e &= w_e + \mu \left\{ [tanh(pw_e x(n)) - w_e x(n)] \right\} x(n) \\
&= w_e + \mu \left\{ [tanh(pw_e as(n)) - w_e as(n)] \right\} as(n).
\end{aligned} \tag{6.36}$$

Manipulating Equation (6.36), we have:

$$\mu a \left\{ tanh\left(pw_e as(n)\right) s(n) - w_e a \right\} = 0. \tag{6.37}$$

Noticing that the hyperbolic tangent is an odd function, we have:

$$tanh(pw_e a) - w_e a = 0 \to tanh(pw_e a) = w_e a. \tag{6.38}$$

Again, the trivial solution $w_e = 0$ is an equilibrium point. The remaining pair can be obtained by a numerical analysis of the intersections between the

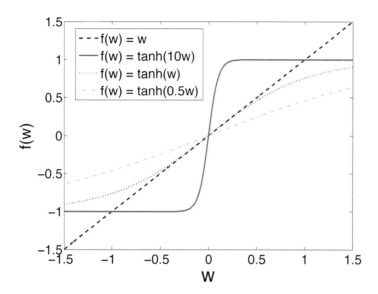

FIGURE 6.8

Graphical solutions — equilibrium points.

graphics of both members of the equation - some examples are shown in Figure 6.8.

These equilibrium points can be analyzed in terms of stability for each particular case, but we will not address this point in more detail here. Instead, let us study the bifurcation diagram for a case for which $p = 5$ and $k = 2$, depicted in Figure 6.9.

An important difference with respect to the diagram in Figure 6.6 is that, analogously to what was verified for the CMA, there is a clear cascade of period-doubling bifurcations leading to chaos and periodicity windows. In Figure 6.10, we present the Lyapunov exponent associated with this diagram.

This analysis clearly shows that the dynamical behavior of the DD algorithm can be significantly distinct depending on whether soft or hard decision is used. It also seems to indicate that the existence of a Feigenbaum scenario for Bussgang algorithms is related to the presence of abrupt transitions in the nonlinear memoryless estimator.

We will now consider the case of a two-tap channel equalized by a two-tap filter. The decision-directed algorithm will be treated in its deterministic version, similarly to what was done in the study of the CMA. The considered channel will be $\mathbf{h} = [1, 0.6]$, and the transmitted signal will be considered to have unit amplitude. In Figure 6.11, we present the bifurcation diagram for an initial condition close to the DD global minimum (which is coincident with the best Wiener solution - $\mathbf{w}_{wiener} = [0.913, -0.403]^T$).

The followed dynamical pattern is, in essence, identical to that verified

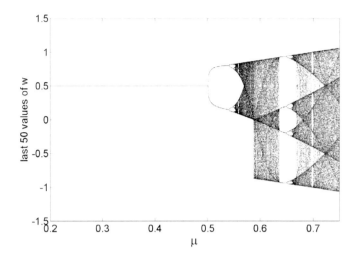

FIGURE 6.9
Bifurcation diagram — modified single-tap DD equalizer.

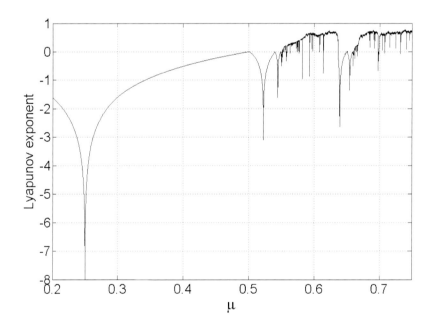

FIGURE 6.10
Lyapunov exponent — modified single-tap DD equalizer.

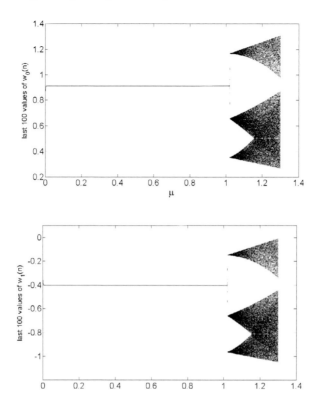

FIGURE 6.11

Bifurcation diagram — two-tap DD equalizer (global minimum).

for the single-tap case - convergence to the best Wiener solution for a certain range of step-size values followed by a direct transition to a chaotic regime (which is followed by divergence towards infinity). In Figure 6.12, the largest Lyapunov is presented for the same step-size values, confirming the qualitative analysis.

The use of a soft decision device leads, also in this case, towards a period-doubling route to chaos, as shown in Figure 6.13 (again, $p = 5$). The main difference is that, when the strange attractor loses its stability, the algorithm converges towards a stable limit cycle in another region of the space, which also follows the pattern of period-doubling and chaos. It seems reasonable to assume that this second sequence of bifurcations is originated by a local minimum.

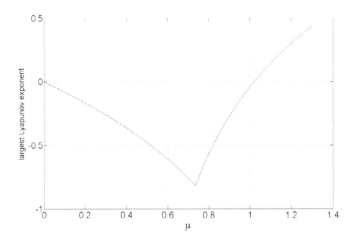

FIGURE 6.12
Largest Lyapunov exponent — two-tap DD equalizer (global minimum).

6.2.3 Some comments on the transition between stochastic and deterministic regimes

In order to deal with cases involving equalizers with multiple taps, we resorted to the expedient of studying deterministic versions of both the CMA and the DD algorithm. This expedient is justifiable from the standpoint of the very distinct types of dynamical analyses performed for deterministic and stochastic systems, but it is also pertinent in view of the fact that the deterministic version of the algorithm can be viewed as a limit case of an algorithm that uses batch gradient estimation for a sufficiently elevated number of employed samples. To illustrate this fact, we present, in Figure 6.13, a bifurcation diagram for the two-tap DD algorithm, in the same scenario analyzed in Figure 6.11, with hard decision and 1200 samples used for gradient calculation.

The diagram indicates that the dynamical behavior of the algorithm is similar to that observed for the deterministic case, although some stochastic fluctuations are noticeable.

6.3 Conclusions

In this work, we perform, along the lines established in [2] and [17], a dynamical analysis of two emblematic unsupervised equalization methods - the decision-directed and constant modulus algorithms. This analysis shows that the variety of potential steady-state behaviors achievable by algorithms based

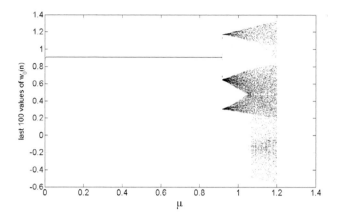

FIGURE 6.13
Bifurcation diagram — two-tap stochastic DD equalizer (global minimum).

on higher-order statistics cannot be properly dealt with within the dual "convergence to a fixed point/divergence" framework that took form in the context of supervised approaches. It is possible that these conclusions also apply insofar as other unsupervised (equalization and source separation) methods are concerned, which may have a tangible impact on studies about the convergence of adaptive techniques, especially when precise gradient estimation is in the order of the day.

6.4   Acknowledgments

The authors thank FAPESP and CNPq for the financial support.

Bibliography

[1] R. Attux, A. Neves, L. T. Duarte, R. Suyama, C. C. Junqueira, L. Rangel, T. Dias, and J. M. Romano. On the relationships between blind equalization and blind source separation - part ii: Relationships. *Journal of Communication and Information Systems*, 22:53–61, 2007.

[2] R. Attux and J. M. T. Romano. Chaotic phenomena in adaptive blind equalisers. *IEE Proceedings – Vision, Image and Signal Processing*, 150(6):360–364, 2003.

[3] C. de Sousa Jr., R. Attux, R. Suyama, and J. M. T. Romano. Lyapunov-based stability analysis of supervised and unsupervised adaptive algorithms. In *Proc. IEEE International Symposium on Circuits and Systems*, Paris, 2010.

[4] Z. Ding and Y. Li. *Blind Equalization and Identification*. Marcel Dekker, New York, 2001.

[5] P. S. R. Diniz. *Adaptive Filtering: Algorithms and Practical Implementation*. Springer, New York, 4th edition, 2012.

[6] M. R. Frater, R. R. Bimead, and C. R. Johnson Jr. Local minima escape transients by stochastic gradient descent algorithms in blind adaptive equalizers. *Automatica*, 31:637–641, 1995.

[7] N. Friedler-Ferrara and C. C. do Prado. *Chaos: An Introduction*. Blücher, São Paulo, Brazil, 1994.

[8] D. Godard. Self-recovering equalization and carrier tracking in two-dimensional data communication systems. *IEEE Transactions on Communications*, 28(11):1867–1875, 1980.

[9] S. Haykin. *Adaptive Filter Theory*. Prentice-Hall, Upper Saddle River, NJ, 4th edition, 2002.

[10] S. Haykin (Ed.). *Unsupervised Adaptive Filtering*. John Wiley & Sons, Inc., New York, 2000.

[11] C. R. Johnson Jr. et al. Blind equalization using the constant modulus criterion: a review. *Proceedings of the IEEE*, 86:1927–1950, October 1998.

[12] O. Macchi and E. Eweda. Convergence analysis of self-adaptive equalizers. *IEEE Transactions on Information Theory*, 30(2):161–176, 1984.

[13] J. E. Mazo. Analysis of decision-directed equalizer convergence. *Bell Syst. Tech. Journal*, 59(10):1857–1877, 1980.

[14] L. H. A. Monteiro. *Sistemas Dinâmicos* (in Portuguese), 3rd edition. Editora Livraria da Física, São Paulo, Brazil, 2011.

[15] P. A Regalia. On the equivalence between the Godard and Shalvi-Weinstein schemes of blind equalization. *Signal Processing*, 73:185–190, 1999.

[16] J. M. T. Romano, R. Attux, C. C. Cavalcante, and R. Suyama. *Unsupervised Signal Processing: Channel Equalization and Source Separation*. CRC Press, Boca, Raton, FL, 2010.

[17] D. C. Soriano, E. Z. Nadalin, R. Suyama, J. M. T. Romano, and R. Attux. Chaotic convergence of the decision-directed blind equalization algorithm. *Communications in Nonlinear Science and Numerical Simulation*, 17:5097–5109, 2012.

[18] J. Treichler and B. Agee. New approach to multipath correction of constant modulus signals. *IEEE Transactions on Acoustics, Speech and Signal Processing*, ASSP-31(2):459–472, 1983.

7

Matched filters for chaotic signals

Jonathan N. Blakely and Ned J. Corron

Charles M. Bowden Laboratory, US Army Aviation & Missile Research, Development, and Engineering Center

CONTENTS

Conventional communication waveforms are composed of a small set of basis functions that are superposed to represent a stream of bits. Optimal detection of such signals in the presence of noise is achieved by a matched filter, i.e., a linear filter whose impulse response is the time-reverse of the basis function. In contrast, signals from well-known chaotic systems cannot be represented as a superposition of a small number of basis functions. Thus, no simple linear filter produces optimal detection through noise. Recently, a new family of hybrid chaotic systems has been identified whose solutions can be written in terms of basis functions. Since matched filtering can be applied to signals from these systems, their discovery suggests that chaotic waveforms may be suitable for carrying information in a practical communication technology. In this chapter, we first review the definition and significance of the matched filter in order to clarify the sense in which it is optimal. Next, we examine two hybrid chaotic systems in detail, and show how simple matched filters can be defined for their oscillations. Finally, we describe an experimental implementation of an electronic chaotic oscillator and its matched filter.

7.1 Introduction

A central concern in the design of a communication system is the problem of noise from extraneous sources interfering with reception of transmitted information. A receiver must decide whether a bit of information in a received waveform originated at the transmitter or is due to entropy-generating external noise sources. An important tool for mitigating errors due to noise is the matched filter. Under certain common conditions, the matched filter is the optimal device for detecting signals in noise [5, 6, 13, 18]. Thus, the absence of matched filters from proposed schemes for chaos-based communications is particularly noteworthy. Familiar chaotic systems such as the Lorenz equations do not have solutions that can be written in the form of a sequence of basis functions like a standard communication waveform [9]. Thus, designers have tended to utilize either digital signal processing to store waveforms and perform correlations, sub-optimal incoherent reception techniques, or nonlinear filters [7, 8, 10, 11, 16]. In this chapter, we report on a new class of chaotic oscillators for which simple implementations of matched filters do indeed exist, and have been experimentally demonstrated [2, 3].

In Section 7.2, we begin with a brief tutorial introduction to matched filters. The aim here is to provide a clear statement of exactly the sense in which a matched filter is optimal, especially for readers with a background in dynamics but not communications. In Section 7.3, we give two examples of chaotic systems that do have matched filters [2, 3]. Topologically, the first bears some resemblance to the famous Lorenz system [12] while the second resembles the Rössler system [14]. However, the systems are hybrid dynamical systems in the sense that they contain both continuous and discrete states. More importantly, they are distinguished from most known chaotic systems by the surprising fact that analytic solutions can be written in terms of individual basis functions. These basis functions can then be matched by linear filters. In Section 7.4, we describe an experimental implementation of a matched filter for a chaotic electronic oscillator. Finally, in Section 7.5 we comment on the outlook for matched filters in chaos communications.

7.2 Matched filters optimize signal to noise

A matched filter is a key component in many communication and radar systems [18]. Specifically, a matched filter is a linear filter whose output maximizes the ratio of signal power to noise power when receiving a known signal (i.e., the signal to which the filter is matched) contaminated with additive white Gaussian noise (AWGN). In communications, typically one of a small set of basis functions is transmitted each cycle, and a receiver containing a matched filter for each basis function must identify the transmitted functions. In radar, a known signal is reflected off a target, and the receiver must determine the time at which the reflected signal arrives at the receiver. In both cases a matched filter is used to determine at some time whether the signal to which it is matched is being received or not. The presence of additive noise picked up in either the channel or the receiver contributes uncertainty to this determination. Thus, it is desirable to maximize signal to noise in order to minimize errors in these decision processes. We now demonstrate that a matched filter is optimum in the sense that it maximizes signal to noise ratio when trying to detect a signal of known shape subject to additive white Gaussian noise [5, 6, 13].

To begin, we model the transmitted known signal by a function $s(t)$. Since realistic communication signals do not diverge, we restrict $s(t)$ to be square integrable. The transmitted signal is received along with additive white Gaussian noise represented by a function $n(t)$. Thus, when the signal $s(t)$ is transmitted, the received signal $r(t)$ is

$$r(t) = s(t) + n(t). \tag{7.1}$$

When no signal is being transmitted, the received signal is simply $r(t) = n(t)$.

We assume the received signal is input to a linear filter with impulse response $h(t)$. The output of the filter $y(t)$ is the convolution of the input with the impulse response function, that is

$$y(t) = \int_{-\infty}^{\infty} h(\tau)s(t-\tau)d\tau \tag{7.2}$$

$$= \int_{-\infty}^{\infty} h(\tau)s(t-\tau)d\tau + \int_{-\infty}^{\infty} h(\tau)n(t-\tau)d\tau \tag{7.3}$$

$$= y_s(t) + y_n(t), \tag{7.4}$$

where $y_s(t)$ is the contribution due to the transmitted signal and $y_n(t)$ is that of the additive noise. We seek a linear filter whose output will be much larger when the known signal $s(t)$ is received than when noise alone is being received. The average signal to noise ratio (conventionally denoted SNR) is defined as

$$\text{SNR} = \frac{y_s^2(t)}{\langle y_n^2(t) \rangle}, \tag{7.5}$$

where $\langle \cdot \rangle$ represents the average over an ensemble of realizations of the noise $n(t)$. The *optimum* linear filter is that filter whose impulse response $h(t)$ maximizes SNR.

In pursuit of this optimum, we first evaluate the denominator of Equation (7.5):

$$\langle y_n^2(t) \rangle = \left\langle \int_{-\infty}^{\infty} h(\tau')n(t-\tau')d\tau' \int_{-\infty}^{\infty} h(\tau)n(t-\tau)d\tau \right\rangle, \tag{7.6}$$

$$= \int_{-\infty}^{\infty} \int_{-\infty}^{\infty} \langle n(t-\tau')n(t-\tau) \rangle h(\tau')h(\tau)d\tau'd\tau. \qquad (7.7)$$

By assumption, the noise is white, meaning samples of the noise waveform at distinct times are on average uncorrelated. Mathematically, this is expressed by

$$\langle n(t)n(t') \rangle = \frac{N_0}{2}\delta(t-t'), \qquad (7.8)$$

where N_0 is the spectral height. Substituting this result into Equation (7.7) gives the average SNR as

$$\text{SNR} = \frac{\left[\int_{-\infty}^{\infty} h(\tau)s(t-\tau)d\tau \right]^2}{N_0/2 \int_{-\infty}^{\infty} h^2(\tau)d\tau}. \qquad (7.9)$$

This expression is of the form of an inner product between two vectors in the Hilbert space of square integrable functions on the interval $(-\infty, \infty)$. The Cauchy-Schwarz inequality states that for two such vectors, $a(t)$ and $b(t)$,

$$\left[\int_{-\infty}^{\infty} a(\tau)b(\tau)d\tau \right]^2 \leq \int_{-\infty}^{\infty} a^2(\tau)d\tau \int_{-\infty}^{\infty} b^2(\tau')d\tau', \qquad (7.10)$$

where the equality is reached if and only if $a(\tau) = b(\tau)$. If we let

$$a(\tau) = \frac{h(\tau)}{\left[N_0/2 \int_{-\infty}^{\infty} h^2(\tau')d\tau' \right]^{1/2}} \qquad (7.11)$$

and

$$b(\tau) = \frac{s(t-\tau)}{\left[N_0/2 \int_{-\infty}^{\infty} h^2(\tau')d\tau' \right]^{1/2}}, \qquad (7.12)$$

then the SNR is bounded according to

$$\text{SNR} = \frac{\left[\int_{-\infty}^{\infty} h(\tau)s(t-\tau)d\tau \right]^2}{\frac{N_0}{2} \int_{-\infty}^{\infty} h^2(\tau')d\tau')} \leq \frac{\int_{-\infty}^{\infty} h^2(\tau)d\tau \int_{-\infty}^{\infty} s^2(\tau)d\tau}{\frac{N_0}{2} \int_{-\infty}^{\infty} h^2(\tau')d\tau'} \qquad (7.13)$$

and the equality is reached only when $h(\tau) = s(t-\tau)$. In general, this condition can only be met for a single time, say $t = 0$. Then $h(\tau) = s(-\tau)$. Thus, the filter that maximizes the SNR at time $t = 0$ is that whose impulse response is the time reverse of the signal to be detected, i.e., the *matched filter*. For this optimal filter, the SNR is simply

$$\text{SNR} = \frac{2}{N_0} \int_{-\infty}^{\infty} s^2(\tau)d\tau. \qquad (7.14)$$

It is important to understand that the matched filter output $y(t)$ is not a reconstruction of the transmitted signal $s(t)$. The shape of the transmitted signal is assumed known, so it carries no new information. The new information lies in the determination of whether the signal is actually received or not. The usefulness of the matched filter is in answering this question, rather than in directly reconstructing the transmitted waveform. It is also important to keep in mind that the matched filter is the optimum *linear* filter only. In general, there may be *nonlinear* filters that could outperform the matched filter. The restriction to linear filters is due to the practical convenience of linear mathematics.

Some examples of conventional modulation schemes and their matched filters will be relevant to our later discussion of chaotic waveforms. One common modulation scheme used in digital communications is binary phase-shift keying (BPSK) in which a single basis function $s(t)$ is transmitted with either positive or negative polarity corresponding to a single binary symbol. A filter matched to $s(t)$ emits a large positive or negative value depending on the transmitted polarity allowing determination of the transmitted symbol. In the next section, we will introduce a Lorenz-like system [12] whose oscillations bear a strong resemblance to a BPSK waveform. Another common modulation scheme is on-off keying (OOK) which also uses a single basis function. In this case, the two possible states of a binary symbol are indicated by whether the basis function is transmitted (on) or not (off). We will introduce a Rössler-like chaotic system [14] whose oscillations resemble OOK.

Another important particular case of a matched filter arises in direct sequence spread spectrum communications as well as in pulse compression radar. In such systems, the transmitted signal is composed of a string of elemental signals evenly spaced in time, for which individual matched filters are known. As a specific example, consider a signal made up of two elemental basis functions $f(t)$ and $-f(t)$. Such a signal can be written as

$$s(t) = \sum_{j=0}^{N-1} a_j f(t-j), \qquad (7.15)$$

where N is a positive integer, and $a_j \in \{-1, 1\}$ for each $j = 1 \ldots N$. The matched filter for a single signal $f(t)$ is $h_f(t) = f(\tau - t)$. The matched filter for the entire composite signal $s(t)$ is

$$h(t) = s(\tau - t) = \sum_{j=0}^{N-1} a_j f(\tau - t - j), \qquad (7.16)$$

$$= \sum_{j=0}^{N-1} a_j h_f(t-j). \qquad (7.17)$$

Thus, the matched filter of the transmitted signal is simply a sum of appropriately weighted and delayed copies of the matched filter of an individual basis function. Since the filter is linear, this sum can be implemented using a multi-tapped delay line to feed a single filter with impulse response $h_f(t)$. In Section 7.3, we will see examples of chaotic oscillators whose waveforms can be written as sums similar to these composite signals. Each oscillation will correspond to a single term in the sum. We will then identify the matched filter $h_f(t)$ and use it to build up the composite matched filter for a given multi-oscillation segment of a waveform.

7.3 Chaotic oscillators with matched filters

It is commonly assumed that the complexity of chaos denies analytic solution, but this is not necessarily true. In this section, we present two examples of low-dimensional exactly solvable chaotic oscillators. Both oscillators can be derived from a single model that was originally proposed by Saito and Fujita [15] as a model of manifold piecewise-linear chaos. These examples are hybrid dynamical systems in that they contain both continuous and discrete state variables. More importantly, a general solution for each system can be written as a linear convolution of binary symbols with a fixed basis function. A linear filter can then be matched to these basis functions. These elemental matched filters can be used to build up matched filters for longer sequences of oscillations.

7.3.1 Exact shift oscillator

Our first example oscillator is a manifold piecewise-linear system that exhibits Lorenz-like chaos [12], which was also previously used for circuit demonstrations of chaos control [17]. We refer to this system as the *exact shift oscillator* for reasons given later. This hybrid dynamical system contains continuous states $u(t) \in \Re$ and $\dot{u}(t) \in \Re$, and a discrete state $s(t)$. The continuous-time dynamics are described by the differential equation

$$\frac{d^2 u}{dt^2} - 2\beta \frac{du}{dt} + \left(\omega^2 + \beta^2\right) \cdot (u - s) = 0, \tag{7.18}$$

where $\omega = 2\pi$ and $0 < \beta < \ln 2$. Transitions in the discrete state are defined by the guard condition

$$\frac{du}{dt} = 0 \quad \Rightarrow \quad s(t) = \text{sgn}\left(u(t)\right), \tag{7.19}$$

meaning $s(t)$ is set to the sign of $u(t)$ whenever its time derivative vanishes, and $s(t)$ maintains this value until the guard condition is next met. Here, we use the right-continuous signum function

$$\text{sgn}(u) = \begin{cases} +1 & \text{if } u \geq 0 \\ -1 & \text{if } u < 0 \end{cases} \tag{7.20}$$

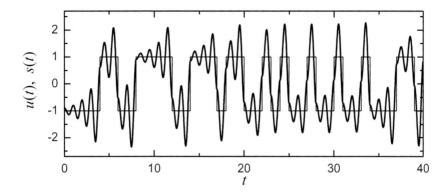

FIGURE 7.1
Typical continuous waveform $u(t)$ and discrete state $s(t)$ from numerical integration of the exact shift oscillator for $\beta = \ln 2$.

so that the discrete state takes the values $s(t) = \pm 1$. The hybrid system admits two fixed point $u(t) = s(t) = \pm 1$, both of which are unstable.

Figure 7.1 shows a typical solution obtained via numerical integration of the exact shift oscillator for $\beta = \ln 2$. In this figure, the state $u(t)$ is the continuous waveform, while $s(t)$ appears as the random square wave. The solution waveform exhibits growing harmonic oscillations combined with random-like switching. Figure 7.2 shows the corresponding $u(t)$ versus $\dot{u}(t)$ phase portrait exhibiting a Lorenz-like dual-lobe structure [12].

To illustrate how this oscillator works, we consider the typical trajectory shown in Figure 7.3. The dynamics of the continuous state $u(t)$ are described by the harmonic oscillator in Equation (7.18), which is unstable with $\beta > 0$. Between switching events, the continuous waveform $u(t)$ exhibits growing oscillations about the fixed level $s(t)$. This growth provides the exponential divergence of initial conditions necessary for chaos. Possible switching events occur whenever the derivative of $u(t)$ is zero, which are indicated by dots in Figure 7.3. The discrete state $s(t)$ switches only if the continuous state $u(t)$ has crossed a threshold at zero and is of opposite sign to $s(t)$. After a switch, the unstable oscillations grow about the new value of $s(t)$, and the process repeats. The nonlinear switching provides a folding action that is essential for chaos. For $\beta \leq \ln 2$, the continuous state at switching times is guaranteed to stay between the set points; thus, oscillations are bounded [3].

Recently, it was shown that an exact, analytic solution for this hybrid system can be found [1]. Without loss of generality, we consider an initial condition at time $t = t_n$ such that $u(t_n) = u_n$, $\dot{u}(t_n) = 0$, and $s(t_n) = s_n$, where $s_n = \text{sgn}(u_n)$. That is, we consider an initial state that coincides with

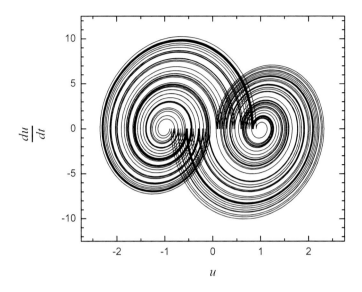

FIGURE 7.2
Phase space projection for a typical continuous waveform in the exact shift
oscillator with $\beta = \ln 2$.

a triggering of the guard condition. The solution to the initial value problem
is then given by the linear convolutions

$$u(t) = \sum_{m=0}^{\infty} s_m \cdot P(t - t_n - m), \tag{7.21}$$

$$s(t) = \sum_{m=0}^{\infty} s_m \cdot \phi(t - t_n - m), \tag{7.22}$$

where $t \geq 0$ and each symbol $s_m = \pm 1$. The fixed basis functions are

$$P(t) = \begin{cases} \left(1 - e^{-\beta}\right) e^{\beta t} \left(\cos \omega t - \frac{\beta}{\omega} \sin \omega t\right), & t < 0 \\ 1 - e^{\beta(t-1)} \left(\cos \omega t - \frac{\beta}{\omega} \sin \omega t\right), & 0 \leq t \leq 1 \\ 0, & 1 \leq t \end{cases} \tag{7.23}$$

and

$$\phi(t) = \begin{cases} 0, & t < 0 \\ 1, & 0 \leq t \leq 1 \\ 0, & 1 \leq t \end{cases} \tag{7.24}$$

which are shown in Figure 7.4 for $\beta = \ln 2$. In Equations 7.21 and 7.22,
each binary symbol s_m modulates the fixed basis functions centered at time

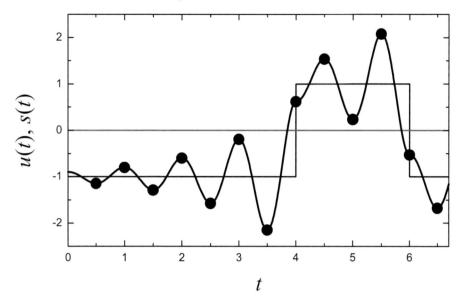

FIGURE 7.3
Exponential growth in the harmonic oscillations $u(t)$ producing transitions in the discrete state $s(t)$ when critical points (dots) cross the threshold at zero.

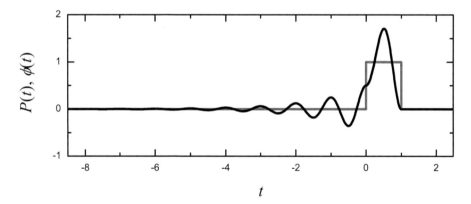

FIGURE 7.4
Continuous basis function $P(t)$ (black) and discrete basis function $\phi(t)$ (gray) for the exact shift oscillator with $\beta = \ln 2$.

$t = t_n + m$. Notably, this chaotic waveform resembles a binary phase-shift keyed communications waveform.

A discrete map can be derived from regular samples of the continuous waveform Equation (7.21). Evaluating Equation (7.21) at the initial time t_n

yields

$$u_n = (1 - e^{-\beta} u_n) \sum_{m=0}^{\infty} s_{m+n} e^{-m\beta}, \qquad (7.25)$$

Then, the state of the system on time unit later satisfies the recurrence relation

$$u_{n+1} = e^{\beta} u_n - (e^{\beta} - 1) \cdot \text{sgn}(u_n), \qquad (7.26)$$

which also coincides with a triggering of the guard condition Equation (7.19). Since the oscillator is autonomous, this map yields an iterated one-dimensional map for successive samples at unit times. As shown in Figure 7.5, the map is piecewise linear and has a slope $e^{\beta} > 1$. As such, the iterated map is a chaotic shift, with Lyapunov exponent $\lambda = \beta$. For $\beta = \ln 2$, the map is a full shift and the grammar is unrestricted, i.e., any infinite sequence of s_m's in Equation (7.21) gives a solution to Equation (7.18). However, for $\beta < \ln 2$, the grammar is restricted and not every sequence of symbols is allowed.[1] The derivation of a shift map from the continuous dynamics constitutes rigorous proof of chaos, and is the reason we chose to refer to the system as the exact shift oscillator.

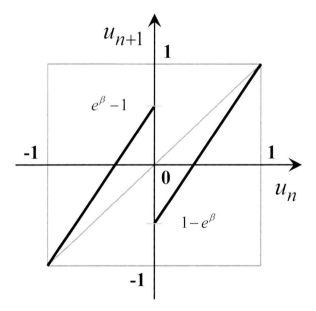

FIGURE 7.5
Return map for continuous state in the exact shift oscillator sampled for each symbol.

---

[1]Since Equation (7.18) is unstable when $\beta > \ln 2$, any practical circuit implementation of this system must necessarily have $\beta < \ln 2$. Therefore, a grammar restriction is inevitable in practice and must be kept in mind in designing a communication system.

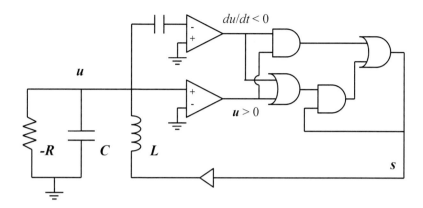

FIGURE 7.6
Circuit realization of exact shift oscillator.

A circuit schematic depicting an implementation of the exact shift oscillator is shown in Figure 7.6. This hybrid circuit includes both analog and digital components. Analog circuitry realizes the continuous state $u(t)$ as the tank voltage in a harmonic LC oscillator in parallel with a negative resistor $-R$ that supplies gain. The digital circuitry realizes the dynamics of the discrete switching state $s(t)$. The circuit does not use an explicit implementation of the guard condition such as shown in a previous report [3]. Instead, the circuit achieves equivalent operation using two threshold comparators to detect the sign of the tank voltage $u(t)$ and its derivative, which are combined in a simple logic circuit to generate the binary logic signal $s(t)$. The simplicity of the circuit suggests that high-frequency implementations of the exact shift oscillator may be easily realized.

It is possible to develop a matched filter for the basis function Equation (7.23). To this end, we note that the time-reversed basis function $P_-(t) = P(-t)$ satisfies the ordinary differential equation

$$\frac{d^2 P_-}{dt^2} + 2\beta \frac{P_-}{dt} + \left(\omega^2 + \beta^2\right) P_- = \left(\omega^2 + \beta^2\right) h(t), \qquad (7.27)$$

where

$$h(t) = \begin{cases} 1, & -1 \leq t < 0 \\ 0, & \text{otherwise} \end{cases} \qquad (7.28)$$

is a square pulse with unit amplitude and duration [4]. By differentiating Equation (7.28), we see that

$$\frac{d}{dt} h(t) = \delta(t+1) - \delta(t), \qquad (7.29)$$

where the impulse $\delta(t)$ has unit area. From these observations, we deduce that the time-reversed basis function is the impulse response of the linear filter

$$\frac{d}{dt}\eta = \nu(t+1) - \nu(t), \tag{7.30}$$

$$\frac{d^2\xi}{dt^2} + 2\beta\frac{d\xi}{dt} + \left(\omega^2 + \beta^2\right)\xi = \left(\omega^2 + \beta^2\right)\eta(t), \tag{7.31}$$

where $\nu(t)$ is the filter input, $\eta(t)$ is an intermediate state, and $\xi(t)$ is the filter output. Consequently, Equations 7.30 and 7.31 are a matched filter for the basis function Equation (7.23).

A finite segment of a waveform, e.g., the segment shown in Figure 7.3, is a finite sum of basis functions of the form

$$u(t) = \sum_{m=0}^{N-1} s_m \cdot P(t - t_n - m) + R(t), \tag{7.32}$$

where N is a positive integer and

$$R(t) = \sum_{m=N}^{\infty} s_m \cdot P(t - t_n - m) \tag{7.33}$$

represents the exponetially small tails of all future basis functions. Following the development of the previous section, a filter matched to such a finite segment (neglecting the contributions of future symbols) has an impulse response of the form

$$h(t) = \sum_{j=0}^{N-1} a_j P(\tau - t - j). \tag{7.34}$$

This filter is of the same form as the direct sequence spread spectrum matched filter in Equation (7.17). Due to linearity, it follows that such a response is obtained from the filter

$$\frac{d\eta}{dt} = \sum_{m=1-N}^{0} s_m \cdot \{\nu(t+1+m) - \nu(t+m)\}, \tag{7.35}$$

$$\frac{d^2\xi}{dt^2} + 2\beta\frac{d\xi}{dt} + \left(\omega^2 + \beta^2\right)\xi = \left(\omega^2 + \beta^2\right)\eta(t), \tag{7.36}$$

where, again, $\nu(t)$ is the filter input, $\eta(t)$ is an intermediate state, and $\xi(t)$ is the filter output.

Remarkably, this compact set of equations (7.35 and 7.36) gives a matched filter for any finite segment of a solution of the exact shift oscillator. Such a result is virtually unprecedented for chaotic waveforms. Moreover, the clear resemblance to direct sequence spread spectrum suggests that the exact shift oscillator may be a candidate for the full range of communications applications to which direct sequence has been applied. Lest one wrongly conclude that this is a unique system, we present another chaotic system that has a matched filter but whose attractor is topologically distinct in the next subsection.

7.3.2 Exact folded band oscillator

As a second example of an exactly solvable oscillator, we consider a manifold piecewise-linear system that exhibits folded-band chaos, which is topologically distinct from the Lorenz-type chaos of the exact shift oscillator [15]. This system is also a hybrid dynamical system, containing continuous states $u(t) \in \Re$ and $\dot{u}(t) \in \Re$, as well as a discrete state $s(t) \in \{0, 1\}$. The continuous-time dynamics are again described by the differential equation

$$\frac{d^2 u}{dt^2} - 2\beta \frac{du}{dt} + (\omega^2 + \beta^2) \cdot (u - s) = 0, \qquad (7.37)$$

where $\omega = 2\pi$ and $\beta > 0$. However, the discrete dynamics are now defined by the guard condition

$$\frac{du}{dt} = 0 \quad \Rightarrow \quad s(t) = H(u(t) - 1) \qquad (7.38)$$

where

$$H(t) = \begin{cases} 1, & x > 0 \\ 0, & x \leq 0 \end{cases} \qquad (7.39)$$

is the left-continuous Heaviside function. We note this hybrid system admits the unique fixed point $u(t) = s(t) = 0$, which is unstable.

Typical waveforms calculated numerically for this hybrid system with $\beta = 0.81 \cdot \ln 2$ are shown in Figure 7.7. In this figure, the continuous state shows intervals of various duration in which oscillations grow about the fixed state $s(t) = 0$. Each of these intervals concludes with a switching event where $s(t) = 1$ for one half time unit, during which the growing oscillation is squelched. A phase-space projection of the continuous state is shown in Figure 7.8. In this projection the solution trajectory spirals outward from the origin, rotating in a clockwise direction with time. Switching events triggered at the guard condition are evident on the positive u axis, folding the outer trajectories

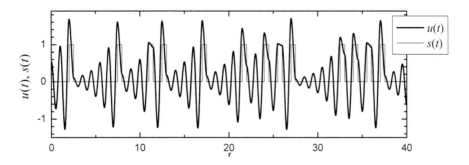

FIGURE 7.7
Typical waveforms calculated numerically for the exact folded-band oscillator with $\beta = 0.81 \ln 2$.

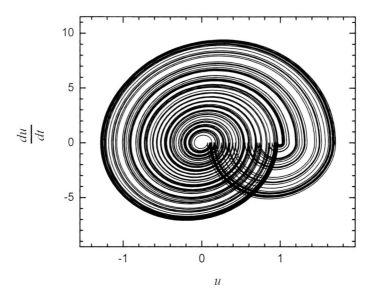

FIGURE 7.8

Phase space projection of a typical waveform generated by the continuous state in the exact folded-band oscillator with $\beta = 0.81 \ln 2$.

back into the center of the spiral. The folded band topology of the attractor resembles that of the famous Rössler oscillator and justifies our designation of this system as the *exact folded-band oscillator* [14].

The operation of the oscillator is better explained using the short segment shown in Figure 7.9. Initially the unstable linear oscillator, Equation (7.37), generates a growing oscillation about the fixed set point $s(t) = 0$. During these oscillations, the guard condition, Equation (7.38) is triggered at each critical point, indicated by dots in the figure. While oscillations are small (amplitude less than one), the discrete state remains unchanged. However, the growing oscillations eventually yield a guard condition with $u(t) > 1$, triggering a transition to $s(t) = 1$. The continuous state now starts to oscillate about the elevated set point. At the next guard condition, which occurs a half cycle later, the state $u(t) < 1$ is again small, resetting $s(t) = 0$ and enabling the growing oscillation process to repeat.

As with the exact shift oscillator, the exact folded-band oscillator also admits an exact analytic solution [2]. Without loss of generality, we again consider an initial condition corresponding to a guard condition at time $t = t_n$ such that $u(t_n) = u_n$, $du/dt(t_n) = 0$, and $s(t_n) = s_n$, where $s_n = H(u_n - 1)$. The solution can be written as the linear convolution

$$u(t) = \sum_{m=0}^{\infty} \sigma_m \cdot Q(t - t_n - \frac{m}{2}), \qquad (7.40)$$

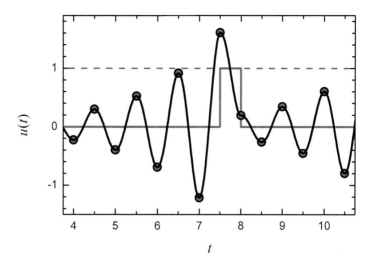

FIGURE 7.9
Waveform segment illustrating the hybrid dynamics in the exact folded-band oscillator.

$$s(t) = \sum_{m=0}^{\infty} \sigma_m \cdot \psi(t - t_n - \frac{m}{2}), \qquad (7.41)$$

where

$$Q(t) = \begin{cases} \left(1 + e^{-\beta/2}\right) e^{\beta t} \left(\cos \omega t - \frac{\beta}{\omega} \sin \omega t\right), & t < 0 \\ 1 + e^{\beta(t-1/2)} \left(\cos \omega t - \frac{\beta}{\omega} \sin \omega t\right), & 0 \le t \le \frac{1}{2} \\ 0, & \frac{1}{2} \le t \end{cases} \qquad (7.42)$$

and

$$\psi(t) = \begin{cases} 1, & 0 \le t < \frac{1}{2} \\ 0, & \text{otherwise} \end{cases} \qquad (7.43)$$

are fixed basis functions for the continuous and discrete states, respectively. The basis functions are plotted in Figure 7.10. Notably, the basis pulses in Equations 7.40 and 7.41 are spaced one half time unit apart, as opposed to the unit spacing of the exact shift oscillator. Each coefficient σ_m is an amplitude that weights the basis functions at the corresponding time lag in the convolution. In particular, the discrete state simply yields $s(t) = \sigma_m$ for $\frac{m}{2} \le t - t_n < \frac{m+1}{2}$; hence, the amplitude values are restricted to the set $\sigma_m \in \{0, 1\}$. This general solution is valid for $t \ge t_n$.

It can be shown that successive maxima of the continuous waveform $u(t)$ are related by the one-dimensional, piecewise-linear map

$$u_{n+1} = f(u_n), \qquad (7.44)$$

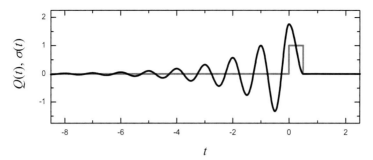

FIGURE 7.10
Basis pulses $Q(t)$ (black) and $\psi(t)$ (gray) for the exact folded-band oscillator with $\beta = 0.81 \ln 2$.

where

$$f(u_n) = \begin{cases} e^\beta u_n, & 0 < u_n \leq 1 \\ -e^{3\beta/2} u_n + \left(e^\beta + e^{3\beta/2}\right), & 1 < u_n \leq 1 + e^{-\beta/2} \\ e^\beta u_n - \left(e^{\beta/2} + e^\beta\right), & u_n > 1 + e^{-\beta/2} \end{cases} \qquad (7.45)$$

which is shown in Figure 7.11 [2]. The map contains three branches labeled with the symbols A, B, and C. For any continuous waveform of the form of Equation (7.40), there is a corresponding symbol sequence $\{S_n\}$ where each $S_n \in \{A, B, C\}$. Importantly, the substitutions

$$\begin{aligned} A &\to 00 \\ B &\to 100 \\ C &\to 10 \end{aligned} \qquad (7.46)$$

provide a mapping from a symbol sequence $\{S_n\}$ to an amplitude sequence $\{\sigma_n\}$, where each amplitude σ_n occurs for a half time unit. As a result, sequences σ_n generated by the free running oscillator must be built up from the segments in Equation (7.46).[2] Importantly, when β is less than a critical value of approximately $0.81 \cdot \ln 2$, chaotic solutions are confined to the first two branches (A and B) of the map and the phase space projection reveals a folded band attractor as in Figure 7.8.

A circuit schematic showing an electronic folded-band oscillator is shown in Figure 7.12. This circuit differs from the shift oscillator circuit of Figure 7.6 only in the digital components that evaluate the guard condition. Its simplicity again suggests that high-frequency implementations of the exact folded-band oscillator may also be easily realized.

By reasoning similar to that of the previous section, it is straightforward

---

[2]The consequence of this grammar restriction on the σ_m's is redundancy in the waveform. Such redundancy typically reduces the bit-error rate of a communication waveform, but may be exploited by a receiver in performing error correction.

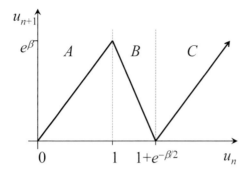

FIGURE 7.11
Successive maxima return map for the exact folded-band oscillator.

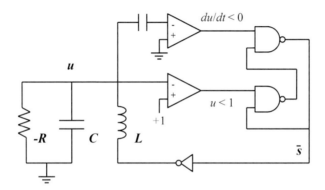

FIGURE 7.12
Circuit realization of exact folded band oscillator.

to find that the matched filter for the basis function $Q(t)$ is

$$\frac{d}{dt}\eta(t) = \nu(t + 1/2) - \nu(t), \tag{7.47}$$

$$\frac{d^2\xi}{dt^2} + 2\beta\frac{d\xi}{dt} + \left(\omega^2 + \beta^2\right)\xi = \left(\omega^2 + \beta^2\right)\eta(t), \tag{7.48}$$

where $\nu(t)$ is the filter input, $\eta(t)$ is an intermediate state, and $\xi(t)$ is the filter output. Remarkably, the only difference between this filter and that of the shift oscillator, Equations 7.30 and 7.31, is the half time unit delay in the first equation. It also follows similarly that a matched filter for segment of a waveform made up of several basis functions is

$$\frac{d\eta}{dt} = \sum_{m=1-N}^{0} s_m \cdot \{\nu(t + 1/2 + m/2) - \nu(t + m/2)\}, \tag{7.49}$$

$$\frac{d^2\xi}{dt^2} + 2\beta\frac{d\xi}{dt} + \left(\omega^2 + \beta^2\right)\xi = \left(\omega^2 + \beta^2\right)\eta(t). \tag{7.50}$$

Again, the only difference between this and the filter for the shift oscillator is the delay time. However, it should be kept in mind that the folded-band oscillator's waveform contains the aforementioned redundancy which is not present in the waveform of the shift oscillator. It then is reasonable to expect that receivers for these two waveforms may be quite different since the folded-band receiver could be designed to exploit the redundancy in an error correction scheme.

7.4 Experimental demonstration of a matched filter for chaos

Having provided examples of chaotic oscillators with matched filters, we now present an experimental demonstration of a matched filter in recent acoustic experiments. For these experiments, an amplified speaker emits an audio-frequency waveform generated by an electronic realization of the exact shift oscillator. A complementary receiver circuit incorporates a matched filter for the chaotic waveform. At repeated intervals, a sequence of symbols detected in the symbolic dynamics of the emitted waveform is captured, thereby defining a transmitted signal. The captured symbol sequence is provided to the receiver, where it defines the matched filter for the transmitted signal. Practically, the symbols define weights applied to elements of a microphone array, the outputs of which are summed and passively filtered. In operation, a peak in the output of the matched filter is observed, which indicates reception of the transmitted symbol sequence. The entire experimental system is realized using simple analog and digital electronic circuit components.

The complete transmitter is shown schematically in Figure 7.13. At the center of the transmitter is the chaotic oscillator, an audio frequency implementation of the exact shift oscillator circuit shown in Figure 7.6. Chaotic oscillations occur at approximately 10 kHz. The transmitted signal is the continuous state of the free-running chaotic oscillator, which is amplified and emitted by a conventional speaker. The additional circuitry at the bottom of the transmitter schematic derives a clock signal from the regular return times of the oscillator. The clock signal drives a binary shift register, which uses the signal $s(t)$ for the data input. A divide-by-N counter circuit provides a signal to alternately enable and disable the shift register. For our experimental system, we typically use the value $N = 1024$.

In operation, the free-running oscillator generates a chaotic waveform that is continuously emitted from the speaker. While the shift register is enabled, the symbolic logic state $s(t)$ is sampled for each return and stored in the shift register. At any time, a fixed number of the most recent values of the

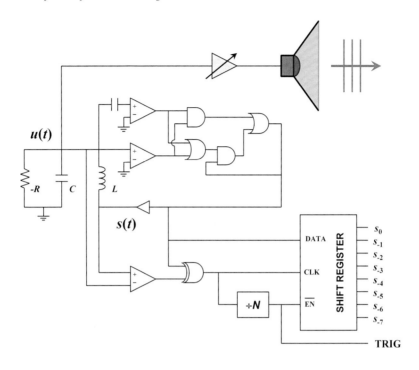

FIGURE 7.13
Acoustic transmitter incorporating the exactly solvable chaotic oscillator.

logic state $s(t)$ are stored, which correspond to a sequence of amplitudes s_m generated by the free-running oscillator. Although the figure only shows an eight-bit shift register, for the acoustic system we used a twelve-bit register.

After shifting N successive returns through the register, the shift register is disabled by the signal from the divide-by-N circuit. When disabled, the contents of the shift register are locked, thereby storing symbols that identify the signal transmitted just prior to the disabling transition. These stored symbols constitute all the information needed to re-generate the transmitted waveform. Compared to the usual Nyquist sampling criteria, this symbolic representation provides at an order of magnitude reduction in the sampling and storage requirements for the reference waveform.

A simulated waveform and shift register content is shown in Figure 7.14. The top plot shows the oscillator waveforms $u(t)$ and $s(t)$. The continuous waveform $u(t)$ is the transmitted waveform emitted by the speaker. The middle plot shows the clock signal that is extracted from the oscillator waveforms and defines the symbol timing. The bottom waveform shows the trigger signal derived by the divide-by-N operation from the clock signal. The shift register is disabled by the low-to-high transition of the trigger signal. The dots in the top plot show the most recent eight symbols captured and stored by the

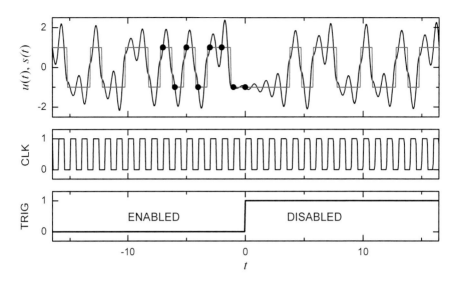

FIGURE 7.14
Simulated typical transmitter waveforms, including the derived clock and trigger signals. Dots shown on the switching waveform indicate the most recent eight symbols stored in the shift register when disabled.

shift register when it is disabled by the trigger. These eight symbols define a reference waveform for the correlation receiver.

Figure 7.15 shows a schematic implementation of the matched filter for the acoustic system. The received waveform impinges on a microphone array, shown at top. The spacing of the microphones in the array is chosen to realize the evenly spaced time delays in the first equation of the matched filter. Differential amplifiers between adjacent microphones provide the difference signal of successively lagged signals, which are multiplied by ±1 according to the symbols defining the reference waveform. The summed differences are integrated and drive the harmonic filter to generate the matched filter output.

Both oscillator and matched filter circuits were constructed using discrete analog and digital components on a solderless breadboard as shown in Figure 7.16. The transmitter and receiver were installed in an acoustically anechoic chamber for demonstration and test. A handheld oscilloscope, triggered on the disable signal to the shift register, was used to monitor the receiver output. The transmitter and matched filter states were also connected to a computer for instrumentation and tuning purposes.

In operation, a consistent spike in the matched filter output was evident at a delay corresponding to the distance from speaker to microphone array. In a single return for one instance of the reference waveform, this peak may be indistinguishable from background noise, intersymbol interference, or wave-

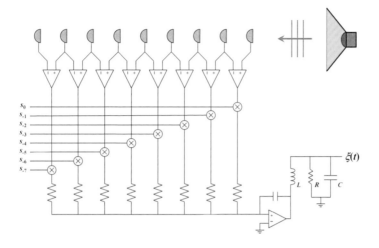

FIGURE 7.15

Acoustic correlation receiver that realizes a matched filter for the solvable oscillator.

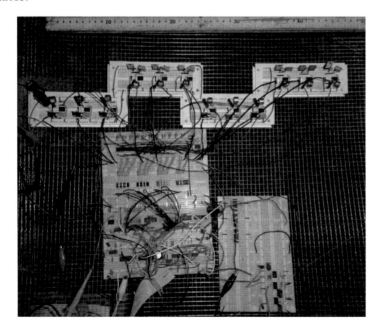

FIGURE 7.16 (SEE COLOR INSERT)

Photo of transmitter and receiver circuits for experimental system.

form sidelobes. However, the consistent peak emerges when multiple receiver outputs are averaged, which was conveniently provided by the oscilloscope.

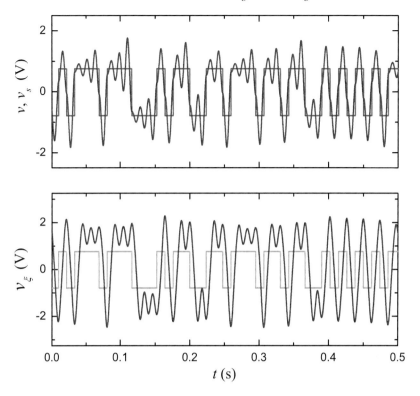

FIGURE 7.17
Typical transmitted waveform and resulting matched filter output observed
in acoustic experiments.

Typical outputs for 64 averaged returns are shown in Figure 7.17, which shows
the detection and correctly estimated range at three speaker positions.

7.5   Outlook

In this chapter, we have reported on the surprising development of chaotic
waveforms with matched filters. Two examples were explored in detail, one
producing a waveform resembling BPSK modulation, and the other displaying
a waveform similar to OOK modulation. Matched filters for arbitrary chaotic
waveforms were defined allowing reception similar to direct sequence spread
spectrum systems. An experimental acoustic demonstration of such operation
was described.

   In fact, the chaotic oscillators detailed here along with their respective matched filters provide a new means of implementing spread spectrum digital communications. The primary expected benefit of this new approach is the reduced cost and complexity due to the fantastic simplicity of circuit implementations. Much experimental and theoretical work remains in order to realize this benefit. For example, high speed implementations of solvable oscillators must be developed and integrated into a larger communication system architecture. Other hybrid chaotic systems should be developed to increase the available waveform diversity. Nonetheless, the considerable progress to date should motivate serious efforts to mature chaos communications technology to a practical level in the near future.

Bibliography

[1] N. J. Corron. An exactly solvable chaotic differential equation. *Dyn. Contin. Discrete Impuls. Syst. A*, 16:777–788, 2009.

[2] N. J. Corron and J. N. Blakely. Exact folded-band chaotic oscillator. *Chaos*, 22:023113, 2012.

[3] N. J. Corron, J. N. Blakely, and M. T. Stahl. A matched filter for chaos. *Chaos*, 20:023123, 2010.

[4] N. J. Corron, S. T. Hayes, S. D. Pethel, and J. N. Blakely. Chaos without nonlinear dynamics. *Phys. Rev. Lett.*, 97:024101, 2006.

[5] R. G. Gallager. *Information theory and reliable communication*. John Wiley & Sons, New York, 1968.

[6] J. C. Hancock and P. A. Wintz. *Signal detection theory*. McGraw-Hill, New York, 1966.

[7] S. Hayes, C. Grebogi, and E. Ott. Communicating with chaos. *Phys. Rev. Lett.*, 70:3031–3034, May 1993.

[8] S. Hayes, C. Grebogi, E. Ott, and A. Mark. Experimental control of chaos for communication. *Phys. Rev. Lett.*, 73:1781–1784, 1994.

[9] S. T. Hayes. Chaos from linear systems: implications for communicating with chaos, and the nature of determinism and randomness. *J. Phys.: Conf Ser.*, 23:215–237, 2005.

[10] M. P. Kennedy and G. Kolumbán. Special issue on noncoherent chaotic communications. *IEEE Transactions on Circuits and Systems I*, 47(12):1661–1662, 2000.

[11] G. Kolumban, M. P. Kennedy, Z. Jako, and G. Kis. Chaotic communications with correlator receivers: theory and performance limits. *Proceedings of the IEEE*, 90(5):711–732, 2002.

[12] E. N. Lorenz. Deterministic nonperiodic flow. *J. Atmosf. Sci.*, 20:130–141, 1963.

[13] A. Papoulis. *Probability, random variables and stochastic processes.* McGraw-Hill, New York, 3rd edition, 1991.

[14] O. E. Rössler. An equation for continuous chaos. *Physics Letters A*, 57:397–398, 1976.

[15] T. Saito and H. Fujita. Chaos in a manifold piecewise linear system. *Electron. Commun. Jpn. 1*, 64(10):9–17, 1981.

[16] M. Sushchik, L. S. Tsimring, and A. R. Volkovskii. Performance analysis of correlation-based communication schemes utilizing chaos. *IEEE Transactions on Circuits and Systems I*, 90(5):711–732, 2000.

[17] T. Tsubone and T. Saito. Stabilizing and destabilizing control for a piece-wise-linear circuit. *IEEE Transactions on Circuits and Systems I*, 45:172–177, 1998.

[18] G. L. Turin. An introduction to matched filters. *IRE Transactions on Information Theory*, 6:311–329, 1960.

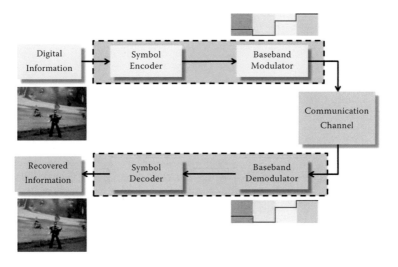

FIGURE 2.14
Baseband digital communication system diagram.

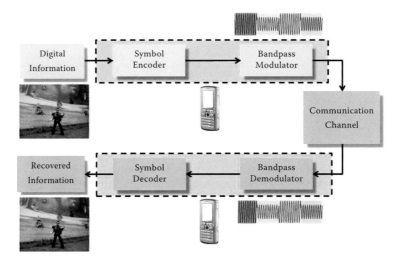

FIGURE 2.15
Bandpass digital communication system diagram.

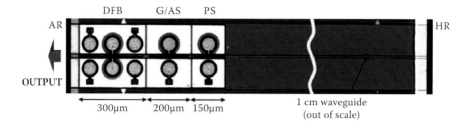

FIGURE 5.6

Photonic integrated circuit with straight waveguide for secure optical communications applications. DFB: Distributed feedback laser, G/AS: Gain - absorption section, PS: Phase section, AR: Anti-reflective coating, HR: Highly reflective coating.

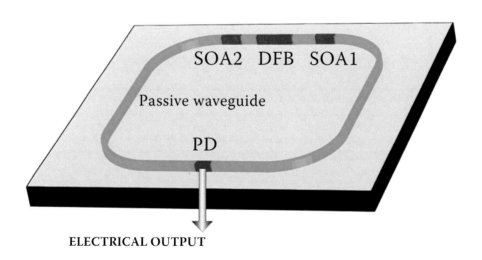

FIGURE 5.7

Photonic integrated circuit with ring waveguide and embedded photodetector for secure optical communications applications. DFB: Distributed feedback laser, SOA: Semiconductor optical amplifier, PD: Photodetector.

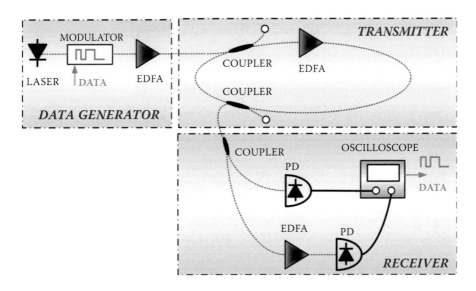

FIGURE 5.9
The optical chaos communication setup proposed by Van Wiggeren and Roy.

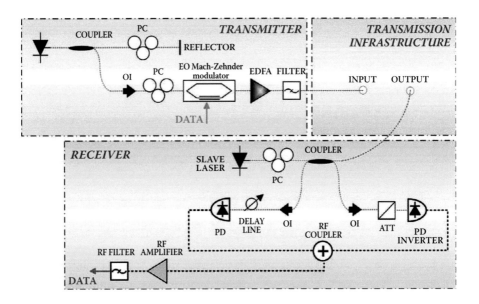

FIGURE 5.10
Experimental setup of an all-optical communication transmission system based on chaotic carriers. PC: polarization controller, OI: optical isolator, PD: photoreceiver, ATT: attenuator.

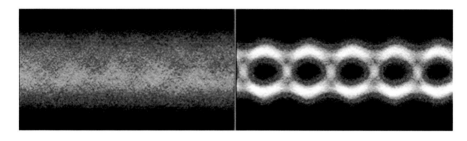

FIGURE 5.11
Eye diagrams for a $2^{31} - 1$, 1-Gb/s pseudorandom encrypted message with BER 0.06 (left) while the decrypted message is recovered with BER 10^{-7} (right).

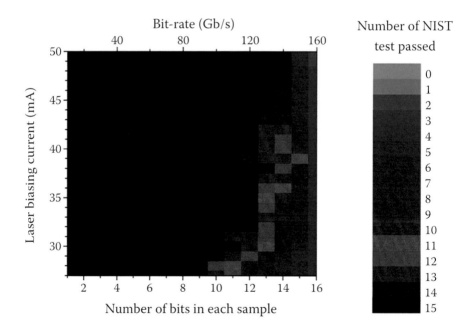

FIGURE 5.19
Mapping of the TRBG performance in terms of bit rate and number of NIST randomness tests passed, versus the laser current and the number of LSBs of each sample included in the output sequence. Black color grade regions designate operating conditions where all NIST randomness tests are successful. (Reproduced from A. Argyris, © OSA, 2010.)

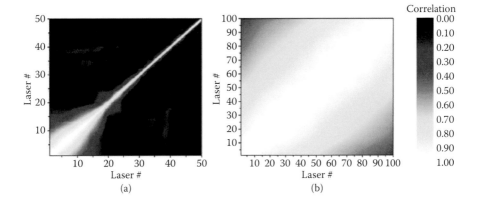

FIGURE 5.21
Zero-lag cross-correlation among the chaotic emitted optical signals from (a) $N = 50$ and (b) $N = 100$ lasers that participate in the star coupling network of Munt et al. configuration. Lasers had identical operational characteristics and a frequency detuning within a linear spread of -40 GHz (laser ♮1), 40 GHz (laser ♮N). The N lasers were biased just above threshold, hub laser was biased below threshold, and all couplings were set to 30 GHz.

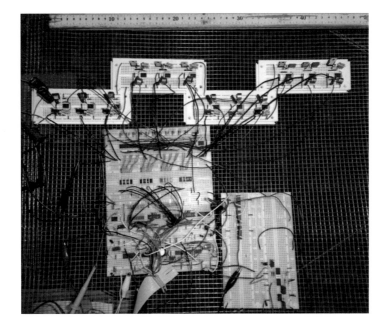

FIGURE 7.16
Photo of transmitter and receiver circuits for experimental system.

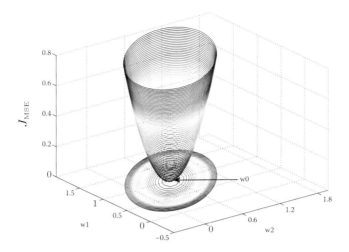

FIGURE 8.6
Mean-square-error cost function, assuming the transmission of a binary sequence $s(n) \in \{-1, +1\}$ through the channel of Table 8.1.

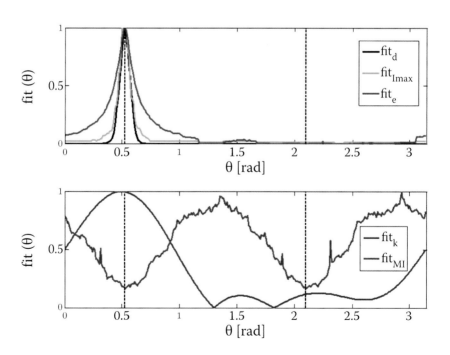

FIGURE 12.5
Exhaustive search in θ parameter for the maximization of statistical independence and deterministic characteristics of the output component y_1. All the cost functions were normalized to the unity. The recurrence parameters $d_e = 3$, $\tau = 6$, ϵ were adjusted to produce a recurrence rate - the density of points in the recurrence plot - of 1%, $N = 2000$ samples.

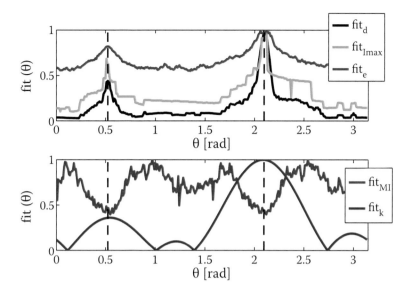

FIGURE 12.7
Exhaustive search in θ parameter that enhances the statistical independence or deterministic characteristics of the output component y_1. All the cost functions were normalized to the unity. Recurrence parameters: $d_e = 3$, $\tau = 6$, $\epsilon = 0.1$, $N = 2000$ samples. The perfect solutions that restore the Lorenz and Rössler signals in the y_1 component are given by, respectively, $\theta = \pi/6$ and $\theta = 2\pi/3$.

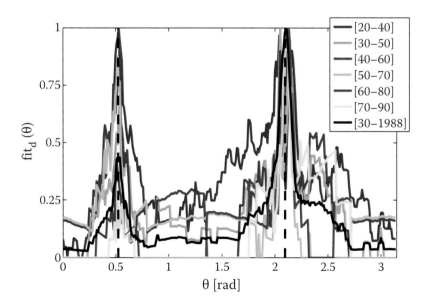

FIGURE 12.8
Normalized $\mathrm{fit_d}$ as a function of θ for different diagonal length intervals ($[a\ b]$ — legend) for the output component y_1. Recurrence parameters: $d_e = 3$, $\tau = 6$, $\epsilon = 0.1$, $N = 2000$ samples. The perfect solutions that restore the Lorenz and Rössler signals in the y_1 component are given by, respectively, $\theta = \pi/6$ and $\theta = 2\pi/3$.

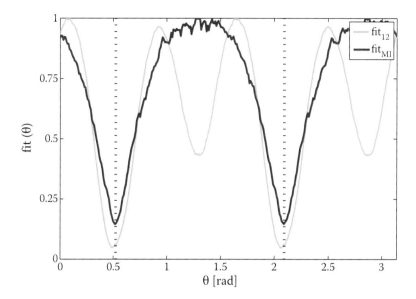

FIGURE 12.9
Estimated mutual information of the output vector for different values of θ using classical histogram — (fit$_{MI}$) and recurrence (fit$_{I2mod}$) — based approaches. All the cost functions were normalized to have maximum value equal to one. The recurrence parameters are: $d_e = 1$, $\epsilon = 0.4$, $N = 6000$ samples. The solutions that perfectly restore the original sources are given by, respectively, $\theta = \pi/6$ and $\theta = 2\pi/3$.

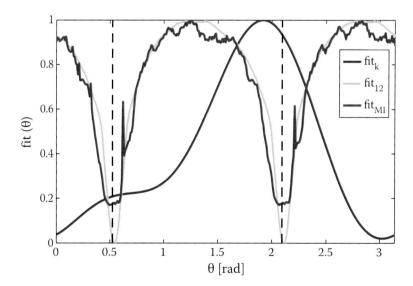

FIGURE 12.10
Normalized fit$_k$, fit$_{I2}$, and fit$_{MI}$ for the y_1 component as a function of θ. Recurrence parameters: $d_e = 3$, $\tau = 6$, $\epsilon = 0.1$, $N = 2000$ samples. Original sources: Lorenz time series and $\sin(5t)$. The solutions that perfectly restore the original sources are given by, respectively, $\theta = \pi/6$ and $\theta = 2\pi/3$.

8

Channel equalization for chaotic communications systems

Renato Candido, Magno T. M. Silva, and Marcio Eisencraft

Escola Politécnica
Universidade de São Paulo

CONTENTS

In this chapter, we propose an adaptive equalization scheme for a chaos-based digital communication system. The studied system is a discrete-time version of the one proposed by Wu and Chua that has recently been considered for optical applications. Due to the nonlinear characteristics of the transmitters and receivers and the lack of robustness of chaos synchronization, even minor channel imperfections are enough to hinder communication. Adaptive equalization is succinctly reviewed by focusing on the normalized least-mean-squares (NLMS) algorithm, which is extended to fit the studied chaotic communication system. Simulation results show that the proposed *chaotic* NLMS can successfully equalize the channel.

8.1 Introduction

In the last years, chaos theory has been studied in several areas of Engineering including signal processing, control, and communications systems (see, e.g., [23]). In communications, nonlinear dynamics has potential applications in several blocks of the system, such as data compression, cryptography, and modulation [11].

Specifically, when it comes to baseband modulation of digital data, many ideas have been proposed [11, 23, 25, 43]. Many papers in the optical communications domain appeared using as chaotic generator the intrinsic nonlinear properties of lasers (see, e.g., [3, 8, 21, 24, 27]). For instance, in [3] a high-speed long-distance communication system based on chaos synchronization over a commercial fibre-optic channel was demonstrated.

Two of the most studied schemes for possible practical optical communication applications are (i) *chaotic masking* [6] and (ii) *chaotic modulation* [44].

In chaotic masking, the transmitter generates a chaotic signal that must be synchronized in the receiver. After an attenuation, the message is added to this signal so that the chaotic signal's characteristics are not significantly changed. After some time, the receiver is supposed to synchronize with the transmitter, which means that the chaotic signal obtained in the receiver is supposed to be equal to the chaotic signal generated in the transmitter. Therefore, it is possible to recover the message signal by calculating the difference between the received signal and the chaotic signal generated in the receiver. It is relevant to note that in this system, the synchronism between master (transmitter) and slave (receiver) is not perfect. This can cause errors when recovering the transmitted message even under ideal conditions, which consists in a serious drawback.

Chaotic modulation, by its turn, avoids synchronization error by injecting the information signal in the transmitter. The message is not simply added to the transmitted chaotic signal. The idea proved to work very well and the performance is acceptable in ideal or *quasi*-ideal situations with a fixed channel that is almost perfectly equalized, as shown in [3].

However, when there is additive noise or if the channel frequency response is not ideal, the chaotic synchronization is simply lost and communication is not possible [10, 43]. Although this fact can be simply checked using computational simulations, it was not deeply investigated yet. Since dispersive channels introduce intersymbol interference, an adaptive equalizer must be included in the system to recover adequately the transmitted message.

Over the last decades, adaptive equalization has been an area of intense research in the signal processing community (see, e.g., [15, 16, 19, 22, 29, 33, 39, 41] and the references therein). In many practical situations, as, e.g., satellite channels, nonlinear equalization is required. In the case of chaotic communications, the nonlinearities of the dynamical system can be constituent of the discrete-time channel model and nonlinear equalization schemes may also be necessary. Therefore, we deduce in this work a nonlinear equalization scheme that can be used to allow a chaos-based digital communication [44] to work in a more practical scenario.

It is important to notice that although the idea of equalization applied to chaotic signals is not new (see, e.g., [7, 26, 45]), we are not aware of works on equalization applied in the discrete-time domain for this particular chaotic modulation that appears in many optical communication system proposals. As digital-implemented equalizers are a common practice, we hope that these results should be relevant in the context of practical chaos-based optical communications.

The chapter is organized as follows. In Section 8.2, we revisit the discrete-time version of the Wu and Chua's chaotic modulation [10]. Section 8.3 brings a succinct review of equalization methods applied in Section 8.4 to chaotic modulation. Finally, Section 8.5 presents simulation results and in Section 8.6, we draft some conclusions.

8.2   Communication system by Wu and Chua

Wu and Chua's synchronization scheme proposed in [44] is a simple way to use chaos for communication. They addressed chaotic system synchronization differently from Pecora and Carroll's seminal paper [32]. Instead of using conditional Lyapunov exponents to check the asymptotic stability of the slave system and hence the possibility of synchronism, Wu and Chua rewrote the master and slave equations in such a way that it is easy to verify the convergence of the synchronization error to zero. Based on this synchronization scheme, a communication system was proposed in [44] and a discrete-time version appeared later in [10]. In this section, we succinctly revise these ideas.

Consider two discrete-time systems defined by

$$\mathbf{x}(n+1) = \mathbf{A}\mathbf{x}(n) + \mathbf{b} + \mathbf{f}(x_1(n)) \tag{8.1}$$

$$\widehat{\mathbf{x}}(n+1) = \mathbf{A}\widehat{\mathbf{x}}(n) + \mathbf{b} + \mathbf{f}(x_1(n)), \tag{8.2}$$

where $n \in \mathbb{N}$ represents time instants, $\mathbf{x}(n)$ and $\widehat{\mathbf{x}}(n)$ are real-valued column vectors of length K, i.e, $\mathbf{x}(n) = [x_1(n)\, x_2(n)\, \cdots \, x_K(n)]^T$ and $\widehat{\mathbf{x}}(n) = [\widehat{x}_1(n)\, \widehat{x}_2(n)\, \cdots \, \widehat{x}_K(n)]^T$, and $(\cdot)^T$ stands for transposition.[1] $\mathbf{A}$ is a square matrix and $\mathbf{b}$ a column vector, both constants, real-valued, and of dimension K. The vector function $\mathbf{f}(\cdot)$: $\mathbb{R} \to \mathbb{R}^K$ is nonlinear in general and is assumed to depend solely on the first component of $\mathbf{x}(n)$, having the form

$$\mathbf{f}(x_1(n)) = [f(x_1(n))\; 0\; 0\; \cdots \; 0]^T, \tag{8.3}$$

---

[1]The vector $\widehat{\mathbf{x}}(n)$ is not always an estimate of $\mathbf{x}(n)$. However, in the case considered in this chapter, $\widehat{\mathbf{x}}(n)$ can be interpreted as an estimate of $\mathbf{x}(n)$, which justifies the notation.

where $f(\cdot)$ is a scalar function. The system described by Equation (8.1) is autonomous and is called *master*, whereas the one described by Equation (8.2) depends on $x_1(n)$ and is called *slave*.

The synchronization error dynamics between the two systems $\mathbf{e}(n) \triangleq \widehat{\mathbf{x}}(n) - \mathbf{x}(n)$ is given by

$$\mathbf{e}(n+1) = \mathbf{A}\mathbf{e}(n). \tag{8.4}$$

Master and slave are said *completely synchronized* if $\mathbf{e}(n) \to \mathbf{0}$ as n grows. Consequently, they synchronize completely if the eigenvalues λ_i of $\mathbf{A}$ satisfy [1]

$$|\lambda_i| < 1, \quad 1 \leq i \leq K. \tag{8.5}$$

Therefore, if a system can be written as Equation (8.1) with the eigenvalues of $\mathbf{A}$ satisfying Equation (8.5), it is easy to set up a slave system that synchronizes with it.

Using this synchronization method, Wu and Chua [44] proposed an information transmission system using chaotic signals that leads to no errors under ideal channel conditions. A block diagram of the discrete-time version of this system is shown in Figure 8.1 [10]. In this scheme, the information signal $m(n)$ is encoded by using the first component of the state vector $\mathbf{x}(n)$ via a coding function

$$s(n) = c\left(x_1(n), m(n)\right), \tag{8.6}$$

so that the information signal can be decoded using the inverse function with respect to $m(n)$, i.e.,

$$m(n) = c^{-1}\left(x_1(n), s(n)\right) = c^{-1}\left(x_1(n), c\left(x_1(n), m(n)\right)\right). \tag{8.7}$$

The equations governing the global system have the same form as Equations (8.1)-(8.2). The only changes are the arguments of $\mathbf{f}(\cdot)$:

$$\mathbf{x}(n+1) = \mathbf{A}\mathbf{x}(n) + \mathbf{b} + \mathbf{f}(s(n)) \tag{8.8}$$

$$\widehat{\mathbf{x}}(n+1) = \mathbf{A}\widehat{\mathbf{x}}(n) + \mathbf{b} + \mathbf{f}(r(n)), \tag{8.9}$$

where $r(n)$ is the signal that the channel delivers to the receiver. Here, the channel is modeled by a finite impulse response (FIR) filter with system function $H(z) = h_0 + h_1 z^{-1} + \cdots + h_{L-1} z^{L-1}$ and additive white Gaussian noise (AWGN), $\eta(n)$, with zero mean and power σ_η^2 [30]. Using this model, $r(n)$ can be written as

$$r(n) = \sum_{k=0}^{L-1} h_k s(n-k) + \eta(n). \tag{8.10}$$

For an ideal channel, considering zero delay for simplicity, i.e., $h_0 = 1$, $h_k = 0$, $k > 0$, and absence of noise ($\sigma_\eta^2 = 0$), $r(n) = s(n)$ and Equations (8.8)-(8.9) can be rewritten as

$$\mathbf{x}(n+1) = \mathbf{A}\mathbf{x}(n) + \mathbf{b} + \mathbf{f}(s(n)) \tag{8.11}$$

$$\widehat{\mathbf{x}}(n+1) = \mathbf{A}\widehat{\mathbf{x}}(n) + \mathbf{b} + \mathbf{f}(s(n)). \tag{8.12}$$

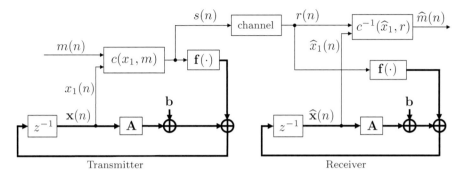

FIGURE 8.1
Block diagram of Wu and Chua's proposal adapted to discrete-time systems [10].

Since the synchronization error dynamics is given again by Equation (8.4) and if Equation (8.5) holds, then $\widehat{\mathbf{x}}(n) \to \mathbf{x}(n)$ and, in particular, $\widehat{x}_1(n) \to x_1(n)$. Thus, using Equation (8.7), we obtain

$$\widehat{m}(n) = c^{-1}\left(\widehat{x}_1(n), s(n)\right) \to c^{-1}\left(x_1(n), s(n)\right) = m(n). \qquad (8.13)$$

Therefore, when transmitter and receiver parameters are perfectly matched over an ideal channel, the message is recovered without degradation at the receiver except for a synchronization transient.

In this chapter, we consider as dynamical system the Hénon map [2, 20]

$$\mathbf{x}(n+1) = \begin{bmatrix} x_1(n+1) \\ x_2(n+1) \end{bmatrix} = \begin{bmatrix} \alpha - x_1^2(n) + \beta x_2(n) \\ x_1(n) \end{bmatrix}, \qquad (8.14)$$

that can be rewritten as Equation (8.1) with $K = 2$, $\mathbf{A} = \begin{bmatrix} 0 & \beta \\ 1 & 0 \end{bmatrix}$, $\mathbf{b} = [\alpha\ 0]^T$, and $\mathbf{f}(x_1(n)) = \left[-x_1^2(n)\ 0\right]^T$. For this choice, the eigenvalues of $\mathbf{A}$ are $\lambda_{1,2} = \pm\sqrt{\beta}$ and, according to Equation (8.5), there is chaotic synchronization for $|\beta| < 1$.

As coding function we have chosen

$$s(n) = c\left(x_1(n), m(n)\right) = m(n)x_1(n), \qquad (8.15)$$

considering a binary message $m(n) \in \{-1, 1\}$, for $n \in \mathbb{N}$. For this particular choice of $c(\cdot\,, \cdot)$, the decoding function can be implemented by

$$\widehat{m}(n) = c^{-1}\left(\widehat{x}_1(n), r(n)\right) = \frac{r(n)}{\widehat{x}_1(n)}. \qquad (8.16)$$

To avoid division by zero in Equation (8.16), we simply make $\widehat{m}(n) = r(n)/\delta$ when $\widehat{x}_1(n) = 0$, where δ is a small positive constant. Simulations have shown

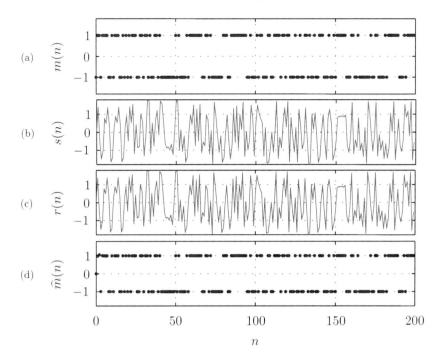

FIGURE 8.2
Simulation of the communication system shown in Figure 8.1 with the Hénon map ($\alpha = 1.4$ and $\beta = 0.3$) for the ideal channel: (a) message $m(n)$; (b) transmitted signal $s(n)$; (c) received signal $r(n)$; and (d) recovered message $\widehat{m}(n)$.

that this choice of $c(\cdot, \cdot)$ presents better bit error rate (BER) performance in an AWGN channel than the additive coding function commonly used in the chaos-communication literature (see, e.g., [10, 14, 35, 44]). With the adopted codification, the signal $s(n)$ may not be chaotic. However, this fact does not affect our results. Verifying if $s(n)$ is in fact chaotic or presents properties generally attributed to signals used in chaos-based communications is out of the scope of this chapter.

Figure 8.2 shows an example of message $m(n)$, transmitted signal $s(n)$ encoded by the Hénon map with $\alpha = 1.4$ and $\beta = 0.3$, received signal $r(n) = s(n)$, and recovered message $\widehat{m}(n)$ for the ideal channel. We can observe in this case that the message is recovered perfectly after a transient, as expected.

However, as pointed out by [10, 35], for a non-ideal channel, synchronism is impaired; consequently, $\widehat{m}(n) \nrightarrow m(n)$. Due to the nonlinear nature of the systems involved, if any spectral component is amiss, all spectral components at the receiver can be affected. Figure 8.3 shows an example of the signals

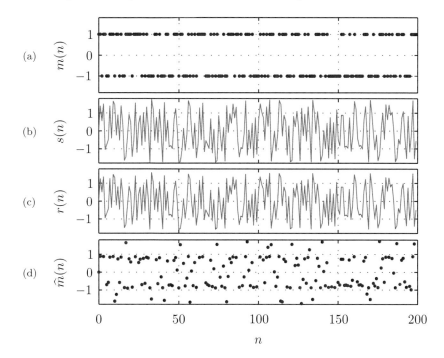

FIGURE 8.3
Simulation of the communication system shown in Figure 8.1 with the Hénon map ($\alpha = 1.4$ and $\beta = 0.3$) for the channel $H(z) = 0.9$ in the absence of noise: (a) message $m(n)$; (b) transmitted signal $s(n)$; (c) received signal $r(n)$; and (d) recovered message $\widehat{m}(n)$.

involved for the channel $H(z) = 0.9$ in the absence of noise. As can be seen, even when the channel introduces a simple attenuation, the message is severely damaged.

A way of circumventing these difficulties for ideal but bandlimited channels is to adjust the spectrum of the transmitted chaotic signal to the channel band using linear filters in the feedback loop of the transmitter and receiver. This idea was followed in [10, 12]. However, it is convenient only for a bandlimiting almost ideal channel. When the frequency response of the channel is not known in advance it is not possible to use this approach.

The idea explored in the next sections is to use an adaptive filter to equalize the channel and to permit the system of Figure 8.1 to work properly for a non-ideal $H(z)$.

8.3   A brief review of adaptive equalization

In digital communication systems, information-bearing signals transmitted between remote locations are affected by intersymbol interference (ISI) and noise introduced by dispersive channels. Examples of dispersive channels include coaxial, fiber optic, or twisted-pair cable in wired communications and the atmosphere or ocean in wireless communications [22]. To remove the effects of channel distortion, it is common to use adaptive equalizers, which attempts to recover the transmitted symbol sequence by counteracting the effects of ISI [9, 19, 22, 33, 41].

A simplified communication system with an adaptive equalizer is depicted in Figure 8.4. We assume that this communication system works in baseband and, therefore, the transmitted message $m(n)$ and the modulated sequence $s(n)$ are the same. Thus, the sequence $m(n) \equiv s(n)$ is transmitted through an unknown channel, whose model was described in Section 8.2. Due to the channel memory, the signal at the receiver contains contributions not only from $s(n)$ but also from the previous symbols $s(n-1)$, $s(n-2)$, ..., $s(n-L+1)$. This can be highlighted by rewriting Equation (8.10) as

$$r(n) = \underbrace{\sum_{k=0}^{\Delta-1} h_k s(n-k)}_{\text{pre-ISI}} + h_\Delta s(n-\Delta) + \underbrace{\sum_{k=\Delta+1}^{L-1} h_k s(n-k)}_{\text{post-ISI}} + \eta(n). \qquad (8.17)$$

Assuming that the overall channel-equalizer system imposes a delay of Δ samples, the adaptive equalizer will try to mitigate the two summations in Equation (8.17), thereby mitigating the ISI and finding an approximation $y(n)$ for $s(n-\Delta)$. The equalizer in the scheme of Figure 8.4 works in the *training mode*, since a delayed version of the transmitted sequence $s(n-\Delta)$ (training sequence) is known in advance at the receiver. During the training mode, the equalizer updates its coefficients using the estimation error $e(n) = s(n-\Delta) - y(n)$ in conjunction with an adaptive algorithm. When information is effectively transmitted, the receiver will not have access to $s(n-\Delta)$. In this case, the training signal $s(n-\Delta)$ is replaced by its estimate $\hat{s}(n-\Delta)$ obtained at the output of the decision device, as shown in Figure 8.5. In this case, the equalizer works in the so-called *decision-directed mode*.

Due to its simplicity, the linear transversal equalizer adapted with stochastic gradient algorithms is the most employed scheme in digital communication systems. However, a linear equalizer may perform poorly in many practical situations, such as channels with long and sparse impulse responses, non-minimum phase, spectral nulls, or nonlinearities (e.g., satellite channels) [4,18,31,34]. In fact, Forney Jr. showed in [15] that the optimum equalizer is nonlinear and can be implemented with a whitening matched filter followed by the Viterbi algorithm. The necessity of estimating the channel and the high

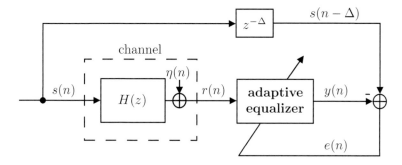

FIGURE 8.4
Simplified communication system with an adaptive equalizer in the training mode.

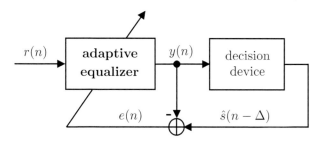

FIGURE 8.5
Adaptive equalizer in the decision-directed mode.

computational cost of the Viterbi algorithm make this solution unfeasible. The same occurs with Bayesian methods for channel equalization which, although optimum, are usually developed from the perspective of known channel characteristics and also present high computational cost [5, 29]. Therefore, suboptimal nonlinear solutions that present a more favorable trade-off between computational cost and efficient behavior, independently of the channel, have been proposed in the literature. Some examples of nonlinear equalizers are: decision feedback equalizers [33,34,42], neural networks [16,29,31,38,39], support vector machines [36], Volterra series [18], and kernel adaptive filters [28].

Another important nonlinear approach for channel equalization is known as blind (or unsupervised) equalization. In this case, the nonlinearity is in estimating the transmitted signal based on its high-order statistics (HOS). Since the transmission of the training sequence is avoided, the available bandwidth is used in a more efficient manner [22]. There are many blind equalization algorithms based on HOS in the literature, but the most popular is the constant modulus algorithm (CMA), proposed independently by Godard [17] and Treichler and Agee [40]. A good review of the main results related to CMA can be found in [22].

In this chapter, to obtain a relatively simple solution to equalize chaotic sequences, we focus only on a supervised scheme based on the NLMS algorithm [19,37]. This algorithm is implemented in a transversal filter, whose input and weight vectors are given, respectively, by

$$\mathbf{r}(n) = [\, r(n) \; r(n-1) \; \cdots \; r(n-M+1) \,]^T \tag{8.18}$$

and

$$\mathbf{w}(n-1) = [\, w_0(n-1) \; w_1(n-1) \; \cdots \; w_{M-1}(n-1) \,]^T, \tag{8.19}$$

and whose output is computed as $y(n) = \mathbf{r}^T(n)\mathbf{w}(n-1)$.

In the sequel, we describe the NLMS algorithm in detail. In order to simplify the arguments, we assume that all the quantities are real.

8.3.1   Linear equalization with NLMS

Stochastic-gradient algorithms seek to minimize the mean-squared error (MSE), defined as $J_{\mathrm{MSE}}(n) \triangleq \mathrm{E}\{e^2(n)\}$, where $\mathrm{E}\{\cdot\}$ stands for the expectation operation and $e(n) = s(n-\Delta) - y(n)$. The MSE is a convex cost function, whose minimum is given by the Wiener solution

$$\mathbf{w}_o = \mathbf{R}^{-1}\mathbf{p}_\Delta, \tag{8.20}$$

in which $\mathbf{R} = \mathrm{E}\{\mathbf{r}(n)\mathbf{r}^T(n)\}$ is the autocorrelation matrix of the input signal and $\mathbf{p}_\Delta = \mathrm{E}\{s(n-\Delta)\mathbf{r}(n)\}$ is the cross-correlation between the input regressor vector and the transmitted signal. The Wiener solution is known as *optimal linear solution* and depends on the delay Δ of the transmitted signal [13,19,37]. Figure 8.6 shows the MSE cost function assuming the transmission of a binary sequence $s(n) \in \{-1, +1\}$ through the channel of Table 8.1 (Channel 1), whose impulse response corresponds to the ten first samples of the impulse response of $H(z) = 1/[1 + 0.6z^{-1}]$. In this case, the minimum MSE occurs for $\mathbf{w}_o \approx [\, 1 \;\; 0.6 \,]^T$ [9] and this solution can approximately eliminate the channel effects, achieving a *quasi*-perfect equalization in the absence of noise with only $M = 2$ coefficients and zero delay.

TABLE 8.1
Coefficients of the impulse response of Channel 1.

k	h_k	k	h_k
0	1.0000	5	−0.0778
1	−0.6000	6	0.0467
2	0.3600	7	−0.0280
3	−0.2160	8	0.0168
4	0.1296	9	−0.0101

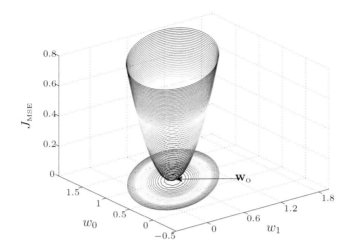

FIGURE 8.6 (SEE COLOR INSERT)
Mean-square-error cost function, assuming the transmission of a binary sequence $s(n) \in \{-1, +1\}$ through the channel of Table 8.1.

The Wiener solution can be achieved by the steepest descent algorithm, which avoids the computation of the inverse of $\mathbf{R}$ by updating $\mathbf{w}(n-1)$ in a direction opposite to that of the gradient of J_{MSE}, i.e.,

$$\mathbf{w}(n) = \mathbf{w}(n-1) + \mu \left[\mathbf{p}_\Delta - \mathbf{R}\mathbf{w}(n-1) \right], \qquad (8.21)$$

where μ is a step size and $\mathbf{w}(-1)$ is an initial guess for the minimum point of J_{MSE}, commonly assumed equal to the null vector ($\mathbf{w}(-1) = \mathbf{0}$). With a proper choice of μ (see [13, 19] for details about stability), this algorithm achieves exactly the Wiener solution. However, the exact gradient requires a prior knowledge of $\mathbf{R}$ and $\mathbf{p}_\Delta$, which is not feasible for equalization. It is important to remark that in practical situations, the channel is time-variant, $r(n)$ is nonstationary, and, therefore, $\mathbf{R}$ and $\mathbf{p}_\Delta$ cannot be exactly estimated at each time instant.

To circumvent this problem, the least-mean-squares (LMS) algorithm was proposed. Instead of using the exact gradient, the LMS algorithm uses an instantaneous approximation to it, i.e.,

$$\nabla_\mathbf{w} J_{\mathrm{MSE}} \approx -2s(n - \Delta)\mathbf{r}(n) + 2\mathbf{r}(n)\mathbf{r}^T(n)\mathbf{w}(n-1) = -2e(n)\mathbf{r}(n), \quad (8.22)$$

which leads to the update equation

$$\mathbf{w}(n) = \mathbf{w}(n-1) + \mu e(n)\mathbf{r}(n), \qquad (8.23)$$

with $\mathbf{w}(-1) = \mathbf{0}$ and μ being a step size. The LMS algorithm is the most popular adaptive filter due to its low computational cost, robustness, and many analytical results. From a second-order analysis, it is possible to show that LMS converges in the mean-square sense to the Wiener solution if [13]

$$0 < \mu < \frac{2}{3\lambda_{\max}} < \frac{2}{3M\sigma_r^2}, \tag{8.24}$$

where $\lambda_{\max}$ is the maximum eigenvalue of $\mathbf{R}$ and σ_r^2 is the power of the input signal $r(n)$.

The maximum value of the step size to ensure the LMS stability depends on the statistics of the input signal. Therefore, when the statistics of the input signal change quickly, the LMS algorithm is not a good solution since a time varying step size must be used. In this context, the normalized LMS (NLMS) algorithm was proposed as a version of LMS with the following step size:

$$\mu(n) = \frac{\tilde{\mu}}{\delta + \|\mathbf{r}(n)\|^2}, \quad \text{with } 0 < \tilde{\mu} < 2, \tag{8.25}$$

where δ is a small positive constant used to avoid division by zero and $\| \cdot \|$ stands for Euclidean norm. Thus, the update equation of NLMS is given by

$$\mathbf{w}(n) = \mathbf{w}(n-1) + \frac{\tilde{\mu}}{\delta + \|\mathbf{r}(n)\|^2} e(n)\mathbf{r}(n). \tag{8.26}$$

The time varying step size $\mu(n)$ depends inversely on the instantaneous power of the input vector $\mathbf{r}(n)$, which enables NLMS to track variations in the signal statistics.

It is important to notice that NLMS can be derived from different manners. The most usual is to derive NLMS by finding the step size $\mu(n)$ that makes the *a posteriori* estimation error

$$e_p(n) = d(n) - \mathbf{r}^T(n)\mathbf{w}(n) \tag{8.27}$$

equal to zero at each iteration [13]. Thus, using Equation (8.23) with $\mu(n)$ instead of μ in Equation (8.27), we obtain

$$
\begin{aligned}
e_p(n) &= d(n) - \mathbf{r}^T(n)\left[\mathbf{w}(n-1) + \mu(n)e(n)\mathbf{r}(n)\right] \\
&= e(n)\left[1 - \mu(n)\mathbf{r}^T(n)\mathbf{r}(n)\right].
\end{aligned} \tag{8.28}
$$

To enforce $e_p(n) = 0$ at each iteration n, we must select

$$\mu(n) = \frac{1}{\|\mathbf{r}(n)\|^2}. \tag{8.29}$$

Finally, to avoid division by zero in Equation (8.29) and to control the rate of convergence, it is common to introduce the regularization factor δ and the step size $\tilde{\mu}$, which leads to Equation (8.25) and to the update Equation (8.26).

Another popular approach interprets NLMS as a *quasi*-Newton algorithm, which explains its faster convergence when compared with LMS [37].

8.3.2 Numerical examples

In order to illustrate the behavior of NLMS, we assume the transmission of a binary sequence $s(n) \in \{-1, +1\}$ through Channel 1 (Table 8.1) with a signal-to-noise ratio (SNR) of 30 dB. The SNR is computed as

$$\text{SNR} = 10 \log_{10} \left(\frac{\sigma_{\check{r}}^2}{\sigma_\eta^2} \right), \tag{8.30}$$

where $\sigma_{\check{r}}^2$ represents the power of the signal $\check{r}(n) = r(n) - \eta(n)$. We also assume zero delay ($\Delta = 0$), an equalizer with $M = 2$ coefficients, $\tilde{\mu} = 0.1$, and $\delta = 0.5$. It is important to notice that to obtain a good behavior of NLMS in this case, we used a relatively large regularization factor δ. This is necessary since $\|\mathbf{r}(n)\|^2$ has a high probability of becoming close to zero when few coefficients are considered, which can de-stabilize the NLMS algorithm.

The output of the NLMS equalizer is shown in Figure 8.7(b); the average of the coefficients and the MSE along the iterations, estimated from the ensemble-average of 1000 independent runs, are shown, respectively, in Figures 8.7(c) and (d). As expected, NLMS obtains a good estimate of the transmitted sequence, converging in the mean to the Wiener solution $\mathbf{w}_o \approx [1 \ 0.6]^T$, whose coefficients are shown as dashed lines in Figure 8.7(c). For comparison, we also show in Figure 8.7(a) the received sequence $r(n)$ and in Figure 8.7(d) the respective MSE. We can observe that the equalizer plays an important role in the communication system, since when it is not considered the sequence is not recovered and the achieved MSE is approximately -3.5 dB. On the other hand, the NLMS equalizer achieves a steady-state MSE of approximately -26 dB, which is sufficient to recover the transmitted sequence after a decision device.

Since Channel 1 models approximately a channel with infinite impulse response (IIR), an FIR filter can achieve perfect equalization in absence of noise. To show the behavior of the NLMS equalizer in a more challenging scenario, we assume now the transmission of a binary sequence $s(n) \in \{-1, +1\}$ through Channel 2, whose system function is given as $H_2(z) = 0.3 + z^{-1} + 0.3z^{-2}$. We also assume an SNR of 30 dB, a delay of $\Delta = 3$ samples, an equalizer with $M = 5$ coefficients, $\tilde{\mu} = 0.1$, and $\delta = 0.1$. In this case, the Wiener solution is given approximately by

$$\mathbf{w}_o \approx [0.1078 \ -0.3971 \ 1.2327 \ -0.3972 \ 0.1078]^T, \tag{8.31}$$

whose coefficients are shown as dashed lines in Figure 8.8(c). As we can observe in Figure 8.8, the NLMS obtains a good estimate of the transmitted sequence, converges in the mean to the Wiener solution, and achieves a steady-state MSE of approximately -23 dB. Again, the equalizer is essential to recover the transmitted message. It is relevant to observe that the received sequence $r(n)$ varies around six distinct levels as shown in Figure 8.8(a). However, the estimated sequence (with no equalizer) after a decision device is very different from the transmitted sequence independently of the delay and the achieved MSE is approximately 3 dB.

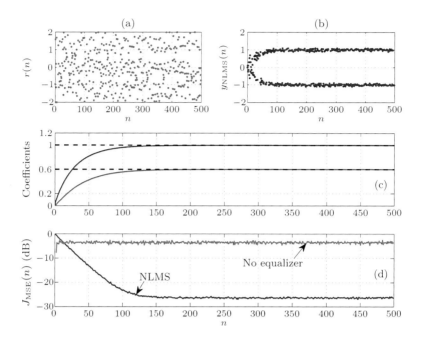

FIGURE 8.7

Estimated sequence considering (a) no equalization and (b) NLMS ($\mu = 0.1$, $\delta = 0.5$); (c) average of the coefficients of NLMS and Wiener solution (dashed lines); (d) estimated MSE; ensemble-average of 1000 independent runs, Channel 1, SNR = 30 dB; $M = 2$; $\Delta = 0$.

8.4   The chaotic NLMS algorithm

Assuming that the communication channel is noisy and dispersive, the chaotic communication system of Figure 8.1 must include an adaptive equalizer, resulting in the scheme depicted in Figure 8.9. In the sequel, we extend the NLMS algorithm of Section 8.3.1 to update the coefficients of the equalizer in this scheme.

To obtain a stochastic gradient algorithm to adapt the equalizer in the scheme of Figure 8.9, we first need to define an MSE-type cost-function. Since our objective is to recover the transmitted message $m(n)$ with some delay Δ, we define the instantaneous *chaotic* MSE cost-function as[2]

$$\hat{J}_{\text{cMSE}}(n) = e_c^2(n) = [m(n - \Delta) - \widehat{m}(n)]^2 . \tag{8.32}$$

---

[2]We use the term *chaotic* for the MSE and the algorithms in this section only for distinguishing the versions derived here from the versions of Section 8.3. The use of this term does not imply a chaotic behavior of the algorithms.

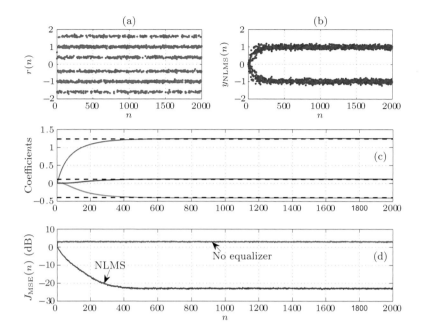

FIGURE 8.8
Estimated sequence considering (a) no equalization and (b) NLMS ($\mu = 0.1$, $\delta = 0.1$); (c) average of the coefficients of NLMS and Wiener solution (dashed lines); (d) estimated MSE; ensemble-average of 1000 independent runs, Channel 2 ($H_2(z) = 0.3 + z^{-1} + 0.3z^{-2}$), SNR = 30 dB; $M = 5$; $\Delta = 3$.

Computing the gradient of $\hat{J}_{\mathrm{cMSE}}(n)$ with respect to the coefficient vector $\mathbf{w}(n-1)$, we obtain

$$\nabla_{\mathbf{w}} \hat{J}_{\mathrm{cMSE}}(n) = 2e_c(n) \frac{\partial e_c(n)}{\partial \mathbf{w}(n-1)} = -2e_c(n) \frac{\partial \widehat{m}(n)}{\partial \mathbf{w}(n-1)}. \tag{8.33}$$

Assuming that $\widehat{x}_1(n) \neq 0$ for all n and taking into account the equalizer in the scheme of Figure 8.9, Equation (8.16) can be rewritten as

$$\widehat{m}(n) = \frac{y(n)}{\widehat{x}_1(n)} = \frac{\mathbf{r}^T(n)\mathbf{w}(n-1)}{\widehat{x}_1(n)}. \tag{8.34}$$

Using Equation (8.34) and recalling that $\widehat{x}_1(n)$ depends only on $\widehat{\mathbf{x}}(n-1)$ and $y(n-1)$, which in turn do not depend on $\mathbf{w}(n-1)$, we arrive at

$$\nabla_{\mathbf{w}} \hat{J}_{\mathrm{cMSE}}(n) = -2\frac{e_c(n)}{\widehat{x}_1(n)} \frac{\partial y(n)}{\partial \mathbf{w}(n)} = -2\frac{e_c(n)}{\widehat{x}_1(n)}\mathbf{r}(n). \tag{8.35}$$

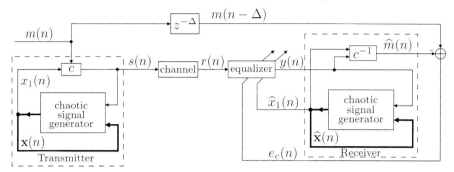

FIGURE 8.9

Chaotic communication system with an adaptive equalizer.

Thus, the update equation of the chaotic LMS (cLMS) algorithm is given by

$$\mathbf{w}(n) = \mathbf{w}(n-1) + \mu \frac{e_c(n)}{\widehat{x}_1(n)} \mathbf{r}(n). \tag{8.36}$$

To obtain a normalized version of cLMS, we first define the *a posteriori* error as

$$e_{c,p}(n) = m(n-\Delta) - \frac{\mathbf{r}^T(n)\mathbf{w}(n)}{\widehat{x}_1(n)}. \tag{8.37}$$

Using Equation (8.36), $e_{c,p}(n)$ can be rewritten as

$$
\begin{aligned}
e_{c,p}(n) &= m(n-\Delta) - \frac{\mathbf{r}^T(n)\left[\mathbf{w}(n-1) + \mu\dfrac{e_c(n)}{\widehat{x}_1(n)}\mathbf{r}(n)\right]}{\widehat{x}_1(n)} \\
&= e_c(n)\left[1 - \mu\frac{\|\mathbf{r}(n)\|^2}{\widehat{x}_1^2(n)}\right].
\end{aligned}
\tag{8.38}
$$

In order to enforce $e_{c,p}(n) = 0$ at each iteration n, we must select

$$\mu_c(n) = \frac{\widehat{x}_1^2(n)}{\|\mathbf{r}(n)\|^2}. \tag{8.39}$$

Introducing a fixed step-size $\tilde{\mu}_c$ to control the rate of convergence and a regularization factor δ to prevent division by zero in $\mu_c(n)$, and replacing the resulting step size in Equation (8.36), we obtain the update equation of the chaotic NLMS (cNLMS) algorithm, i.e.,

$$\mathbf{w}(n) = \mathbf{w}(n-1) + \frac{\tilde{\mu}_c}{\delta + \|\mathbf{r}(n)\|^2} \widehat{x}_1(n) e_c(n) \mathbf{r}(n). \tag{8.40}$$

Different from the version of Section 8.3.1, cNLMS depends not only on the estimation error $e_c(n)$, but also on $\hat{x}_1(n)$. Since $\hat{x}_1(n)$ depends nonlinearly on $y(n)$, cNLMS is a nonlinear version of NLMS. Moreover, the synchronization between master and slave in chaotic communication system depends on the mitigation of the intersymbol interference, which is the role played by the equalizer.

It is important to notice that we prevent division by zero in the computation of the estimate $\hat{m}(n)$, by making $\hat{m}(n) = y(n)/\delta$ when $\hat{x}_1(n) = 0$. However, $\hat{x}_1(n)$ can assume values too close to zero and in this case, the estimate $\hat{m}(n)$ and the error $e_c(n)$ acquire large values. Apparently, this does not affect the stability of the algorithm since the error $e_c(n)$ is multiplied by $\hat{x}_1(n)$ in the update Equation (8.40), but large values of $\hat{m}(n)$ are not desirable since the transmitted message is assumed to be binary. Therefore, we restricted the estimate $\hat{m}(n)$ to the interval $[-3, +3]$. More restricted intervals can affect the convergence of the algorithm and consequently the synchronization between master and slave. The proposed algorithm is summarized in Table 8.2, where

$$\text{sign}[x] = \begin{cases} -1, & x < 0 \\ 0, & x = 0 \\ 1, & x > 0 \end{cases}$$

is the sign function. Note that the constant δ is used to avoid division by zero in the update equation of the weights, as well as in the computation of the estimate $\hat{m}(n)$. An analysis of cNLMS is necessary to obtain the interval of $\tilde{\mu}_c$ that ensures its stability. We intend to pursue this matter in a future work. It is important to mention that we did not observe divergence in our preliminary simulations for $0 < \tilde{\mu}_c < 0.2$.

8.5 Simulation results

In this section, we show some simulation results for the chaotic communication system of Figure 8.9. We consider the Hénon map of Equation (8.14) with $\alpha = 1.4$ and $\beta = 0.3$, which was proposed in [20] as an example of a dynamical system with a fractal attractor. The dynamical system of Equation (8.14) was initialized in the transmitter and in the receiver as $\mathbf{x}(0) = \mathbf{0}$ and $\hat{\mathbf{x}}(0) = \mathbf{0}$, respectively. In all simulations, we assume the transmission of a binary sequence $m(n) \in \{-1, 1\}$ and equalizers initialized as $\mathbf{w}(0) = \mathbf{0}$. Note that, different from the numerical examples of Section 8.3.2, $m(n) \neq s(n)$ due to the chaotic signal generator. For comparison, we also consider the system of Figure 8.9 without equalizer, in which $y(n) = r(n)$.

We first assume that the encoded sequence $s(n)$ is transmitted through Channel 1 (Table 8.1) with SNR $= 30$ dB, and $\Delta = 0$. The output of the cNLMS equalizer is shown in Figure 8.10(b), the average of the two coefficients and the cMSE along the iterations, estimated by an ensemble-average of 1000

TABLE 8.2
Summary of the cNLMS algorithm applied to the Hénon map with $K = 2$.

Initialize the algorithm by setting:
$$\mathbf{w}(0) = \mathbf{0}, \quad \widehat{\mathbf{x}}(0) = [0\ \ 0]^T$$

$$\mathbf{A} = \begin{bmatrix} 0 & \beta \\ 1 & 0 \end{bmatrix}, \quad \mathbf{b} = [\alpha\ \ 0]^T$$

δ: small positive constant

For $n = 1, 2, 3 \ldots$, compute:

$$y(n) = \mathbf{r}^T(n)\mathbf{w}(n-1)$$

$$\widehat{\mathbf{x}}(n+1) = \mathbf{A}\widehat{\mathbf{x}}(n) + \mathbf{b} + \begin{bmatrix} -y^2(n) & 0 \end{bmatrix}^T$$

if $\ \widehat{x}_1(n) = 0$

$$\widehat{m}(n) = \frac{y(n)}{\delta}$$

else

$$\widehat{m}(n) = \frac{y(n)}{\widehat{x}_1(n)}$$

end

if $\ |\widehat{m}(n)| > 3$

$$\widehat{m}(n) = 3\,\mathrm{sign}[\widehat{m}(n)]$$

end

$$e_c(n) = m(n - \Delta) - \widehat{m}(n)$$

$$\mathbf{w}(n) = \mathbf{w}(n-1) + \frac{\tilde{\mu}_c}{\delta + \|\mathbf{r}(n)\|^2}\widehat{x}_1(n)e_c(n)\mathbf{r}(n)$$

end

---

independent runs and filtered by a moving average filter with 16 coefficients to facilitate the visualization, are shown, respectively, in Figures 8.10(c) and (d). We can observe that the cNLMS converges in the mean to the Wiener solution $\mathbf{w}_o \approx [1 \quad 0.6]^T$, whose coefficients are shown as dashed lines in Figure 8.10(c). Therefore, the equalizer is working as expected since this solution mitigates the intersymbol interference. The received sequence $r(n)$ is shown in Figure 8.10(a) and the respective cMSE in Figure 8.10(d). Comparing to Figure 8.7, we can observe that, although cNLMS converges to the Wiener solution, the steady-state cMSE is approximately -13 dB, being superior than that of the case with no chaos. The effect of this superior error can be noticed in the estimated sequence $\widehat{m}(n)$. In Figure 8.10(b), this estimate

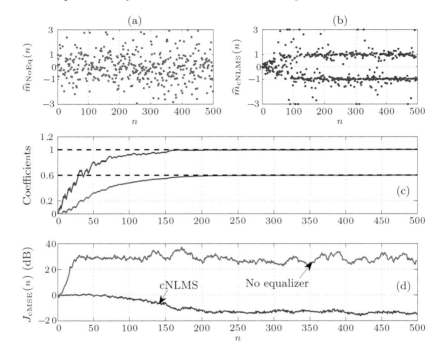

FIGURE 8.10
Estimated sequence considering (a) no equalization and (b) cNLMS ($\mu = 0.1$, $\delta = 0.5$); (c) average of the coefficients of cNLMS and Wiener solution (dashed lines); (d) estimated cMSE; ensemble-average of 1000 independent runs, Hénon map ($\alpha = 1.4$, $\beta = 0.3$), Channel 1, SNR = 30 dB; $M = 2$; $\Delta = 0$.

varies in the interval $[-3, +3]$, whereas in the case of Figure 8.7(b) the output of the equalizer varies little around -1 and $+1$. Again, the communication is completely lost in the case with no equalizer.

Assuming now that the encoded sequence $s(n)$ is transmitted through Channel 2 ($H_2(z) = 0.3 + z^{-1} + 0.3z^{-2}$) with SNR = 30 dB and $\Delta = 3$, we obtain the results shown in Figure 8.11. cNLMS converges again to the Wiener solution given by Equation (8.31) and the transmitted message $m(n - \Delta)$ can be recovered with a relatively low bit error rate (BER) at the output of a decision device by using the estimated sequence $\widehat{m}(n)$. Comparing to the results of Figure 8.8, the chaotic communication system introduces more errors in the recovery of the transmitted message, since the cMSE is higher than the MSE and the estimate sequence varies in the interval $[-3, +3]$. Additionally, the importance of restricting $\widehat{m}(n)$ in that interval becomes more clear in this case. Again, the equalizer plays an important role to mitigate the intersymbol interference since the performance of the system without equalizer is much worse.

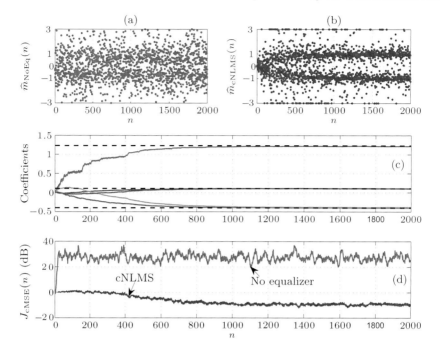

FIGURE 8.11
Estimated sequence considering (a) no equalization and (b) cNLMS ($\mu = 0.1$, $\delta = 0.1$); (c) average of the coefficients of cNLMS and Wiener solution (dashed lines); (d) estimated cMSE; ensemble-average of 1000 independent runs, Hénon map ($\alpha = 1.4$, $\beta = 0.3$), Channel 2 ($H_2(z) = 0.3 + z^{-1} + 0.3z^{-2}$), SNR = 30 dB; $M = 5$; $\Delta = 3$.

Considering SNR = 30 dB, $M = 5$, $\Delta = 3$, and the channel $H_3(z) = h_0 + z^{-1} + h_0 z^{-2}$, $0 \leq h_0 \leq 0.5$, we obtain BER curves as a function of h_0, as shown in Figure 8.12. The BER was estimated after the convergence of the algorithms and counting the number of errors when comparing the transmitted sequence with the sequence obtained at the output of a decision device with zero threshold. We disregarded 8×10^4 bits due to the initial convergence and used 8×10^5 bits in the computation of the BER. Furthermore, it is important to remark that in the case with no equalizer, the delay is due only to the channel. Therefore, we compared the recovered sequence with the transmitted one, assuming $\Delta = 1$ in this case. The smaller the value of h_0 the lower the intersymbol interference introduced by the channel. The communication system (with no chaos) of Figure 8.4 with the equalizer coefficients adapted with NLMS is also considered here for comparison. In this case, $m(n) \equiv s(n)$ and the estimate of the transmitted sequence is obtained by passing the output of the equalizer $y(n)$ through a decision device. We can observe that cNLMS outperforms the case with no equalizer for $h_0 \geq 0.05$, providing a reasonable

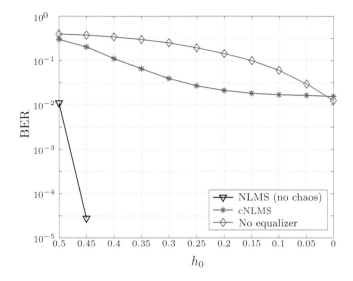

FIGURE 8.12
Bit error rate as a function of the channel $H_3(z) = h_0 + z^{-1} + h_0 z^{-2}$ with SNR = 30 dB; cNLMS and NLMS ($\mu = 0.1$, $\delta = 0.5$); Hénon map ($\alpha = 1.4$, $\beta = -0.3$); $M = 5$; $\Delta = 3$.

BER. The case with no equalizer achieves a BER slighter lower than that of cNLMS only for the ideal channel ($h_0 = 0$). On the other hand, the system with no chaos presents the lowest BER in this case. Although the chaotic communication system presents a BER higher than that of the system with no chaos, it has some intrinsic advantages of chaotic signals as spread spectrum and the possibility of a substantial security level (see, e.g., [3,8,23,25]). M-ary transmission with chaos can also be an alternative to improve efficiency of chaos-based communications in dispersive channels.

8.6 Conclusion and perspectives

The chaos-based communication system proposed by Wu and Chua [44] has been considered for practical optical communication systems. However, its performance in non-ideal channels is prohibitive: even minor channel imperfections are enough to hinder communication if no channel equalization is inserted. Thus, it can perform with acceptable BER only in almost ideal and very controlled environments, as the optical link of [3].

In this chapter, we have extended the NLMS algorithm so that it can be applied to this problem. Preliminary simulations show that the proposed algorithm can successfully permit chaotic communications. As far as we are concerned this is the first adaptive scheme proposed for this chaos-based chaotic system. As to current and future works, the influence of the maps, and their parameters in the success of equalization, other techniques and more challenging channels will be investigated.

Bibliography

[1] R. P. Agarwal. *Difference equations and inequalities*, volume 155 of *Monographs and Textbooks in Pure and Applied Mathematics*. Marcel Dekker Inc., New York, 1992. Theory, methods, and applications.

[2] K. T. Alligood, T. Sauer, and J. A. Yorke. *Chaos: An Introduction to Dynamical Systems*. Textbooks in Mathematical Sciences. Springer, New York, 1996.

[3] A. Argyris, D. Kanakidis, and A. Bogris. Spectral synchronization in chaotic optical communication systems. *IEEE Journal of Quantum Electronics*, 41:892–897, 2005.

[4] S. Bouchired, D. Roviras, and F. Castani. Equalisation of satellite mobile channels with neural network techniques. *Space Communications*, 15:209–220, 1999.

[5] S. Chen, B. Mulgrew, and S. McLaughlin. Adaptive Bayesian equalizer with decision feedback. *IEEE Transactions on Signal Processing*, 41(9):2918–2927, September 1993.

[6] K. M. Cuomo and A. V. Oppenheim. Circuit implementation of synchronized chaos with applications to communications. *Phys. Rev. Lett.*, 71(1):65–68, July 1993.

[7] K. M. Cuomo, A. V. Oppenheim, and R. J. Barron. Channel equalization for self-synchronizing chaotic systems. In *Proc. of IEEE Int. Conf. on Acoustics, Speech, and Signal Processing (ICASSP)*, volume 3, pages 1605–1608, May 1996.

[8] T. Deng, G.-Q. Xia, L.-P. Cao, J.-G. Chen, X.-D. Lin, and Z.-M. Wu. Bidirectional chaos synchronization and communication in semiconductor lasers with optoelectronic feedback. *Optics Communications*, 282(11):2243–2249, June 2009.

[9] Z. Ding and Y. Li. *Blind Equalization and Identification*. Marcel Dekker, New York, 2001.

[10] M. Eisencraft, R. D. Fanganiello, and L. A. Baccala. Synchronization of discrete-time chaotic systems in bandlimited channels. *Mathematical Problems in Engineering*, 2009:207971, May 2009.

[11] M. Eisencraft, R. D. Fanganiello, J. M. V. Grzybowski, D. C. Soriano, R. R. F. Attux, A. M. Batista, E. E. N. Macau, L. H. A. Monteiro, J. M. T. Romano, R. Suyama, and T. Yoneyama. Chaos-based communication systems in non-ideal channels. *Communications in Nonlinear Science Numerical Simulation*, 17(12):4707–4718, 2012.

[12] M. Eisencraft, R. D. Fanganiello, and L. H. A. Monteiro. Chaotic synchronization in discrete-time systems connected by bandlimited channels. *IEEE Communications Letters*, 15(6):671–673, 2011.

[13] B. Farhang-Boroujeny. *Adaptive Filters - Theory and Applications*. John Wiley & Sons, West Sussex, 1998.

[14] M. Feki, B. Robert, G. Gelle, and M. Colas. Secure digital communication using discrete-time chaos synchronization. *Chaos, Solitons & Fractals*, 18(4):881–890, 2003.

[15] G. D. Forney Jr. Maximum-likelihood sequence estimation of digital sequences in the presence of intersymbol interference. *IEEE Transactions on Information Theory*, IT-18, May 1972.

[16] G. J. Gibson, S. Siu, and C. F. N. Cowan. The application of nonlinear structures to the reconstruction of binary signals. *IEEE Transactions on Signal Processing*, 39(8):1877–1884, August 1991.

[17] D. Godard. Self-recovering equalization and carrier tracking in two-dimensional data communication systems. *IEEE Transactions on Communications*, 28(11):1867–1875, 1980.

[18] A. Gutierrez and W. E. Ryan. Performance of volterra and MLSD receivers for nonlinear band-limited satellite systems. *IEEE Transactions on Communications*, 48:1171–1177, July 2000.

[19] S. Haykin. *Adaptive Filter Theory*. Prentice-Hall, Upper Saddle River, 4th edition, 2002.

[20] M. Hénon. A two-dimensional mapping with a strange attractor. *Communications in Mathematical Physics*, 50:69–77, 1976.

[21] L. Illing. Digital communication using chaos and nonlinear dynamics. *Nonlinear Analysis: Theory, Methods & Applications*, 71(12):E2958–E2964, December 2009.

[22] C. R. Johnson Jr. et al. Blind equalization using the constant modulus criterion: a review. *Proceedings of the IEEE*, 86:1927–1950, October 1998.

[23] M. P. Kennedy and G. Kolumbán. Digital communications using chaos. *Signal Processing*, 80(7):1307–1320, 2000.

[24] A. A. Koronovskii, O. I. Moskalenko, and A. E. Hramov. On the use of chaotic synchronization for secure communication. *Phys.-Usp.*, 52(12):1213–1238, 2009.

[25] F. C. M. Lau and C. K. Tse. *Chaos-Based Digital Communication Systems*. Springer, New York, 2003.

[26] H Leung. System identification using chaos with application to equalization of a chaotic modulation system. *IEEE Transactions on Circuits and Systems I*, 45(3):314–320, March 1998.

[27] N. Li, W. Pan, B. Luo, L. Yan, X. Zou, N. Jiang, and S. Xiang. High bit rate fiber-optic transmission using a four-chaotic-semiconductor-laser scheme. *IEEE Photonics Technology Letters*, 24(12):1072–1074, June 2012.

[28] W. Liu, J. C. Príncipe, and S. Haykin. *Kernel Adaptive Filtering: A Comprehensive Introduction*. Wiley, New York, 2010.

[29] B. Mulgrew. Applying radial basis function. *IEEE Signal Processing Magazine*, 13:50–65, March 1996.

[30] A. V. Oppenheim and A. S. Willsky. *Signals and Systems*. Prentice-Hall, Hoboken, NJ, 2nd edition, 1996.

[31] D.-C. Park and T.-K. J. Jeong. Complex-bilinear recurrent neural network for equalization of a digital satellite channel. *IEEE Transactions on Neural Networks*, 13(3):711–725, May 2002.

[32] L. M. Pecora and T. L. Carroll. Synchronization in chaotic systems. *Phys. Rev. Lett.*, 64:821–824, February 1990.

[33] S. U. H. Qureshi. Adaptive equalization. *Proceedings of the IEEE*, 73(9):1349–1387, September 1985.

[34] A. A. Rontogiannis and K. Berberidis. Efficient decision feedback equalization for sparce wireless channel. *IEEE Transactions on Wireless Communications*, 2:570–581, May 2003.

[35] N. F. Rulkov and L. S. Tsimring. Synchronization methods for communication with chaos over band-limited channels. *International Journal of Circuit Theory and Applications*, 27(6):555–567, 1999.

[36] I. Santamaria, C. Pantaleon, L. Vielva, and J. Ibanez. Blind equalization of constant modulus signals using support vector machines. *IEEE Transactions on Signal Processing*, 52(6):1773–1782, June 2004.

[37] A. H. Sayed. *Adaptive Filters*. John Wiley & Sons, Hoboken, NJ, 2008.

[38] C. Z. W. H. Sweatman, B. Mulgrew, and G. J. Gibson. Two algorithms for neural-network design and training with application to channel equalization. *IEEE Transactions on Neural Networks*, 9(3):533–543, May 1998.

[39] S. Theodoridis, C. F. N. Cowan, C. P. Callender, and C. M. S. See. Schemes for equalisation of communication channels with nonlinear impairments. *IEE Proceedings - Communications*, 142(3):165–171, June 1995.

[40] J. Treichler and B. Agee. New approach to multipath correction of constant modulus signals. *IEEE Transactions on Acoustics, Speech and Signal Processing*, ASSP-31(2):459–472, 1983.

[41] J. R. Treichler, I. Fijalkow, and C. R. Johnson Jr. Fractionally spaced equalizers. *IEEE Signal Processing Magazine*, 13:65–81, May 1996.

[42] J. Tsimbinos and L.B. White. Error propagation and recovery in decision-feedback equalizers for nonlinear channels. *IEEE Transactions on Communications*, 49:239–242, February 2001.

[43] C. Williams. Chaotic communications over radio channels. *IEEE Transactions on Circuit and Systems I*, 48(12):1394–1404, December 2001.

[44] C. W. Wu and L. O. Chua. A simple way to synchronize chaotic systems with applications to secure communication systems. *International Journal of Bifurcation and Chaos*, 3(6):1619–1627, December 1993.

[45] Z. Zhu and H. Leung. Adaptive blind equalization for chaotic communication systems using extended-kalman filter. *IEEE Transactions on Circuit and Systems I*, 48(8):979–989, August 2001.

9

Chaotic communications in bandlimited channels

Renato D. Fanganiello

Escola de Engenharia da Universidade Presbiteriana Mackenzie

Rodrigo T. Fontes and Marcio Eisencraft

Escola Politécnica, Universidade de São Paulo

Luiz H. A. Monteiro

Escola de Engenharia da Universidade Presbiteriana Mackenzie
Escola Politécnica da Universidade de São Paulo

CONTENTS

In chaos-based communications, it is essential to know and to control the spectral features of the chaotic signals to be transmitted, since every communication channel is bandlimited to some extent. In this chapter, we review a proposal of discrete-time communication system using chaotic synchronization in which the spectrum of the transmitted signals can be moderately shaped. Numerical studies on the Lyapunov exponents of the generated signals are presented and related to the performance of this chaotic communication system.

9.1 Introduction

One of the possible motivations for using chaos in communication systems is the fact that these signals can be broadband [5,10]. In systems in which chaotic signals modulate independent narrowband sources, as the ones proposed in [1,13], the resulting signals usually have a bandwidth larger than what would be considered necessary to their transmission. Although this can look like a waste in a first approach, the use of a larger bandwidth is a well-known technique in Telecommunications called *Spread Spectrum* (SS). An SS system must satisfy the following criteria [6,12]:

- the bandwidth of the transmitted signal must be much larger than the minimum amount necessary for transmitting the corresponding information;

- this bandwidth is determined from a spreading signal that is independent of the information;

- in the receiver, the original data are recovered from the correlation between the spreading signal and a synchronized local replica of such a signal.

Thus, the communication systems based on the Wu and Chua's synchronization scheme [1, 13] can be classified as an SS system. Therefore, these systems present the same advantages of the conventional SS systems.

The SS modulation was first developed for military applications, in which the resistance to intentional perturbations is of utmost importance. There are also civilian applications that can be benefited from the unique characteristics of SS modulation. For instance, it can be used to mitigate multipath interferences in wireless channels and to increase the system capacity in terms of number of users [6,12].

When it comes to practical applications, it is not enough to qualify a chaotic signal as "broadband" as almost all chaos-based communication texts do. Because every communication channel is bandlimited, it is fundamental to quantify the effective bandwidth [9] of the transmitted signals. Just a few papers on the bandwidth characteristics of chaotic signals have been published so far (see, e.g., [3,4,8]) and the spectrum characterization of chaos is still an open problem.

In [1], a technique for band limiting discrete-time chaotic signals before transmission in a bandlimited channel was proposed. The main idea is to use identical digital filters in the feedback loops of the transmitter and the receiver so that the synchronization is not affected. However, the Lyapunov exponent of the generated signals can become negative; consequently, the generated signals can cease to be chaotic [2].

In this chapter, we review this technique and present numerical studies on the behavior of the generated signals when the values of filter parameters are varied.

The chapter is organized as follows. In Section 9.2, we review the chaos-based communication system for bandlimited channels proposed in [1]. Numerical calculations of the largest Lyapunov exponents of the involved signals and associated discussions are presented in Section 9.3. Conclusions and perspectives are drafted in Section 9.4.

9.2 Chaos-based communications in bandlimited channels

Chapter 8 presented, among other important concepts, the communication system proposed by Wu and Chua [13], Equations (8.8)-(8.9), which define the transmitter and the receiver systems, and the condition which makes the synchronization between the two systems possible, given by Equation (8.5). It was also shown that, in a non-ideal channel, communication is impaired and the original message can be no longer recovered.

In the particular case of an ideal but bandlimited channel, a way to circumvent the signal degradations imposed by the bandwidth limitations in the communication channel is to insert filters $H_S(\omega)$ in the feedback loops of both transmitter and receiver, so that the transmitted signal power is contained within the channel bandwidth [1].

Figure 9.1 shows a block diagram of the communication system proposed in [1]. By considering that $H_S(\omega)$ represents a finite impulse response (FIR)

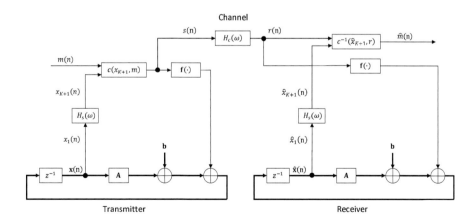

FIGURE 9.1
Block diagram of the communication system proposed for bandlimited channels.

filter of order N_S, its output is given by

$$x_{K+1}(n) = c_0 x_1(n) + c_1 x_1(n-1) + \ldots + c_{N_S} x_1(n - N_S), \qquad (9.1)$$

where $c_0, c_1, \ldots, c_{N_S}$ are the filter coefficients. Hence, the dimension of both the transmitter and the receiver systems is $K + N_S$, where K is the dimension of these systems in the absence of filters in their feedback loops.

Figure 9.1 suggests that when $H_S(\omega) = 1$ for every ω, the transmitter and the receiver systems will synchronize as described in Chapter 8, Section 8.2, because $x_{K+1}(n) = x_1(n)$. However, when $H_S(\omega) \neq 1$, $x_{K+1}(n)$ is the map component that is actually combined with a sample of the message to be transmitted and then fed back into the chaotic signal generation stage of the transmitter. So it is relevant to perform an analysis on how the filter $H_S(\omega)$ affects the characteristics of the transmitted signal and the synchronization. In other words, it is important to evaluate if the presence of the filter $H_S(\omega)$ can impair the synchronization, and which combinations of filter order and cutoff frequency allow the signal $x_{K+1}(n)$ to be chaotic. This latter goal is presented in Section 9.3 of this chapter.

Now, we perform an analytic evaluation of the synchronization of this communication system by using the Hénon map [7] given by

$$\begin{cases} x_1(n) = 1 - \alpha x_1^2(n-1) + x_2(n-1) \\ x_2(n) = \beta x_1(n-1). \end{cases} \qquad (9.2)$$

When the feedback loop filter rule Equation (9.1) is included, the transmitter is given by

$$\begin{cases} x_1(n) = 1 - \alpha x_3^2(n-1) + x_2(n-1) \\ x_2(n) = \beta x_1(n-1) \\ x_3(n) = c_0 x_1(n) + c_1 x_1(n-1) + \ldots + c_{N_S} x_1(n - N_S). \end{cases} \qquad (9.3)$$

We can rewrite $x_3(n)$ as a function of samples at the instant $n-1$ using the following set of auxiliary equations:

$$\begin{cases} x_4(n) = x_1(n-1) \\ x_5(n) = x_4(n-1) \\ \vdots \\ x_N(n) = x_{N-1}(n-1), \end{cases} \qquad (9.4)$$

where $N = N_S + 2$ is the dimension of the new system which is formed after inserting the filter $H_S(\omega)$ in the transmitter feedback loop. The same holds for the receiver equations.

In this case, the system of equations related to the transmitter can be rewritten in the form of Equation (8.8) with $\mathbf{A}_{N \times N}$ as

$$
\mathbf{A} = \begin{bmatrix}
0 & 1 & 0 & 0 & & 0 & 0 \\
\beta & 0 & 0 & 0 & & 0 & 0 \\
c_1 & c_0 & 0 & c_2 & \cdots & c_{N_S-1} & c_{N_S} \\
1 & 0 & 0 & 0 & & 0 & 0 \\
0 & 0 & 0 & 1 & & 0 & 0 \\
& \vdots & & & \ddots & \vdots & \\
0 & 0 & 0 & 0 & \cdots & 1 & 0
\end{bmatrix}. \tag{9.5}
$$

The eigenvalues λ of this matrix are

$$
\begin{cases}
\lambda_{1,2} & = \pm\sqrt{\beta} \\
\lambda_{3,\dots,N} & = 0.
\end{cases} \tag{9.6}
$$

Therefore, chaotic synchronization occurs if $|\beta| < 1$, *independently of the filter coefficients* and of the message $m(n)$ [2].

Even though the coefficients and the order of the filter $H_S(\omega)$ (inserted in the transmitter and in the receiver systems) do not affect the synchronization of this communication system, there are no guarantees that the signal $x_{K+1}(n)$, which corresponds to the transmitter filter output, is chaotic. In principle, this signal could either be chaotic, or periodic, or even diverge toward infinity.

In Section 9.3, this issue is addressed and simulation results regarding the largest Lyapunov exponent of the generated signals are presented. Numerical studies performed in order to determine the conditions for which obtaining a chaotic $x_{K+1}(n)$ is possible are also shown.

9.3 Numerical studies

As pointed out in the previous section, the synchronization in the communication system remains immune to the presence of the filters $H_S(\omega)$ inserted in the transmitter and in the receiver, but no conclusion is straightforward regarding the features of the filter output signal $x_{K+1}(n)$.

In order to investigate the influence of a linear time-invariant low-pass FIR filter on the signal $x_{K+1}(n)$, the Hénon map with parameter values $\alpha = 1.4$ and $\beta = 0.3$ was used, which are numbers commonly found in the literature [7].

Computer simulations were performed so that the variation of the largest Lyapunov exponent h of the map of the transmitter could be analyzed as a function of the filter order N_S and its cutoff frequency ω_c. The filters were projected by employing the classical method of windowed linear-phase FIR digital filter design and Hamming window [11]. The results are exhibited in Figure 9.2.

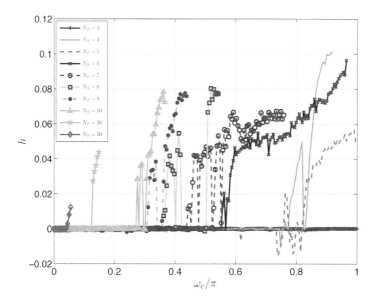

FIGURE 9.2
Largest Lyapunov exponent h of the transmitter system as a function of the
cutoff frequency ω_c for $\alpha = 1.4$, $\beta = 0.3$, and different filter orders N_S.

In this figure, the vertical axis presents the largest values of Lyapunov
exponent of the transmitter system, while the horizontal axis denotes the
normalized cutoff frequency of the filters that were chosen in the simulations.
The largest value for the normalized cutoff frequency, $\omega_c/\pi = 1$, is equivalent
to half the sampling rate used in the communication system. For values of
ω_c for which no Lyapunov exponent is indicated, the map dynamics diverged
towards infinity.

By analyzing Figure 9.2, we can realize that it becomes increasingly difficult to obtain a positive Lyapunov exponent as the order of the filter grows.
In other words, it becomes increasingly difficult to obtain a chaotic signal that
can be used in the proposed communication system.

In view of this fact, we realized the possibility of another value existing
for the α parameter of the Hénon map that would allow generating chaotic
signals in the presence of high-order FIR filters. To answer this question, a
new set of computer simulations was performed maintaining the parameter
$\beta = 0.3$, choosing a fixed order N_S and cutoff frequencies ω_c, and numerically
evaluating h as a function of α. Thus, it became possible to verify which values
of α generate chaotic signals.

Figure 9.3 shows the results obtained from computer simulations in which
the filter order was kept constant at $N_S = 20$ and three cutoff frequencies were
taken: $\omega_c = 0.3\pi$, $\omega_c = 0.5\pi$, and $\omega_c = 0.8\pi$. Again, for values of α for which
the Lyapunov exponent is not indicated, the map dynamics diverged towards

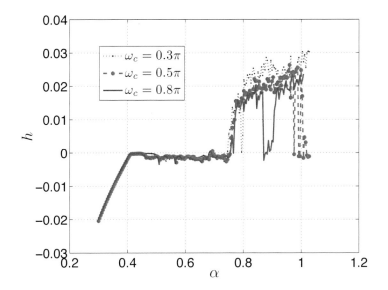

FIGURE 9.3
Largest Lyapunov exponent h of the transmitter system as a function of α, for $\beta = 0.3$, filter order $N_S = 20$, and different cutoff frequencies.

infinity. In light of these results, we chose $\alpha = 0.9$ to carry on the simulations. So now we have a pair of parameter values of the Hénon map which may not necessarily lead to chaotic signals in the absence of the filter $H_S(\omega)$, but that certainly will lead to chaotic signals in the presence of a low-pass FIR filter of order $N_S = 20$.

Next, we performed more simulations for evaluating the variation of h in terms of N_S and ω_c to find out for which other low-pass FIR filters it is possible to obtain chaotic signals. Figure 9.4 shows the results we obtained for the filter order varying from $N_S = 10$ up to $N_S = 100$. It is important to notice that, in this case, there is no value of cutoff frequency causing the map implemented in the transmitter to diverge; and, therefore, communication between transmitter and receiver will always be possible, even though the generated signals may not always be chaotic. The information depicted in Figure 9.4 is summarized in Figure 9.5 in a binary representation of the Lyapunov exponents calculated for each low-pass filter used in the simulations, in which light and dark gray tones represent positive and negative Lyapunov exponents, respectively [2].

In Figure 9.5, we observe large ranges of cutoff frequencies that allow to obtain chaotic signals, even for high-order filters, such as $N_S = 80$ or $N_S = 100$.

In order to verify that these results are well grounded, we implemented a communication system with $\alpha = 0.9$ and $\beta = 0.3$ for the Hénon map, and we inserted low-pass filters of order $N_S = 30$ and cutoff frequency $\omega_c = 0.5\pi$

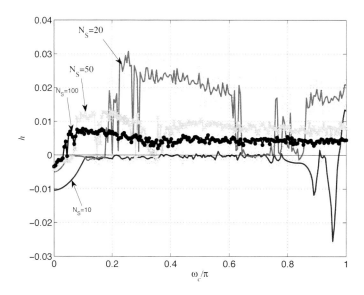

FIGURE 9.4
Largest Lyapunov exponent h of the transmitter system as a function of the normalized cutoff frequency ω_c of the transmitter filter for $\alpha = 0.9$, $\beta = 0.3$, and different filter orders N_S. The solid lines represent the average values; the dotted lines, the intervals determined from the standard deviations.

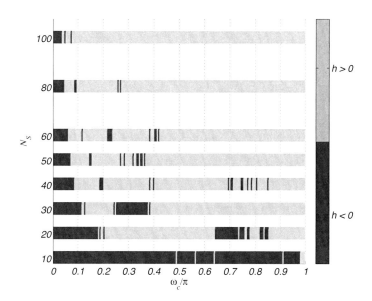

FIGURE 9.5
Chaotic (light gray) and non-chaotic (dark gray) regions for different cutoff frequencies ω_c and filter orders N_S.

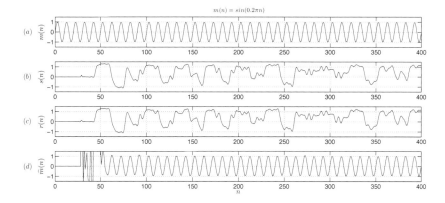

FIGURE 9.6
(a) Message $m(n)$; (b) transmitted signal $s(n)$; (c) received signal $r(n)$; (d) recovered message $\widehat{m}(n)$. A FIR filter $H_S(\omega)$ with $N_S = 30$ and $\omega_c = 0.5\pi$ is used (chaotic condition) and the cutoff frequency of the channel is $\omega_c = 0.8\pi$.

in the transmitter and in the receiver. Notice that these values are in the light gray region of Figure 9.5 leading to chaotic signals. For modeling the communication channel, we used a low-pass FIR filter of order $N_C = 50$ and cutoff frequency $\omega_c = 0.8\pi$. The message to be transmitted was chosen to be the sinusoidal signal

$$m(n) = \sin(0.2\pi n), \tag{9.7}$$

and the coding function was given by

$$s(n) = c(x_1(n), m(n)) = x_1(n) + 0.01m(n), \tag{9.8}$$

so the original message can be recovered by using its inverse function; that is

$$\widehat{m}(n) = c^{-1}(\widehat{x}_1(n), r(n)) = 100(r(n) - \widehat{x}_1(n)), \tag{9.9}$$

based on Equations (8.6)-(8.7), explained in Chapter 8.

Figures 9.6 and 9.7 present the time-domain and frequency-domain representations of the signals involved in this simulation. The frequency response of the communication channel and of the filters inserted are shown in Figures 9.7(b) and (c) by dashed lines. The transmitted and received signals exhibit the aperiodic and broadband features of chaos.

Figures 9.8 and 9.9 reveal the consequence of choosing a cutoff frequency in the negative maximum Lyapunov exponent region of Figure 9.5. Here we took $\omega_c = 0.35\pi$ and $N_S = 30$. As shown in time and frequency domains, the message is correctly recovered but the transmitted signals are not chaotic anymore.

The complex behavior glimpsed in Figure 9.5 must be further analytically studied. For the chosen value of α, it seems that the higher the filter order,

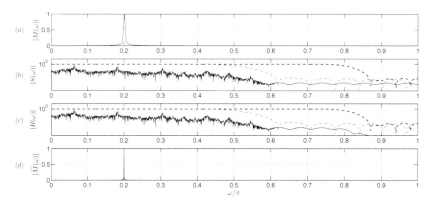

FIGURE 9.7

Normalized power spectrum density representation of the signals in Figure 9.6:
(a) message; (b) transmitted signal; (c) received signal; (d) recovered message.

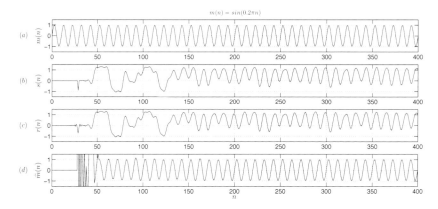

FIGURE 9.8

(a) Message $m(n)$; (b) transmitted signal $s(n)$; (c) received signal $r(n)$; (d)
recovered message $\widehat{m}(n)$. A FIR filter $H_S(\omega)$ with $N_S = 30$ and $\omega_c = 0.35\pi$ is
used (periodic condition) and the cutoff frequency of the channel is $\omega_c = 0.8\pi$.

the larger the range of ω_c that provides chaotic behavior. This inference leads
to the idea of using infinite impulse response filters instead of FIR filters in
the feedback loops. These are research lines currently being followed.

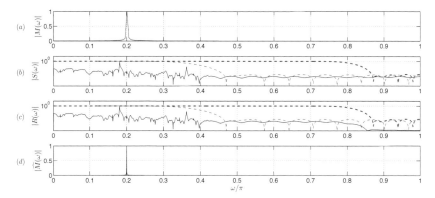

FIGURE 9.9
Normalized power spectrum density representation of the signals illustrated in Figure 9.8: (a) message; (b) transmitted signal; (c) received signal; (d) recovered message.

9.4 Conclusion and perspectives

In this chapter, we explored the idea of using discrete-time linear time-invariant FIR filters to shape the spectrum properties of the chaotic signals generated in the Wu and Chua's communication system [1, 2]. Using these filters does not change the chaotic synchronization, but it can make the involved signals become periodic or even diverge. Numerical experiments showed that the behavior of the largest Lyapunov exponent of the orbits is erratic with the order and the cutoff frequency of the filter. However, when convenient parameter values of the map are chosen, it is possible to obtain chaos for a large range of orders and cutoff frequencies. To analytically address these results is a current work of some of the authors. We also look for more general results not dependent on the used map.

9.5 Acknowledgments

M. Eisencraft and L. H. A. Monteiro are partially supported by CNPq.

Bibliography

[1] M. Eisencraft, R. D. Fanganiello, and L. Baccalá. Synchronization of discrete-time chaotic systems in bandlimited channels. *Mathematical Problems in Engineering*, 2009:1–12, 2009.

[2] M. Eisencraft, R. D. Fanganiello, and L. H. A. Monteiro. Chaotic synchronization in discrete-time systems connected by bandlimited channels. *IEEE Communications Letters*, 15(6):671–673, June 2011.

[3] M. Eisencraft and D. M. Kato. Spectral properties of chaotic signals with applications in communications. *Nonlinear Analysis: Theory, Methods & Applications*, 71(12):e2592–e2599, 2009.

[4] M. Eisencraft, D. M. Kato, and L. H. A. Monteiro. Spectral properties of chaotic signals generated by the skew tent map. *Signal Process.*, 90:385–390, 2010.

[5] J. Grzybowski, M. Eisencraft, and E. Macau. Chaos-based communication systems: current trends and challenges. In S. Banerjee, M. Mitra, and L. Rondoni, editors, *Applications of Chaos and Nonlinear Dynamics in Engineering - Vol. 1*, volume 71 of *Understanding Complex Systems*, pages 203–230. Springer, Berlin/Heidelberg, 2011.

[6] S. Haykin. *Communication Systems*. Wiley, New York, 4th edition, 2000.

[7] M. Hénon. A two-dimensional mapping with a strange attractor. *Communications in Mathematical Physics*, 50:69–77, 1976.

[8] S. H. Isabelle and G. W. Wornell. Statistical analysis and spectral estimation techniques for one-dimensional chaotic signals. *IEEE Transactions on Signal Processing*, 45(6):1495–1506, June 1997.

[9] B. P. Lathi. *Modern Digital and Analog Communication Systems*. Oxford University Press, New York, 1998.

[10] F. C. M. Lau and C. K. Tse. *Chaos-Based Digital Communication Systems*. Springer, Berlin, Heidelberg, 2003.

[11] A. V. Oppenheim and R. W. Schafer. *Discrete-Time Signal Processing*. Prentice-Hall signal processing series. Prentice-Hall, Upper Saddle River, NJ, 2010.

[12] B. Sklar. *Digital Communications*. Prentice-Hall P T R, Upper Saddle River, NJ, 2nd edition, 2004.

[13] C. W. Wu and L. O. Chua. A simple way to synchronize chaotic systems with applications to secure communication systems. *International Journal of Bifurcation and Chaos*, 3(6):1619–1627, December 1993.

10

Using coupled maps to improve synchronization performance

Greta A. Abib

Centro de Engenharia, Modelagem e Ciências Sociais Aplicadas (CECS)
Universidade Federal do ABC (UFABC)

Marcio Eisencraft

Escola Politécnica, Universidade de São Paulo

Antonio M. Batista

Departamento de Matemática e Estatística
Universidade Estadual de Ponta Grossa

CONTENTS

10.1 Introduction

Many of the possible applications of chaotic systems in communications rely on master-slave chaotic synchronization (see, e.g., [14, 15, 17]). This way, to study the robustness of chaotic synchronization under additive noise is fundamental when it comes to practical applications of chaos in communication channels where noise is always present.

In this chapter we access the performance of a chaos-based digital communication system proposed in the literature [1, 7] under additive white Gaussian noise (AWGN). As the results are far from that obtained in conventional digital communication systems, we review a way to increase robustness of chaotic synchronization using map lattices. The presented results can increase the possibility of applying chaos in practical environments.

The chapter is organized as follows: in Section 10.2 a digital chaos-based communication system [7] is studied in terms of bit error rate (BER) performance and synchronization time under AWGN. In Section 10.3, coupled maps lattices and chaotic synchronization are reviewed and in 10.4 its performance under AWGN is studied and compared with using a single map [6]. Finally, some conclusions and perspectives are drafted in Section 10.5.

10.2 Performance of a digital chaos-based communication system

Due to the frequent appearance of new challenges in Telecommunications, studying new techniques and ideas that do not fit into current commercial systems is relevant as a research subject.

Bearing this in mind, chaotic signals have been proposed as broadband information transmitters with the potential to provide a high level of robustness and privacy on data transmission [3]. However, chaos-based communication systems do not reach the performance of conventional systems in terms of BER [16].

In this section, we evaluate the performance of a binary communication system using chaotic synchronization based on the one proposed in [7] when the communication channel introduces AWGN. We tested various maps and encoding and decoding functions and the results are measured in terms of the transient time and BER.

10.2.1 The communication system studied

The digital chaos-based communication system studied here is the same as Section 8.2 of Chapter 8 in this book. This way, we will not enter in the details here and we will use exactly the same notation employed there.

In that system, it is assumed that the master can be expressed as

$$\mathbf{x}(n+1) = \mathbf{A}\mathbf{x}(n) + \mathbf{b} + \mathbf{f}(\mathbf{x}(n)), \qquad (10.1)$$

while the slave system, which depends on $\mathbf{x}(n)$, is expressed as

$$\mathbf{y}(n+1) = \mathbf{A}\mathbf{y}(n) + \mathbf{b} + \mathbf{f}(\mathbf{x}(n)), \qquad (10.2)$$

with $n = 0, 1, 2, 3 \ldots$, $\{\mathbf{x}(n), \mathbf{y}(n)\} \subset \mathbb{R}^K$, $\mathbf{x}(n) = [x_1(n), \ldots, x_k(n)]^T$, $\mathbf{y}(n) = [y_1(n), \ldots, y_k(n)]^T$. The matrix $\mathbf{A}_{K \times K}$ and the vector $\mathbf{b}_{K \times 1}$ are constants. The function $\mathbf{f}(\cdot)$ from $\mathbb{R}^K \to \mathbb{R}^K$ is nonlinear in general.

The master and slave systems synchronize when the synchronism error,

$\mathbf{e}(n) \triangleq \mathbf{y}(n) - \mathbf{x}(n)$, tends to zero with the evolution of n; in other words, $\lim_{n\to\infty} \mathbf{e}(n) = \mathbf{0}$. Using Equations (10.1) and (10.2),

$$\mathbf{e}(n+1) = \mathbf{y}(n+1) - \mathbf{x}(n+1) = \mathbf{A}(\mathbf{y}(n) - \mathbf{x}(n)) = \mathbf{A}\mathbf{e}(n). \quad (10.3)$$

Thus, in order for the synchronism to occur, it is sufficient that the eigenvalues λ_i of $\mathbf{A}$ satisfy [7]

$$|\lambda_i| < 1, \ 1 \le i \le K. \quad (10.4)$$

The communication system considered is shown in Figure 10.1. In the analysis and simulations we consider a memoryless channel with AWGN so that the received signal is given by

$$r(n) = s(n) + \eta(n), \quad (10.5)$$

where $\eta(n)$ is zero-mean AWGN with power σ_η^2.

We considered two pairs of coding and decoding functions $c^{(1)}, d^{(1)}$ and $c^{(2)}, d^{(2)}$, given by

$$\begin{cases} c^{(1)}(x_1, m) = (1 - \gamma)x_1 + \gamma m \\ d^{(1)}(y_1, r) = \frac{r - (1-\gamma)y_1}{\gamma} \end{cases} \quad (10.6)$$

with $\gamma = 0.001$ and

$$\begin{cases} c^{(2)}(x_1, m) = x_1 \cdot m \\ d^{(2)}(y_1, r) = r/y_1 \end{cases} . \quad (10.7)$$

In the simulations, the decoding function $d^{(2)}(y_1, r)$ has been implemented as the product $d^{(2)}(y_1, r) = y_1 \cdot r$ to avoid numerical problems. In this case, the use of multiplication is equivalent to dividing because we considered that $m(n) \in \{-1, 1\}$, as Chapter 8 does.

For the sake of BER calculations we added a threshold device in the receiver described in [7], as shown in Figure 10.1. For each n, if the signal at the input of the device is positive, the decision is $\hat{m} = +1$, and if it is negative, the decision is $\hat{m} = -1$.

10.2.2 Chaotic maps employed

For the simulations, we have chosen classical maps of different dimensions. To be employed, a map must meet the following prerequisites:

1. It must be possible to rewrite it as Equation (10.1);

2. It must lead to the master-slave synchronization of Equations (10.1) and (10.2). This way, the eigenvalues of the associated matrix $\mathbf{A}$ must satisfy the Inequality (10.4);

3. It must generate chaotic signals. Here, we considered that the generated signals are chaotic if during the interval of simulation they present aperiodic bounded behavior and the largest associated Lyapunov exponent is positive [2].

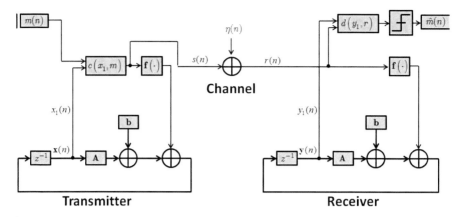

FIGURE 10.1
Communication system studied: the block diagram.

A. Quadratic map

The one-dimensional quadratic map is defined by [5]

$$x_1(n+1) = -2x_1^2(n) + 1, \tag{10.8}$$

where $x_1(0) \in [-1, 1]$. This map can be written in the form of Equation (10.1) with $K = 1$, $\mathbf{A} = 0$, $\mathbf{b} = [1]$, and $\mathbf{f}(\mathbf{x}(n)) = \left[-2x_1^2(n)\right]$. Thus, for this system, the eigenvalue of the matrix $\mathbf{A}$ is $\lambda_1 = 0$, and it satisfies the condition of Inequality (10.4). Therefore, a master-slave system formed by using this map will reach synchronism.

The Lyapunov exponent for the orbits of quadratic map is $h_1 = \ln(2) > 0$ [5] and aperiodic signals generated by the quadratic map are chaotic.F

B. Skew tent map

The skew tent map is a piecewise-linear map composed of two linear parts with inclinations of different signals. It is given by

$$x_1(n+1) = f_I(x_1(n)) = \begin{cases} \frac{2}{\alpha+1}x_1(n) + \frac{1-\alpha}{\alpha+1}, & -1 < x < \alpha \\ \\ \frac{2}{\alpha-1}x_1(n) - \frac{\alpha+1}{\alpha-1}, & \alpha \leqslant x < 1 \end{cases}, \tag{10.9}$$

with $\{\alpha, x(0)\} \subset U = (-1, 1)$. Parameter α determines the coordinate of the tent peak. This map can be written in the form of Equation (10.1) with $K = 1$, $\mathbf{A} = 0$, $\mathbf{b} = 0$, and $\mathbf{f}(\mathbf{x}(n)) = [f_I(x_1(n))]$. For this system, the eigenvalue of the matrix $\mathbf{A}$ is $\lambda_1 = 0$, and it satisfies the condition of Inequality (10.4). Therefore, a master-slave system formed by using this map will reach synchronism. For the simulations it was considered $\alpha = 0.1$.

The Lyapunov exponent for the orbits of $f_I(\cdot)$ depends only on the parameter α and is given by [8]

$$h_1 = \frac{\alpha+1}{2}\ln\left(\frac{2}{\alpha+1}\right) + \frac{1-\alpha}{2}\ln\left(\frac{2}{1-\alpha}\right). \qquad (10.10)$$

For all values of α contained in the range $(-1,1)$, $h_1 > 0$. In particular, for $\alpha = 0.1$, $h_1 \approx 0.688$.

C. Hénon map

The Hénon map is given by [10]

$$\mathbf{x}(n+1) = \begin{bmatrix} x_1(n+1) \\ x_2(n+1) \end{bmatrix} = \begin{bmatrix} 1 - \alpha x_1^2(n) + x_2(n) \\ \beta x_1(n) \end{bmatrix}. \qquad (10.11)$$

This map can be written in the form of Equation (10.1) with $K = 2$, $\mathbf{A} = \begin{bmatrix} 0 & 1 \\ \beta & 0 \end{bmatrix}$, $\mathbf{b} = \begin{bmatrix} 1, & 0 \end{bmatrix}^T$, and $\mathbf{f}(\mathbf{x}(n)) = \begin{bmatrix} -\alpha x_1^2(n) & 0 \end{bmatrix}^T$.

We considered $\alpha = 1.4$ e $\beta = 0.3$. For this system, the eigenvalues of the matrix $\mathbf{A}$ are $\lambda_{1,2} = \pm\sqrt{\beta}$, satisfying the condition of Inequality (10.4). The largest Lyapunov exponent for the chosen parameters is $h_1 \approx 0.42 > 0$ [2].

D. Three-dimensional Hénon map

The three-dimensional Hénon map is given by [11]

$$\mathbf{x}(n+1) = \begin{bmatrix} x_1(n+1) \\ x_2(n+1) \\ x_3(n+1) \end{bmatrix} = \begin{bmatrix} -\alpha x_1(n)^2 + x_3(n) + 1 \\ -\beta x_1(n) \\ \beta x_1(n) + x_2(n) \end{bmatrix}. \qquad (10.12)$$

This map can be written in the form of Equation (10.1) with

$$\mathbf{A} = \begin{bmatrix} 0 & 0 & 1 \\ -\beta & 0 & 0 \\ \beta & 1 & 0 \end{bmatrix}, \qquad (10.13)$$

$\mathbf{b} = \begin{bmatrix} 1, & 0, & 0 \end{bmatrix}^T$, and $\mathbf{f}(x(n)) = \begin{bmatrix} \alpha x_1^2(n), & 0, & 0 \end{bmatrix}^T$.

For the simulations, we have used $\alpha = 1.07$ and $\beta = 0.3$. For this system, the eigenvalues of the matrix $\mathbf{A}$ are [7]

$$\lambda_1 = 0.4084 + 0.4477j,$$
$$\lambda_2 = 0.4084 - 0.4477j,$$
$$\lambda_3 = -0.8169. \qquad (10.14)$$

As $|\lambda_i| < 1, i = 1, 2, 3$, the Inequality (10.4) is satisfied. The Lyapunov exponent is $h_1 \approx 0.23 > 0$ [7], therefore aperiodic signals generated by this map are chaotic.

Figure 10.2 shows examples of transmitted signals $s(n)$ for the considered maps using $c^{(1)}$, Equation (10.6), as coding function. Figure 10.3 shows the transmitted signals for the same maps but using $c^{(2)}$, Equation (10.7), instead.

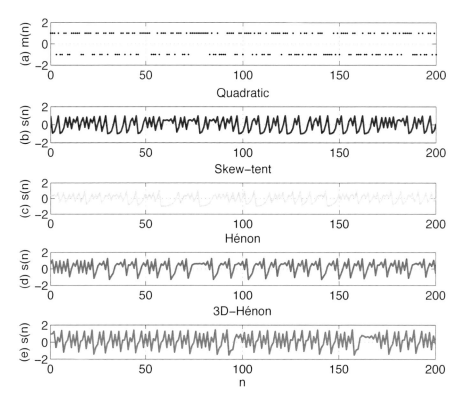

FIGURE 10.2
Transmitted signals for the system in Figure 10.1 using $c^{(1)}$, message (a) $m(n)$, (b) quadratic, (c) skew tent, (d) Hénon, and (e) 3D-Hénon maps.

10.2.3   Computer simulations

In this section we show some simulations to numerically address the performance of the chaos-based digital communication system of Figure 10.1 in an AWGN channel.

The signal-to-noise ratio (SNR) is used as a measure of the quality of the received signal and it is defined here as

$$\text{SNR} = \frac{E_b}{N_0}, \tag{10.15}$$

where E_b is the average energy per bit and $\frac{N_0}{2}$ is the power spectral density of the AWGN in the channel [9].

It is usual to express the SNR in decibels, so that

$$\text{SNR}_{\text{dB}} = 10 \log_{10}(\text{SNR}). \tag{10.16}$$

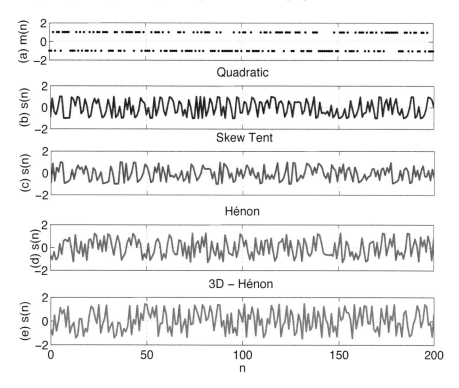

FIGURE 10.3
Transmitted signals for the system in Figure 10.1 using $c^{(2)}$, message (a) $m(n)$, (b) quadratic, (c) skew tent, (d) Hénon, and (e) 3D-Hénon maps.

As an example, Figure 10.4 shows the results of a simulation using the skew tent map and the pair $c^{(1)}, d^{(1)}$, with an $\text{SNR}_{\text{dB}} = 60$ dB. As we can see, even at this high SNR level there are already errors, as signalized in the figure. Specifically, in this case BER $= 0.1370$.

Figure 10.5 shows BER curves as a function of SNR_{dB} for the maps and encoding functions tested. For comparison we also show the optimal result of the conventional phase shift keying (PSK) [9].

For each value of SNR_{dB} the transmission of 10^7 bits was simulated. To eliminate the effect of the synchronization transient, the first 100 bits were not considered.

Analyzing the BER curves in Figure 10.5, we notice that encoding using $c^{(2)}$ presents better results than using $c^{(1)}$. Thus, for digital communications, using $c^{(2)}$ seems more appropriate than $c^{(1)}$ generally used in the literature (see, e.g., [7, 17]).

Table 10.1 presents the average number of samples required for synchro-

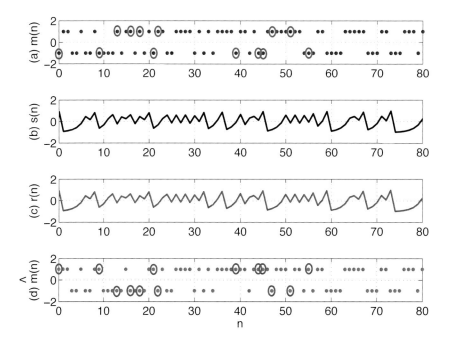

FIGURE 10.4
Simulation of the communication system for $\mathrm{SNR}_{\mathrm{dB}} = 60$ dB: (a) message $m(n)$, (b) transmitted signal $s(n)$, (c) received signal $r(n)$, and (d) retrieved message $\hat{m}(n)$. Bit errors are marked in $m(n)$ and $\hat{m}(n)$.

nization for the tested maps and coding functions. In order to calculate the mean and standard deviation of the transient time we considered a noiseless channel and 1000 repetitions were performed. We considered that systems were synchronized when $|\mathbf{e(n)}| < 10^{-2}$.

There are numerical evidences that the transient is related to the dimension of the adopted map. Maps with lower dimension present lower transients.

In summary, we realize that using the encoding-decoding $c^{(2)}, d^{(2)}$ functions and a unidimensional map is the combination that gives the best results in terms of BER and transient time. However, there is much to progress in order for these systems to reach BER levels comparable to conventional systems, such as PSK.

One possible alternative to obtain better results is to work with chaotic coupled maps lattices as chaos generators, as in [6]. In this work, it was noticed that increasing the number of maps in the lattice can make synchronization more robust to additive noise. In the next sections we review this idea bearing in mind possible applications in the system of Figure 10.1.

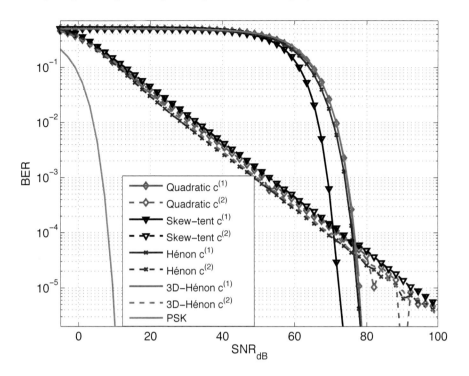

FIGURE 10.5
BER as a function of SNR_{dB} for different maps and coding functions.

10.3 Coupled maps lattices and synchronization

One of the reasons for the popularity exhibited by maps is associated with the use of less computer time. Moreover, there are many maps that present chaotic behavior. Chaos occurs when the system displays exponentially sensitive dependence on initial conditions; for example, the logistic map $x(n+1) = f(x(n)) = \mu x(n)(1 - x(n))$, where $x(n) \in [0, 1]$ and μ is a control parameter. For most values $3.57 < \mu \leq 4$ the logistic map exhibits chaotic behavior.

In Figure 10.6 we show two types of dynamical regimes, where we plot the time series of the state variable. We can observe a periodic regime of period four for $\mu = 3.5$ (Figure 10.6(a)), as well as chaotic regime considering $\mu = 4.0$ (Figure 10.6(b)).

Coupled maps lattices are models of spatially extended dynamical systems. As a matter of fact these lattices have discrete space and time, as well as a continuous state variable [13]. They have been considered for studies of spatio-

TABLE 10.1
Average number of samples required for synchronization for the tested maps and coding functions.

Map	Encoding Function	Transient (samples)
Quadratic	$c^{(1)}$	1.99 ± 0.10
Quadratic	$c^{(2)}$	1.99 ± 0.10
Skew Tent	$c^{(1)}$	1.95 ± 0.22
Skew Tent	$c^{(2)}$	1.98 ± 0.14
Hénon	$c^{(1)}$	7.41 ± 1.71
Hénon	$c^{(2)}$	7.15 ± 2.16
3D-Hénon	$c^{(1)}$	15.14 ± 3.86
3D-Hénon	$c^{(2)}$	15.14 ± 4.68

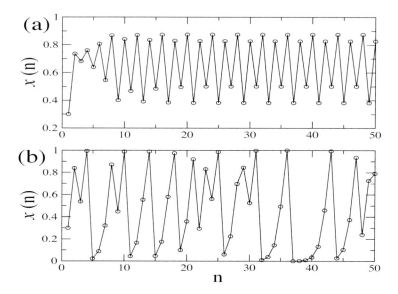

FIGURE 10.6
Time evolution of the state variable for the logistic map with (a) $\mu = 3.5$ and (b) $\mu = 4.0$.

temporal phenomena, like pattern formation, biological networks, lattices of technology, and synchronization of chaos.

The synchronization is a collective phenomenon that has been studied due to the possibility that chaotic systems may synchronize their trajectories [15]. In the synchronization process, the competition occurs between the nonlinear behavior of each system unit and the diffusive effect due to coupling. There are many types of synchronized behavior, whereas we are interested in the completely synchronized states, where all maps have equal amplitude at all times.

Let us consider a one-dimensional chain of N coupled logistic maps, where the state variable is $x_i(n) \in [0, 1]$ for the site i ($i = 1, 2, ..., N$) at time n. We couple the logistic maps according to a global mean-field prescription with periodic boundary conditions and random initial conditions are assumed. The coupled maps lattice is given by

$$x_i(n+1) = (1 - \varepsilon)f(x_i(n)) + \frac{1}{N}\sum_{j=1}^{N} f(x_j(n)), \qquad (10.17)$$

where ε is the coupling strength.

Figure 10.7 shows the time evolution of the state variables using the coupling in Equation (10.17). We can see in Figure 10.7(a) that when the logistic maps are uncoupled ($\varepsilon = 0$) the site amplitudes are so spatially uncorrelated. After a transient the dynamical variables can present a completely synchronized state (Figure 10.7(b)) due to coupling strength. In other words, the state variables can have the same value at all times $x_1(n) = x_2(n) = \cdots = x_N(n)$. This synchronized behavior depends on the coupling strength ε, the control parameter, and the topology.

We use the dispersion in order to verify the synchronization of the lattice according to the coupling strength. Therefore, we have numerically computed the dispersion of the map amplitudes with respect to their space average at a given time,

$$\gamma(n)^2 = \frac{1}{N-1}\sum_{j=1}^{N}(x_j(n) - \langle x(n) \rangle)^2, \qquad (10.18)$$

where $\langle x(n) \rangle = (1/N)\sum_{j=1}^{N} x_j(n)$. In Figure 10.8 we plot the temporal mean of the dispersion,

$$\Gamma = \frac{1}{n_2 - n_1 + 1}\sum_{n=n_1}^{n_2} \gamma(n), \qquad (10.19)$$

versus the coupling strength in the lattice of form in Equation (10.17). For ε close to 0.5 this dispersion is about zero, indicating the synchronization. Moreover, when the coupling strength increases we can observe a transition between nonsynchronized and synchronized states.

10.4   Master-slave synchronization with additive noise

Chaos synchronization in a master-slave structure has been extensively studied as a model of the communication system, due to chaotic signals presenting

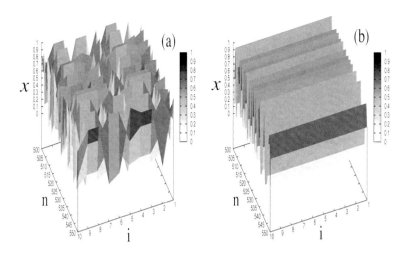

FIGURE 10.7
Time evolution of the state variables for a lattice of 10 logistic maps, where
we consider (a) $\varepsilon = 0.0$ and (b) $\varepsilon = 0.6$.

sensitive dependence on initial conditions. In a master-slave structure the
slave system is driven by a signal derived from the master [4].

First of all, we consider two coupled 3D-Hénon maps in a master-slave
configuration with additive noise

$$x(n+1) = -\alpha x(n)^2 + z(n) + 1, \tag{10.20}$$
$$y(n+1) = -\beta x(n), \tag{10.21}$$
$$z(n+1) = \beta x(n) + y(n), \tag{10.22}$$

$$X(n+1) = (1-\varepsilon)(-\alpha X(n)^2 + Z(n) + 1) + \varepsilon_I I(n), \tag{10.23}$$
$$Y(n+1) = -\beta X(n), \tag{10.24}$$
$$Z(n+1) = \beta X(n) + Y(n), \tag{10.25}$$

$$I(n) = f(x(n), z(n)) + \eta(n), \tag{10.26}$$

where the lower-case letters represent the master states and capital letters the
slave states, $f(x, z) = -\alpha x^2 + z + 1$, $0 \leq \varepsilon_I \leq 1$ is the coupling strength, $\eta(n)$
is zero-mean AWGN with variance σ_η^2, and $n = 0, 1, 2...$ is the discrete time.
When $\alpha = 1.07$ and $\beta = 0.3$ the 3D-Hénon presents chaotic orbits for almost
all initial conditions in the unity sphere [11, 12].

As diagnostic for the noise intensity in the channel we use the SNR defined

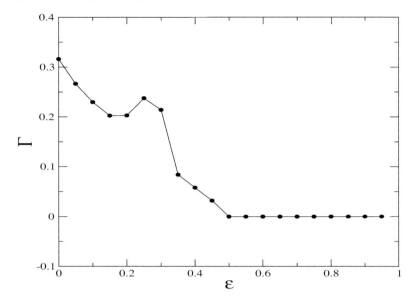

FIGURE 10.8
Temporal mean of the dispersion varying the coupling strength for a lattice of 10 logistic maps, $n_1 = 10000$ and $n_2 = 20000$.

as the mean power of the transmitted signal $I(n)$ divided by σ_η^2. The higher the SNR, the lower the relevance of the noise in the channel. In an ideal situation, SNR $= \infty$. For this reason, we consider that the synchronization error dynamics between the two systems is

$$\delta(n) = |x(n) - X(n)|. \qquad (10.27)$$

We choose parameters in that $\delta(n) \to 0$ when $n \to \infty$ if $\sigma_\eta = 0$ [7]. Moreover, we consider that the system synchronizes for $\delta_\eta(n) < 10^{-3}$.

Figure 10.9(a) exhibits the temporal evolution of the state variable of the master (circles) and slave (squares) map for $\varepsilon_I = 0.3$ and no noise ($\sigma_\eta^2 = 0$). In Figure 10.9(c) we can see the synchronization error and in this case the two 3D-Hénon synchronize after a transient time. Similarly, Figures 10.9(b) and (d) show the synchronization performance at SNR $= 8$, where the synchronization is destroyed by the channel noise.

For different values of the SNR it is possible to observe the behavior of the synchronization error (Figure 10.10). Decreasing SNR the maps pass from synchronized to nonsynchronized behavior exhibiting values of the synchronization error with irregular oscillations.

Coupled maps lattices have been used with the aim of increasing the robustness of synchronization with respect to channel noise. We consider two unidirectional lattices coupled where each one presents a global coupling pre-

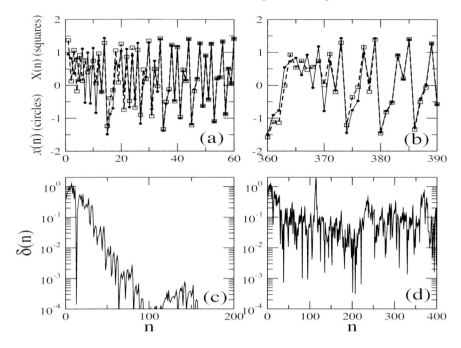

FIGURE 10.9
Master-slave system with $\varepsilon_I = 0.3$, (a) and (c) without noise, (b) and (d) SNR
= 8.

scription [6]. The coupling is given by

$$x_i(n+1) \quad = \quad (1 - \varepsilon_1)(-\alpha x_i(n)^2 + z_i(n) + 1) + \qquad (10.28)$$

$$\frac{\varepsilon_1}{N} \sum_{i=1}^{N} f(x_i(n), z_i(n)), \qquad (10.29)$$

$$y_i(n+1) \quad = \quad -\beta x_i(n), \qquad (10.30)$$

$$z_i(n+1) \quad = \quad \beta x_i(n) + y_i(n), \qquad (10.31)$$

$$X_i(n+1) \quad = \quad (1 - \varepsilon_2)(-\alpha X_i(n)^2 + Z_i(n) + 1) + \qquad (10.32)$$

$$\frac{\varepsilon_2}{N} \sum_{i=1}^{N} f(X_i(n), Z_i(n)) + \frac{\varepsilon_I}{N} I(n), \qquad (10.33)$$

$$Y_i(n+1) \quad = \quad -\beta X_i(n), \qquad (10.34)$$

$$Z_i(n+1) \quad = \quad \beta X_i(n) + Y_i(n). \qquad (10.35)$$

$$I(n) = \sum_{i=1}^{N} f(x_i(n), z_i(n)) - \sum_{i=1}^{N} f(X_i(n), Z_i(n)) + \eta(n), \qquad (10.36)$$

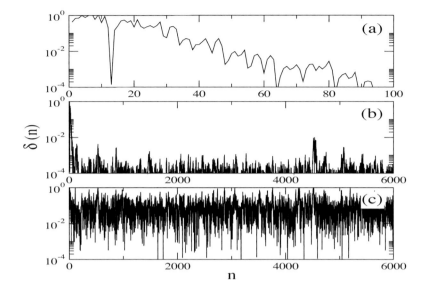

FIGURE 10.10
Master-slave system considering $\varepsilon_I = 0.3$, (a) without noise, (b) SNR = 8000, and (c) SNR = 8.

where index $i = 1, ..., N$ identifies a particular map in the N-maps lattice. Therefore, we define the synchronization error for the two coupled lattices as

$$\Delta(n) = \frac{1}{N} \left| \sum_{i=1}^{N} x_i(n) - \sum_{i=1}^{N} X_i(n) \right|, \qquad (10.37)$$

and there is synchronization if $\Delta(n) < 10^{-3}$.

With regard to synchronization, the error increases when the SNR decreases in the coupled map lattice (Figure 10.11). However, the oscillations of the error present amplitudes no greater than the values obtained in the single map case for the same parameters, shown in Figure 10.10. With this in mind, coupled lattices are more robust to noise than two unidirectionally coupled maps.

Figure 10.12 illustrates the dependence on the coupling strength ε_I, the SNR, and the number of maps in the lattices. The gray region corresponds to the nonsynchronized states when in a time interval $(5000 \leq n \leq 50000)$ we have 5% or more of the values with (a) $\delta(n) > 10^{-3}$ for single 3D-Hénon coupled maps, (b) $\Delta(n) > 10^{-3}$ for two lattices with 10 3D-Hénon each one, and (c) $\Delta(n) > 10^{-3}$ for two lattices with 100 3D-Hénon each one; whereas in the white region there are synchronized states.

To study as the synchronization depends on the lattice size we define a

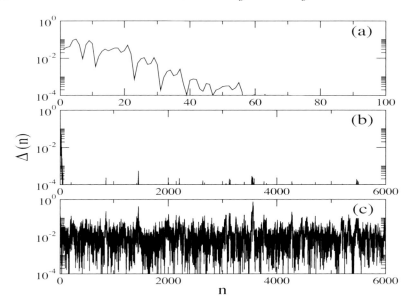

FIGURE 10.11
Master-slave system considering unidirectional coupling between two global
coupled map lattices for $N = 10$, $\varepsilon_1 = \varepsilon_2 = 0.7$, $\varepsilon_I = 0.3$, (a) without noise,
(b) SNR $= 8000$, and (c) SNR $= 8$.

critical SNR$_c$, that is, the maximum value of the SNR which allows synchronization.

In Figure 10.13 we have plotted the SNR$_c$ as a function of the lattice size
N for $\varepsilon_I = 0.3$ (triangles), $\varepsilon_I = 0.5$ (circles), and $\varepsilon_I = 1.0$ (squares). Not
only do we observe that the value of SNR$_c$ decreases following a power-law
with slope ≈ -1, but also we can infer from this figure that for $\varepsilon_I \geq 0.5$ the
coupling strength almost does not influence SNR$_c$.

All things considered, we have observed that the synchronization is more
robust to noise when the number of maps in the lattice increases, whereas
the critical coupling strength does not depend on the lattice size. Moreover,
the dependence of the critical SNR with the lattice size can be fitted by a
power-law, where the slope is approximately equal to -1.

Therefore, these results may imply that using lattices instead of single
maps can be a way to improve performance of chaos-based communication
systems in more realistic environments.

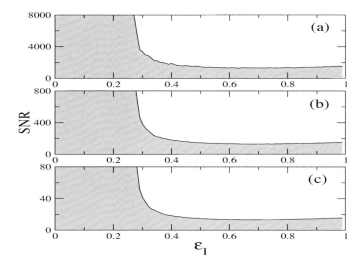

FIGURE 10.12
Gray region represents the values of SNR and the coupling strength in that the system does not present synchronization, while in the white region there is synchronization. (a) Two 3D-Hénon, (b) two lattices with 10 3D-Hénon each one, and (c) two lattices with 100 3D-Hénon each one.

10.5 Conclusions

Chaotic signals are deterministic, aperiodic, and sensitive to initial conditions. They are suited for use in applications that require security due to their difficulty of prediction and because they can be mistaken for noise in channel. In the last decades many works describing communication schemes based on chaotic signals have been published. However, the performance of theses systems under non-ideal conditions is much less studied. In this chapter we studied some aspects of the synchronization performance under noise.

Simulations performed using different maps and two pair of coding and decoding functions were considered. When using product as coding function, the chaos-based communication system performance is closer to the optimal performance of the PSK.

We compared a single map with a coupled maps lattice in terms of a master-slave interaction. Our results have confirmed that the use of the coupled lattices may be a way to improve robustness to noise in chaos-based communication systems.

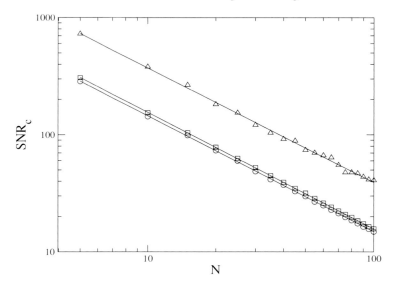

FIGURE 10.13
Vertical length of the nonsynchronization region versus lattice size, for $\varepsilon_I = 0.3$ (triangles), $\varepsilon_I = 0.5$ (circles), and $\varepsilon_I = 1.0$ (squares). We can observe a power-law of SNR_c as N grows with a slope ≈ -1.

Bibliography

[1] G. A. Abib and M. Eisencraft. Comportamento de sistemas de comunicação digital usando sinais caóticos em canal com ruído. *Revista de Tecnologia da Informação e Comunicação*, 1(2):27–32, April 2012.

[2] K. T. Alligood, T. Sauer, and J. A. Yorke. *Chaos: An Introduction to Dynamical Systems*. Textbooks in Mathematical Sciences. Springer, New York, 1996.

[3] A. Argyris, D. Kanakidis, and A. Bogris. Spectral synchronization in chaotic optical communication systems. *IEEE Journal of Quantum Electronics*, 41:892–897, 2005.

[4] T. L. Carroll and L. M. Pecora. Synchronizing chaotic circuits. *IEEE Transactions on Circuits and Systems*, 38(4):453–456, 1991.

[5] M. Eisencraft and L. A. Baccalá. The Cramer-Rao Bound for initial conditions estimation of chaotic orbits. *Chaos Solitons & Fractals*, 38(1):132–139, October 2008.

[6] M. Eisencraft and A. M. Batista. Discrete-time chaotic systems synchronization performance under additive noise. *Signal Processing*, 91(8):2127–2131, 2011.

[7] M. Eisencraft, R. D. Fanganiello, and L. Baccalá. Synchronization of discrete-time chaotic systems in bandlimited channels. *Mathematical Problems in Engineering*, 2009:1–12, 2009.

[8] M. Eisencraft and D. M. Kato. Spectral properties of chaotic signals with applications in communications. *Nonlinear Analysis: Theory, Methods & Applications*, 71(12):e2592–e2599, 2009.

[9] S. Haykin. *Communication Systems*. Wiley, New York, 4th edition, 2000.

[10] M. Hénon. A two-dimensional mapping with a strange attractor. *Communications in Mathematical Physics*, 50:69–77, 1976.

[11] D. L. Hitzl and F. Zele. An exploration of the Henon quadratic map. *Physica D: Nonlinear Phenomena*, 14(3):305–326, 1985.

[12] K. Stefanski. Modelling chaos and hyperchaos with 3-d maps. *Chaos, Solitons & Fractals*, 9(1):83–93, 1998.

[13] K. Kaneko. Overview of coupled map lattices. *Chaos: An Interdisciplinary Journal of Nonlinear Science*, 2(3):279–282, 1992.

[14] M. P. Kennedy, G. Setti, and R. Rovatti, editors. *Chaotic Electronics in Telecommunications*. CRC Press, Inc., Boca Raton, FL, USA, 2000.

[15] L. M. Pecora and T. L. Carroll. Synchronization in chaotic systems. *Phys. Rev. Lett.*, 64:821–824, February 1990.

[16] C. Williams. Chaotic communications over radio channels. *IEEE Transactions on Circuit and Systems I*, 48(12):1394–1404, December 2001.

[17] C. W. Wu and L. O. Chua. A simple way to synchronize chaotic systems with applications to secure communication systems. *International Journal of Bifurcation and Chaos*, 3(6):1619–1627, December 1993.

11

Asymptotically optimal estimators for chaotic digital communications

David Luengo

Universidad Politécnica de Madrid

Ignacio Santamaría

Universidad de Cantabria

CONTENTS

Chaotic signals and systems offer the potential of increased security in digital communications. However, most of the proposed approaches either lack robustness at low signal-to-noise ratios (SNRs) (due to the difficulty of synchronizing chaotic signals) or provide a much worse performance than classical techniques based on sinusoidal carrier functions. In this chapter we show how asymptotically optimal estimators, developed for the estimation of chaotic signals generated by discrete chaotic maps and corrupted by additive white Gaussian noise (AWGN), can be applied to improve the performance of digital chaotic communication schemes. First of all, after a brief review of discrete-time chaotic maps and sequences, we derive the optimal maximum likelihood (ML) estimator for this problem. Unfortunately, its computational cost grows exponentially with the length of the chaotic sequence, thus rendering it unfeasible for moderate/large sequences. Therefore, asymptotically optimal estimators, based on well-known signal processing techniques, such as censoring approaches or the Viterbi algorithm (VA), with a reduced computational cost, are developed. Finally, we show how these methods can be applied to improve the performance of digital chaotic communications schemes based on the iteration of discrete-time chaotic maps, focusing on a recently proposed symbolic coding technique based on backward iteration.

11.1   Introduction: chaotic maps and sequences

Chaotic sequences are signals generated by purely deterministic systems that possess features typical of random signals. The dual deterministic/random nature of these types of signals renders them very interesting for a wide range of engineering applications: communications, time series modeling, criptography, watermarking, pseudorandom number generation, etc.

In this chapter we focus on chaotic sequences generated by the discrete-time iteration of unidimensional (1D) piecewise linear (PWL) chaotic maps. Although this choice may look too restrictive, one-dimensional discrete-time chaotic maps, described by a non-linear difference equation, seem to possess all the interesting features of higher-dimensional continuous-time systems, defined through non-linear differential equations [3, 5, 37]. Furthermore, it has been shown that 1D PWL maps exhibit the same types of dynamic behaviors as any other one-dimensional chaotic maps, and a particular type of 1D PWL maps (Markov PWL maps) can approximate a wide range of non-PWL maps with an arbitrary accuracy [9, 10]. Hence, many works focus on the one-dimensional case, which is much easier to analyze, and has been exploited also in many practical applications. In the remaining of this section we briefly review the type of chaotic signals and systems that we consider in this chapter: discrete-time chaotic sequences obtained iterating piecewise linear maps.

11.1.1   One-dimensional piecewise linear chaotic maps

We define a *unidimensional chaotic map* as a non-linear application from an interval I onto the same interval,[1] $f : I \to I$, that fulfills the following three conditions [6]:

1. f has *sensitive dependence on initial conditions*, i.e., there exists a $\delta > 0$ such that for every $x \in I$ and neighborhood of x, $N_\delta(x) = \{x' : d(x, x') \le \delta\}$ with $d(\cdot, \cdot)$ denoting any appropriate distance function, there is some $x' \in N_\delta(x)$ and $n > 0$ fulfilling that $f^n(x') \notin N_\delta(f^n(x))$.

2. f is *topologically transitive*, i.e., for any pair of open sets $U, V \in I$ there exists an $n > 0$ such that $f^n(U) \cap V \ne \emptyset$.

3. The *periodic points* of f are *dense* in I, i.e., given the set of periodic points of f inside I, $P = \{x : f^n(x) = x \text{ for } n = 1, 2, 3, \ldots\}$, then $\overline{P} = I$ with $\overline{P}$ denoting the closure of P.[2]

A *piecewise linear (PWL) map* is a particular class of chaotic map which is defined by a different affine transformation inside each of the M intervals into which I is subdivided. Mathematically,

$$y = f(x) = \sum_{i=1}^{M} (a_i x + b_i) \chi_{E_i}(x), \tag{11.1}$$

where $I = [e_0, e_M]$ is the domain of the map (usually $I = [-1, 1]$ in our case), $E_i = [e_{i-1}, e_i)$ for $1 \le i \le M - 1$ and $E_M = [e_{M-1}, e_M]$ (with $e_0 < e_1 < \ldots < e_M$) are the M linear intervals of the map, a_i and b_i are the slope and offset of the line that characterize the map inside the i-th interval, respectively, and

$$\chi_R(x) = \begin{cases} 1, & x \in R; \\ 0, & x \notin R; \end{cases} \tag{11.2}$$

is the *characteristic function* of region R, that indicates whether a point x belongs to it or not.

    PWL maps are probably the class of 1D chaotic maps most widely used, due to their simplicity and mathematical tractability. As a first example of a PWL map, we consider the skew tent-map (SK-TM), which is one of the most popular 1D chaotic maps, having been used for cryptography [21], watermarking [22], digital communications [7], etc. The SK-TM has a single parameter, $0 < c < 1$, a phase space $I = [0, 1]$, and two intervals: $E_1 = [0, c)$ with $a_1 = 1/c$ and $b_1 = 0$, and $E_2 = [c, 1]$ with $a_2 = -1/(1 - c)$ and $b_2 = 1/(1 - c)$.

---

[1] A function $f : I \to I$ is *onto* if for every $y \in I$ there is an $x \in I$ such that $f(x) = y$ [6].

[2] Let P be an open subset of I. Its closure, $\overline{P}$, is defined as the set containing all the points in P altogether with all the limit points of P [6].

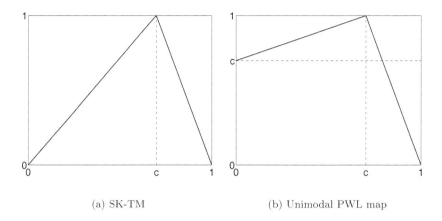

(a) SK-TM　　　　　　　　　　(b) Unimodal PWL map

FIGURE 11.1
Examples of PWL maps: skew tent-map (SK-TM) and unimodal PWL map introduced in [29].

Mathematically, the equation defining the SK-TM can be expressed as

$$f(x) = \begin{cases} \frac{x}{c}, & 0 \le x < c; \\ \frac{1-x}{1-c}, & c \le x \le 1. \end{cases} \qquad (11.3)$$

The shape of the SK-TM is shown in Figure 11.1(a). As a second example, we consider the following unimodal map introduced in [29]:

$$f(x) = \begin{cases} \frac{1-c}{c}x + c, & 0 \le x < c; \\ \frac{1-x}{1-c}, & c \le x \le 1. \end{cases} \qquad (11.4)$$

This map still has a single parameter, $0 < c < 1$, a phase space $I = [0, 1]$, and two intervals: $E_1 = [0, c)$ with $a_1 = (1-c)/c$ and $b_1 = c$, and $E_2 = [c, 1]$ with $a_2 = -1/(1-c)$ and $b_2 = 1/(1-c)$. The shape of this unimodal map is shown in Figure 11.1(b). We note that the second interval is identical to that of the SK-TM, but the line in the first one has a non-zero offset and a smaller slope than the corresponding line in the SK-TM.

11.1.2　Chaotic sequences: forward and backward iteration

Chaotic sequences are obtained by the repeated application of the non-linear function $f(x)$ on the output of the previous iteration. Hence, the n-th sample of the chaotic sequence is given by

$$x[n] = f(x[n-1]; \boldsymbol{\theta}) = f^2(x[n-2]; \boldsymbol{\theta}) = \ldots = f^n(x[0]; \boldsymbol{\theta}), \qquad (11.5)$$

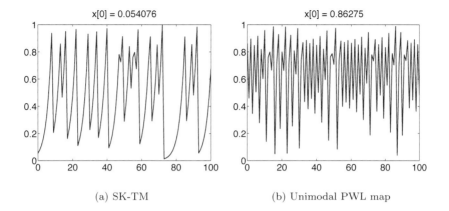

(a) SK-TM   (b) Unimodal PWL map

FIGURE 11.2
Examples of chaotic orbits generated by two PWL maps: skew tent-map (SK-TM) and unimodal PWL map introduced in [29].

where $\boldsymbol{\theta}$ is the vector of parameters that characterizes the chaotic map, $x[0] \in I$ is the initial condition that defines the evolution of the whole sequence, and

$$f^k(x) = \underbrace{f \circ f \circ \cdots \circ f}_{k \text{ times}}(x) = \underbrace{f(f(\cdots f(f(x))\cdots))}_{k \text{ times}} \qquad (11.6)$$

denotes the k-th functional composition of $f(x)$, with $f^0(x) = x$. Chaotic sequences generated using Equations (11.5) and (11.6) are said to be obtained through *forward iteration* starting from an initial condition $x[0] \in I$. Now, let us suppose that we want to generate N samples from a chaotic map starting from an initial condition $x[0] \in I$. Making use of Equation (11.5), the *forward orbit* or *trajectory* of length N of $x[0]$ is the ordered set of $N + 1$ points generated including $x[0]$, i.e.,[3]

$$O_{f,N}^{+}(x[0]) = \{x[0], \ f(x[0]), \ f^2(x[0]), \ \ldots, \ f^N(x[0])\}. \qquad (11.7)$$

Examples of chaotic orbits generated by the two maps displayed in Figure 11.1 are shown in Figure 11.2, for $c = 0.7$ and different values of $x[0]$. The irregular aspect of both signals can be clearly appreciated, although each one shows different characteristic patterns that repeat themselves in an approximate way.

Alternatively, a length $N + 1$ chaotic sequence may be generated through *backward iteration* starting from a final condition $x[N]$. However, since 1D chaotic maps must be non-invertible,[4] in order to define the backward iteration

---

[3]For the sake of simplicity, in the sequel we often remove the dependence on $\boldsymbol{\theta}$ of $f(x)$.
[4]Invertible one-dimensional maps lead only to simple dynamical behaviors and can never produce chaotic dynamics [6].

of the map we must introduce first the concept of symbolic sequences. Let us consider a 1D chaotic map $f(x)$ and a *partition* of its domain or phase space, $\mathcal{P} = \{J_1, J_2, \cdots, J_P\}$, such that $\cup_{i=1}^{P} J_i = I$ and $J_i \cup J_j = \emptyset$ for $1 \leq i, j \leq P$ and $i \neq j$. Now, given an initial condition $x[0] \in I$, we may define the length N *symbolic sequence* associated to $x[0]$ as

$$S_{f,N}^{+}(x[0]) = \{s[0], s[1], \ldots, s[N-1]\} \tag{11.8}$$

$$= \{\sigma(x[0]), \sigma(x[1]), \ldots, \sigma(x[N-1])\}, \tag{11.9}$$

where $s[n] = \sigma(x[n])$ denotes the n-th symbol for $0 \leq n \leq N-1$, and $\sigma : I \to \Sigma$ is the function that relates the phase space of the chaotic map, $I = [e_0, e_M]$, to its symbolic space, $\Sigma = \{1, 2, \ldots, P\}$, which is known as a *symbolic dynamics*, and can be expressed as

$$s[n] = \sigma(x[n]) = \sum_{i=1}^{P} i \cdot \chi_{J_i}(x[n]). \tag{11.10}$$

Obviously, infinitely many different symbolic dynamics can be defined for a given 1D chaotic map. The most frequently used partition is the so-called *natural partition*, $\mathcal{P}_N$, which is composed of the minimum number of regions where $f(x)$ is monotonic and continuous, and thus also invertible. Using the natural partition, any 1D chaotic map can be expressed as

$$y = f(x) = \sum_{i=1}^{P} f_i(x)\chi_{J_i}(x), \tag{11.11}$$

where $f_i(x)$ is the monotonic and continuous function that describes the map inside the i-th interval of the natural partition. For a given chaotic map, the natural partition is unique, can be easily obtained, and allows us to define a piecewise inverse function as

$$x = f^{-1}(y) = \sum_{i=1}^{P} f_i^{-1}(y)\chi_{J_i}(x), \tag{11.12}$$

with $f_i^{-1}(y)$ denoting the inverse of $f_i(x)$ inside the i-th interval of $\mathcal{P}_N$. For the particular class of chaotic maps considered here, PWL maps defined by Equation (11.1), the natural partition is clearly given by $\mathcal{P}_N = \{E_1, E_2, \ldots, E_M\}$, and the one-step backward iteration can be expressed as

$$x = f^{-1}(y) = \sum_{i=1}^{M} \frac{y - b_i}{a_i}\chi_{E_i}(x), \tag{11.13}$$

which corresponds to another PWL map with slopes $1/a_i$ and offsets $-b_i/a_i$. Note that this backward iteration requires prior knowledge of the region of the natural partition to which the generated sample belongs. Denoting $s = \sigma(x)$, Equation (11.13) can be expressed more compactly as

$$x = f_s^{-1}(y) = \frac{y - b_s}{a_s}. \tag{11.14}$$

Unfortunately, the natural partition is too restrictive for the digital communications application considered in this chapter, as will be shown in Section 11.3. Consequently, we define a generalization of the natural partition, that we call the *invertible partition* of the map, $\mathcal{P}_I$, which is composed of the minimum number of connected intervals inside which $f(x)$ is invertible. Although for many chaotic maps (e.g., continuous maps) the invertible partition is identical to the natural partition, it allows us to consider partitions with regions inside which the map is not continuous or even not monotonic as long as the map is still invertible. We remark that, even though the invertible partition cannot be expressed as compactly as the natural partition for a generic PWL map, it can also be easily found, since each interval will be formed by the union of one or more consecutive intervals of the PWL map. For example, for the second map used in Section 11.3, given by Equation (11.45), the natural partition is composed of $M = 5$ intervals, $\mathcal{P}_N = \{E_1, E_2, E_3, E_4, E_5\}$, whereas the invertible partion contains only three intervals, $\mathcal{P}_I = \{E_1 \cup E_2, E_3, E_4 \cup E_5\}$.

Using the symbolic sequence induced either by the natural or the invertible partition, the n-th sample of a chaotic sequence is alternatively given by

$$x[n] = f_{s[n]}^{-1}(x[n+1]; \boldsymbol{\theta}) = f_{s[n],s[n+1]}^{-2}(x[n+2]; \boldsymbol{\theta}) = \ldots =$$
$$= f_{s[n],\ldots,s[N-1]}^{-(N-n)}(x[N]; \boldsymbol{\theta}), \tag{11.15}$$

where the subindex of f indicates the portion of the symbolic sequence required for the backward iteration and $f^{-k}(x)$ denotes the k-th functional composition of $f^{-1}(x)$. Similarly, making use of Equation (11.15) we can define the *backward orbit* or *trajectory* of length N of $x[N]$ as the ordered set of $N + 1$ points generated including $x[N]$, i.e.,

$$O_{f,N}^-(x[N]) = \{x[N], f_{s[N-1]}^{-1}(x[N]), \ldots, f_{s[0],\ldots,s[N-1]}^{-N}(x[N])\}, \tag{11.16}$$

which corresponds to the forward trajectory given by Equation (11.7) in reverse order.

Finally, we remark again that the symbolic sequence must also be specified when determining the backward orbit of length N of $x[N]$. Therefore, a PWL map with M intervals may have up to M^N backward orbits for a given value of $x[N]$. In fact, the number of backward orbits can actually be much less than M^N, since some symbolic sequences may be invalid. For instance, for the chaotic map shown in Figure 11.1(b), since $f(E_1) = E_2$ (i.e., the first interval is mapped onto the second) we must necessarily have $s[n + 1] = 2$ whenever $s[n] = 1$, implying that all the symbolic sequences containing $\{\ldots, 1, 1, \ldots\}$ will be invalid. Denoting the set of valid symbolic sequences of length N for a given map as $\mathcal{S}_N$, the number of possible backward orbits for a given value of $x[N]$ is $\Gamma_s(N) = |\mathcal{S}_N|$, with $|\mathcal{A}|$ indicating the cardinality of set $\mathcal{A}$.

11.2   Optimal and suboptimal estimation of chaotic sequences

11.2.1   Problem formulation

The problem considered in this section is estimating a sequence of samples generated by iterating a 1D PWL chaotic map, given observations corrupted by AWGN. Mathematically, let us consider an $N+1$ column vector containing the $N+1$ samples of the forward orbit of length N of $x[0]$, or equivalently the $N+1$ samples of the backward orbit of $x[N]$ in reverse order,[5]

$$
\begin{aligned}
\boldsymbol{x} &= [x[0],\ x[1],\ \ldots,\ x[N-1],\ x[N]]^\top \\
&= [x[0],\ f(x[0]),\ f^2(x[0]),\ \ldots,\ f^{N-1}(x[0]),\ f^N(x[0])]^\top \\
&= [f_{\boldsymbol{s}}^{-N}(x[N]),\ f_{\boldsymbol{s}}^{-(N-1)}(x[N]),\ \ldots,\ f_{s[N-1]}^{-1}(x[N]),\ x[N]]^\top, \qquad (11.17)
\end{aligned}
$$

where $\boldsymbol{s} = [s[0],\ s[1],\ \ldots,\ s[N-1]]^\top$ is the length N column vector with the symbolic sequence associated to the orbit of length N of $x[N]$,[6] and $\boldsymbol{s}_{n:N} = [s[n-1],\ s[n],\ \ldots,\ s[N-1]]^\top$ for $1 \le n \le N$ is the column vector containing the last $N - n + 1$ samples of the symbolic sequence. The observed sequence is $\boldsymbol{y} = [y[0],\ y[1],\ \ldots,\ y[N-1],\ y[N]]^\top$ with $y[n] = x[n] + w[n]$ and $w[n] \sim \mathcal{N}(0, \sigma^2)$ for $0 \le n \le N$.[7] The goal is obtaining an accurate and efficient estimator of the original chaotic sequence, i.e., an unbiased estimator whose variance decreases with N as fast as possible.

In the following section we develop the optimal maximum likelihood estimator (MLE) for this problem.[8] Unfortunately, the computational cost of the MLE increases exponentially with the length of the sequence. Consequently, in Section 11.2.3 we describe two simple and asymptotically efficient estimators that attain the Cramer-Rao lower bound (CRLB) as the SNR tends to infinity. The CRLB, described in the Appendix, provides us with a lower bound on the variance of any unbiased estimator, allowing us to quantify precisely the concept of an efficient estimator as the one attaining the CRLB [11]. Hence, the two estimators described in Section 11.2.3 can be considered asymptotically optimal in the sense of making the most efficient use of all the information available, at least for large values of SNR.

---

[5] As shown in Reference [14], any point in the chaotic sequence can be used as a reference, mixing forward and backward iteration to obtain the remaining samples. However, in order to simplify the discussion, here we only consider the first and the last samples of the chaotic sequence, $x[0]$ and $x[N]$, respectively.

[6] Note that $s[N]$ is not included in $\boldsymbol{s}$ since it is not required for the backward iteration.

[7] We use the notation $w[n] \sim \mathcal{N}(\mu, \sigma^2)$ to indicate that $w[n]$ is a sample from a Gaussian distribution with mean μ and variance σ^2.

[8] Bayesian estimators have also been developed for this problem [14, 25, 29], but they will not be discussed here, since they do not provide any advantage for the chaotic communications approach described in this chapter.

11.2.2   Maximum likelihood estimator

The MLE of chaotic sequences corrupted by AWGN was formulated originally in [23], where two suboptimal approaches for finding the MLE, based on dynamic programming and the Kalman filter, respectively, were proposed. Then, Papadopoulos and Wornell were the first ones to provide an algorithm that achieved the exact MLE for a particular chaotic map, the tent-map (TM) with $\beta = 2$ [30, 31]. Unfortunately, although this method can be easily applied to the TM with other values of β or to other chaotic maps, in general the estimator obtained will not be the MLE. The exact MLE for generic PWL maps was developed independently by Schimming et al. [35, 36] and Pantaleón et al. [26]. Finally, an algorithm to attain the MLE for non-PWL maps based on Markov chain Monte Carlo (MCMC) methods was proposed in [16]. In this section we develop the exact MLE following the description of [14, 26].

11.2.2.1   Standard problem formulation

Formally, the MLE is obtained solving the following maximization problem [11, 38]:

$$\hat{x}_{\mathrm{ML}} = \arg\max_{x} p(y; x),$$

$$\text{s.t.} \quad x[0] \in [e_0, e_M], \ x[n] = f(x[n-1]) \quad \text{for } 1 \leq n \leq N, \qquad (11.18)$$

where the constraints are imposed by the nature of the chaotic sequence, and $p(y; x)$ is the *likelihood*, i.e., the probability density function (PDF) of the observations conditioned on the parameters to be estimated, in this case the underlying noiseless chaotic sequence. Since the noise samples are Gaussian and independent, the likelihood is a multivariate Gaussian PDF,

$$p(y; x) = (2\pi\sigma^2)^{-(N+1)/2} \exp\left(-\frac{1}{2\sigma^2}(y-x)^\top(y-x)\right). \qquad (11.19)$$

Moreover, since all the samples of the chaotic sequence can be obtained from $x[0]$ through forward iteration, it is straightforward to show that the MLE of the whole sequence can be reformulated in terms of the MLE of the initial condition,

$$\hat{x}_{\mathrm{ML}}[0] = \arg\min_{x[0]} J(x[0]),$$

$$\text{s.t.} \quad x[0] \in [e_0, e_M], \qquad (11.20)$$

with

$$J(x[0]) = \sum_{n=0}^{N} (y[n] - f^n(x[0]))^2 \qquad (11.21)$$

indicating the quadratic error between the observations and the chaotic sequence. Hence, in this case the MLE of $x[0]$ and its least squares (LS) estimator are equivalent, a result which is well known for parameters observed in AWGN. The remaining samples of the chaotic sequence can be obtained by forward iteration from $\hat{x}_{\mathrm{ML}}[0]$, thanks to the invariance property of the MLE [11], resulting in an ML estimate of the whole sequence

$$\hat{\boldsymbol{x}}_{\mathrm{ML}} = [\hat{x}_{\mathrm{ML}}[0],\ f(\hat{x}_{\mathrm{ML}}[0]),\ \ldots,\ f^{N-1}(\hat{x}_{\mathrm{ML}}[0]),\ f^{N}(\hat{x}_{\mathrm{ML}}[0])]^{\top}. \quad (11.22)$$

Unfortunately, although the formulation of this problem looks deceptively simple, solving it efficiently is extremely complicated, due to the highly non-linear dependence of the samples of the chaotic sequence on the initial condition. In fact, the cost function given by Equation (11.21) is extremely rugged, with fractal characteristics (see [14] for a more detailed discussion of this issue) and multiple minima and maxima. As an example, Figure 11.3 shows the cost function for the SK-TM with $N = 10$ and $N = 100$ and two randomly selected values of $x[0]$. Obviously, given the shape of the cost function, any iterative or grid search approach is doomed to get stuck in a local minimum, especially for large values of N. Furthermore, although the exact MLE of $x[0]$ can be obtained as shown in [25, 26], it requires searching for a set of symbolic regions, $R_{\boldsymbol{s}_i}$, associated to the portion of the phase space where the initial condition must lie in order to have $\boldsymbol{s} = \boldsymbol{s}_i$. Finding these regions becomes more involved as N grows, since their size decreases exponentially and the forward iteration of chaotic maps is numerically unstable. Hence, in the following section we introduce an alternative formulation based on backward iteration that avoids all these problems: the cost function is quadratic in $x[N]$, backward iteration is numerically stable, and, whenever $\Gamma_s(N) = M^N$ as it happens for most of the maps used in practice, the symbolic regions agree with the phase space of the chaotic map.

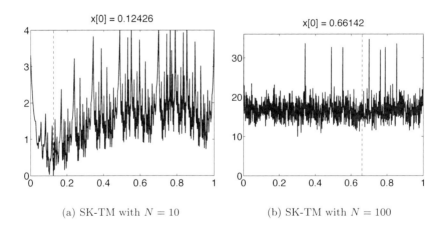

(a) SK-TM with $N = 10$ (b) SK-TM with $N = 100$

FIGURE 11.3
Examples of $J(x[0])$ for the SK-TM with $N = 10$ and $N = 100$. The dashed line indicates the true value of $x[0]$.

11.2.2.2   Alternative problem formulation

For chaotic sequences, since all the samples are related to $x[N]$ through backward iteration with the appropriate symbolic sequence, the MLE can be reformulated in a more practical way:

$$[\hat{x}_{\mathrm{ML}}[N],\ \hat{s}_{\mathrm{ML}}]^{\top} = \underset{x[N],s}{\arg\min}\ J(x[N], s),$$

$$\text{s.t.}\quad x[N] \in [e_0, e_M],\ s \in \mathcal{S}_N, \tag{11.23}$$

where $\mathcal{S}_N$ is again the set of valid symbolic sequences of length N, and

$$J(x[N], s) = \sum_{n=0}^{N} \left(y[N-n] - f_s^{-n}(x[N])\right)^2. \tag{11.24}$$

Hence, the MLE of the chaotic sequence can actually be obtained by estimating the last sample of the sequence, $x[N]$, and the N samples of the symbolic sequence: $s[0]$, $s[1]$, ..., $s[N-2]$, $s[N-1]$.

Note that, although this formulation may look more complicated than the standard one, for PWL maps it leads to much simpler expressions for the MLE. Indeed, for PWL maps it can be shown that the n-th iteration backwards starting from $x[N]$ for $1 \le n \le N$ is given by [14, 15][9]

$$x[N-n] = f_s^{-n}(x[N]) = B_{s_{N-n+1:N}}^{1,n} x[N] - \sum_{m=1}^{n} B_{s_{N-n+1:N-m+1}}^{m,n} b_{s[N-m]},$$
$$\tag{11.25}$$

with $B_s^{1,0} = 1$ and

$$B_{s_{N-n+1:N}}^{m,n} = \prod_{\ell=m}^{n} a_{s[N-\ell]}^{-1}, \tag{11.26}$$

for $1 \le n \le N$ and $1 \le m \le n$. Inserting Equation (11.26) into (11.24), the cost function for the MLE of a PWL map finally becomes

$$J(x[N], s) = \sum_{n=0}^{N} \left(\gamma_s[N-n] - B_s^{1,n} x[N]\right)^2, \tag{11.27}$$

where, instead of the precise samples of the symbolic sequence involved, we have used s in all cases to simplify the notation, and

$$\gamma_s[N-n] = y[N-n] + \sum_{m=1}^{n} B_s^{m,n} b_{s[N-m]}, \qquad \text{for}\quad 0 \le n \le N. \tag{11.28}$$

---

[9]Closed-form equations for the forward iteration of the TM were developed originally in [25], and extended to a generic PWL map in [26]. Regarding backward iteration, analytical expressions were developed first for the polar SK-TM in [27] and extended to generic PWL maps in [29]. All these formulas for the forward and backward iteration of a generic PWL map starting from an arbitrary reference sample, $x[n]$, have been compiled in [14]. Furthermore, efficient algorithms for their implementation are also provided in [14].

In the next section we show how, using this alternative formulation and the analytical expressions for the backward iteration of PWL maps, we can get rid of the fractal cost functions that appear in the forward iteration and obtain a consistent estimator of $x[N]$ that can be expressed in a closed-form.

11.2.2.3   Exact maximum likelihood estimator

Looking back at the cost function of the maximum likelihood estimator, we notice that, for a given symbolic sequence, $s = s_i \in S$, Equation (11.24) is quadratic in $x[N]$. Hence, taking the derivative of Equation (11.24) w.r.t. $x[N]$ and equating it to zero we can easily obtain the estimate of $x[N]$ associated to $s = s_i$:

$$\hat{x}_i[N] = \frac{\sum_{n=0}^{N} B_{s_i}^{1,n} \gamma_{s_i}[N-n]}{\sum_{n=0}^{N} \left(B_{s_i}^{1,n}\right)^2}. \tag{11.29}$$

Note that Equation (11.29) is not the ML estimate of $x[N]$ for the i-th symbolic sequence yet, since there is no guarantee that $\hat{x}_i[N]$ belongs to $I = [e_0, e_M]$. However, assuming that all the symbolic sequences are valid (i.e., $\Gamma_s(N) = |S_N| = M^N$), the ML estimator of $x[N]$ corresponding to the i-th symbolic sequence is obtained simply by thresholding Equation (11.29):[10]

$$\hat{x}_{\text{ML}}^i[N] = \begin{cases} e_0, & \hat{x}_i[N] < e_0; \\ \hat{x}_i[N], & e_0 \leq \hat{x}_i[N] \leq e_M; \\ e_M, & \hat{x}_i[N] > e_M. \end{cases} \tag{11.30}$$

Finally, thanks to the invariance property of the ML estimator [11], the ML estimate of the rest of the sequence for the i-th symbolic sequence can be obtained simply by iterating backwards from $\hat{x}_{\text{ML}}^i[N]$ using the i-th symbolic sequence, s_i, i.e.,

$$\hat{x}_{\text{ML}}^i[N-n] = f_{s_i}^{-n}(\hat{x}_{\text{ML}}^i[N]), \qquad \text{for} \quad 1 \leq n \leq N. \tag{11.31}$$

Unfortunately, the estimation of the optimal symbolic sequence is a nondeterministic polynomial (NP) time hard problem in the general case, implying that algorithms for obtaining the exact MLE in polynomial time cannot be developed except for some particular cases, such as the TM with $\beta = 2$ [31]. Therefore, the only solution that guarantees that the MLE of the symbolic sequence is achieved for a generic PWL map is an *exhaustive search* or *brute force approach*: testing all the valid symbolic sequences and selecting the one that minimizes the cost function. Mathematically, the ML estimators of $x[N]$ and the symbolic sequence are $\hat{x}_{\text{ML}}[N] = \hat{x}_{\text{ML}}^r[N]$ and $\hat{s}_{\text{ML}} = s_r$, respectively, with $\hat{x}_{\text{ML}}^r[N]$ being the ML estimator associated to the r-th valid symbolic

---

[10]When some symbolic sequences are invalid the same approach can be followed, but the limits for the thresholding operation may depend on the symbolic sequence [14].

sequence, s_r, as given by Equation (11.30), and

$$r = \arg \min_i J(\hat{x}^i_{\mathrm{ML}}[N], s_i) \tag{11.32}$$

is the index to the MLE of the symbolic sequence. The MLE of the remaining samples of the chaotic sequence can be obtained again through backwards iteration, resulting in the following MLE of the whole sequence:

$$\hat{x}_{\mathrm{ML}} = [f^{-N}_{\hat{s}_{\mathrm{ML}}}(\hat{x}_{\mathrm{ML}}[N]), \ f^{-(N-1)}_{\hat{s}_{\mathrm{ML}}}(\hat{x}_{\mathrm{ML}}[N]), \ \ldots, \ f^{-1}_{\hat{s}_{\mathrm{ML}}}(\hat{x}_{\mathrm{ML}}[N]),$$
$$\hat{x}_{\mathrm{ML}}[N]]^\top. \tag{11.33}$$

11.2.3 Asymptotically optimal estimators

Due to the computational complexity of the MLE, many suboptimal algorithms have been proposed for estimating chaotic signals corrupted by AWGN. As already mentioned before, Myers et al. were the first ones to develop suboptimal algorithms to approximate the MLE [23]. A method for modeling chaotic systems based on hidden Markov models (HMMs) that could be applied to obtain approximate ML and Bayesian estimators was introduced in [24, 33]. Kay also proposed two suboptimal estimators based on topological conjugacy (the halving method) and dynamic programming with good asymptotic performance [13]. During the following years many other authors developed several simple and suboptimal estimators based on the symbolic sequence that attained the CRLB asymptotically [4, 26, 39]. Iterative approaches based either on the E-M algorithm [28] or the Viterbi algorithm [1, 2, 18–20] have also been proposed. In this section we review the simple hard-censoring approach proposed in [26] and the more elaborate algorithm based on the Viterbi algorithm as described in [18].

11.2.3.1 Hard censoring estimator

The hard censoring maximum likelihood (HC-ML) estimator, proposed in [26], is probably the simplest approximate MLE. The idea behind the HC-ML is simply applying a threshold to the noisy observations to obtain an estimate of the symbolic sequence and using it to compute the MLE for that particular symbolic sequence. Mathematically, the HC-ML estimator is given by

$$\hat{x}_{\mathrm{HC\text{-}ML}}[N] = \hat{x}^r_{\mathrm{ML}}[N], \tag{11.34}$$

where $\hat{x}^r_{\mathrm{ML}}[N]$ is the MLE associated to the r-th symbolic sequence, given by Equation (11.30) for $s_r = \hat{s} = [\hat{s}[0], \ \hat{s}[1], \ \ldots, \ \hat{s}[N-1], \ \hat{s}[N]]^\top$, with the symbols of this symbolic sequence being obtained as

$$\hat{s}[n] = \begin{cases} 1, & y[n] < e_0; \\ \sigma(y[n]), & e_0 \le y[n] \le e_M; \\ M, & y[n] > e_M; \end{cases} \tag{11.35}$$

where $\sigma(y[n])$ is the symbolic dynamics associated to the natural partition of the map, given by Equation (11.10), applied to the noisy observations. The remaining samples of the chaotic sequence are obtained through backward iteration using Equation (11.31) for the r-th MLE.

This HC-ML estimator, denoted as HC-ML(0) in [14], can be improved by locating the k symbols ($1 \leq k \leq N$) most likely to be erroneous (i.e., those associated to observations closer to the borders separating the regions of the natural partition), changing them, and checking whether the modified symbolic sequence provides better results or not [14]. Obviously, for $k = 0$ we get the basic HC-ML estimator as described in [26], whereas for $k = N$ we obtain the exact MLE.

11.2.3.2 Estimator based on the Viterbi algorithm

A more sophisticated approximate MLE than the HC-ML can be obtained using the Viterbi algorithm (VA) as a computationally efficient estimator of s. Note that, unlike the well-known cases of decoding of convolutional codes and detection in channels with intersymbol interference (ISI), where the VA provides us with the exact MLE [32], here the VA is a suboptimal estimator that will not achieve the MLE in general. In fact, the VA is able to obtain the exact ML estimator, but that would require a trellis with $\Gamma_s(N)$ states, thus not providing any computational advantage w.r.t. the exact MLE implementation based on an exhaustive search. Hence, following the approach of [14,18], we propose to use a trellis with a reduced number of states, $R = M^r$, as an approximate MLE. In order to describe the algorithm, we focus on the simplest case: $M = 2$ and $r = 1$. Using $r = 1$ is equivalent to assuming that the next output depends only on the current symbol of the itinerary and the estimated value of $x[n]$. For $M = 2$ and $r = 1$ the basic butterfly of the trellis, shown in Figure 11.4, only has two states.

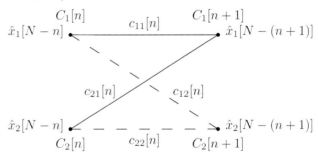

FIGURE 11.4
Basic butterfly for the VA using only two states per iteration of the chaotic sequence and a chaotic map with $M = 2$.

The transition cost for each branch is given, as usual, by the difference between the observations and the expected signals,

$$c_{ij}[n] = |y[N - n] - \hat{x}_{ij}[n]|, \qquad (11.36)$$

with $\hat{x}_{ij}[n] = f_{s_j[n]}^{-1}(\hat{x}_i[N - (n-1)])$ for $i, j \in \{1, 2\}$, and $1 \le n \le N$. The cost of each node is the minimum cost accumulated in all the branches arriving to it,

$$C_j[n] = C_r[n-1] + c_{rj}[n], \qquad (11.37)$$

with

$$r = \arg\min_{i \in \{1, 2\}} \{C_i[n-1] + c_{ij}[n]\}, \qquad (11.38)$$

and the current estimate of $x[n]$ for the j-th node is

$$\hat{x}_j[n] = \hat{x}_{rj}[n] = f_{s_j[n]}^{-1}(\hat{x}_r[n-1]). \qquad (11.39)$$

Finally, the initial cost for each state is $C_j[0] = |y[0] - \hat{x}_j[0]|$, where $\hat{x}_j[0]$ is the estimate of $x[N]$. When $x[N]$ is known in the receiver, then $\hat{x}_j[0] = x[N]$. Otherwise, the initial two samples required to start the recursion can be obtained applying a threshold to $y[0]$:

$$\hat{x}_j[0] = \max(e_{j-1}, \min(y[0], e_j)). \qquad (11.40)$$

Hence, we guarantee that $\hat{x}_1[0] \in E_1 = [e_0, e_1]$ and $\hat{x}_2[0] \in E_2 = [e_1, e_2]$.

Extending this construction to higher values of r is straightforward, as shown in [14], and a slight improvement in the performance of the estimator is observed for low signal to noise ratios (SNRs), as shown in the following section. However, although for $r = 1$ there is a drastic reduction in the number of states of the trellis, the performance of the VA is very close to that of the actual ML estimator [18]. The reason is simple: backward iteration of two different final conditions using the same itinerary leads to very similar trajectories. As a result, in every iteration we only need to store one estimate for each possible symbol, since in general all the other paths do not provide very different chaotic sequences in the long term. In [17] it was proved, using map 1, that the distance between the orbits of two different initial states with the same itinerary decreases by a factor $2/(1-c)$ per iteration. Following the same reasoning, it is easy to show that, using map 2, this factor becomes 2 for any value of c, i.e. different initial states converge even faster than for map 1.

11.2.4   Comparison of optimal and suboptimal estimators

In this section we present some performance results of the exact ML estimator and the two suboptimal estimators described (HC-ML and VA) for two chaotic maps and short sequences. The first map considered is the SK-TM, whereas the second one is another popular chaotic map, the binary shift map (BSM), which is given by

$$f(x) = \begin{cases} 2x, & 0 \le x < 0.5; \\ 2x - 1, & 0.5 \le x \le 1. \end{cases} \tag{11.41}$$

The reference sample used in all cases is $x[N]$, but results are presented for the estimation of $x[0]$, since this is the hardest sample of the sequence to be estimated.[11] The performance measure used is the mean square error (MSE) of the estimate of $x[0]$, $\hat{M}_0(dB) = -10 \log_{10} MSE(\hat{x}[0])$, which is in fact equivalent to the variance, since all the estimators considered are unbiased.

First of all, Figure 11.5 shows the value of $\hat{M}_0(dB)$ obtained for both maps and different values of SNR using the HC-ML, the VA, and the exact MLE for $N = 4$. It can be appreciated that all the estimators attain the CRLB for an SNR above a threshold that depends on the map (e.g., a much larger SNR is required to attain the CRLB for the SK-TM than for the BSM) and on the estimator. On the one hand we observe the excellent performance of the VA, which attains the CRLB at the same SNR than the exact MLE and provides very similar MSE values with only a fraction of its computational cost. On the other hand, the HC-ML provides a worse performance, requiring a much larger SNR than the VA and exact MLE to attain the CRLB for the BSM. In Table 11.1 we compare the average performance of the exact MLE, the VA (with $r = 1$, 2, and 3), and the HC-ML for several values of SNR. Each input in the table has been obtained averaging the results of 1000 simulations with a randomly chosen initial condition. Note how the VA provides very good results, achieving virtually the same performance as the MLE for $r = 3$.

TABLE 11.1
Comparison of MSE obtained by the HC-ML(0), the VA (with $r = 1$, 2, and 3), and the exact MLE for the BSM with $N = 4$.

SNR (dB)	HC-ML	$\hat{M}_0(dB)$			ML	CRLB
		VA $(r = 1)$	VA $(r = 2)$	VA $(r = 3)$		
0	10.5	10.9	11.3	11.4	11.6	20.6
5	14.2	14.2	15.1	15.4	15.4	25.6
10	19.8	19.5	20.6	21.0	21.0	30.6
15	24.7	30.0	30.9	31.2	31.2	35.6
20	31.9	40.6	40.6	40.6	40.6	40.6
25	45.5	45.5	45.5	45.5	45.5	45.6
60	80.4	80.4	80.4	80.4	80.4	80.6

---

[11]In fact, the MLE of $x[N]$ always attains the CRLB, whereas the MLE of $x[0]$ needs a minimum SNR to attain it [14].

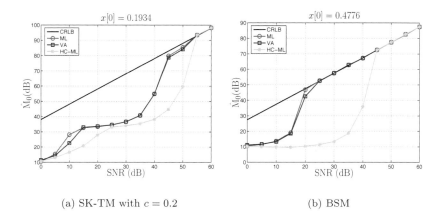

(a) SK-TM with $c = 0.2$ (b) BSM

FIGURE 11.5
Comparison of the MSE obtained with the HC-ML, the VA, and the exact MLE for $N = 4$.

11.3   Inverse symbolic chaotic encoding

In this section we show how the asymptotically optimal estimators developed in the previous section can be applied to a chaotic digital communications system. Although all of these estimators can be applied to the well-known chaos shift keying (CSK) schemes, as shown in [14], in this section we focus on their application to the inverse symbolic chaotic encoding proposed in [17].

11.3.1   Symbolic encoder and decoder

Let us consider the transmission of a vector of N information bits, $\boldsymbol{b} = [b[0], \ldots, b[N-1]]^T$. The basic idea of the inverse symbolic chaotic encoding scheme proposed in [17] is iterating backwards from a known final condition, $x[N]$, using those information bits to construct the symbolic sequence. The structure of the chaotic encoder is shown in Figure 11.6. First of all, a one-to-one correspondence between the sequence of N information bits, and the symbols of the itinerary, $\tilde{s}[n] = s[N - n] = g(b[n])$, is established:

$$\tilde{s}[n] = s[N - n] = 1 + 2b[n], \tag{11.42}$$

for $1 \leq n \leq N$. Then, $\tilde{s}[n]$ is used to generate the chaotic sequence according to Equation (11.15), starting from a given $\tilde{x}[0] = x[N]$ previously fixed or

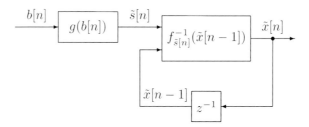

FIGURE 11.6
Block diagram of the inverse symbolic chaotic encoder for a generic map.

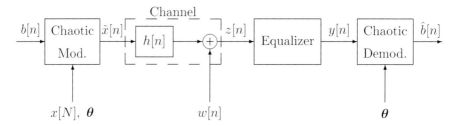

FIGURE 11.7
Baseband chaotic communications system with parameters $x[N]$ and $\boldsymbol{\theta}$: modulator, channel, and receiver.

randomly chosen, i.e.,

$$\tilde{x}[n] = x[N - n] = f^{-1}_{\tilde{s}[n]}(\tilde{x}[n - 1]) = f^{-1}_{s[N-n]}(x[N - n])$$
$$== f^{-1}_{1+2b[n]}(x[N - n]) \tag{11.43}$$

for $1 \leq n \leq N - 1$. Finally, these samples can be directly transmitted through the channel in baseband after a digital to analog conversion (DAC), as shown in Figure 11.7, or up converted to the desired frequency band. In the receiver, a standard equalizer is required to compensate the effect of the channel prior to the chaotic demodulation using the Viterbi algorithm as described in Section 11.2.3.2.

The choice of the chaotic map is crucial to achieve a good performance with this scheme. In [17] a PWL map with $M = 3$ regions (*map 1*) was proposed:

$$f(x) = \begin{cases} \frac{2x+(1+c)}{1-c}, & -1 \leq x \leq -c; \\ \frac{x}{c}, & -c < x < c; \\ \frac{2x-(1+c)}{1-c}, & c \leq x \leq 1. \end{cases} \tag{11.44}$$

The two outer intervals, $E_1 = [-1, -c]$ and $E_3 = [c, 1]$ (with E_1 associated to $b[n] = 0$ and E_2 to $b[n] = 1$), are used for coding, whereas the inner one, $E_2 = (-c, c)$, acts as a guard interval, ensuring a minimum distance between the signals associated to $b[n] = 0$ and $b[n] = 1$. This map shows a good performance in terms of bit error rate (BER), but it has an important drawback: $\tilde{x}[n]$ cannot belong to any region of the state space that maps into E_2 after $1 \leq k \leq n+1$ iterations. This creates exclusion regions for $\tilde{x}[n]$ and a clustering of the samples of the chaotic sequence (see [14] for a detailed description of this issue), thus reducing the security of the system. In order to solve this problem, an alternative map (*map 2*) with $M = 5$ intervals has been introduced in [14], which avoids the innermost interval, $E_3 = (-c, c)$, which is used again as a guard region:

$$f(x) = \begin{cases} 2x + 1, & -1 \leq x \leq -(1+c)/2; \\ 2x + (1 + 2c), & -(1+c)/2 \leq x \leq -c; \\ x/c, & -c < x < c; \\ 2x - (1 + 2c), & c \leq x \leq (1+c)/2; \\ 2x - 1, & (1+c)/2 \leq x \leq 1. \end{cases} \quad (11.45)$$

The shape of these two maps is plotted in Figure 11.8. Note that in both cases we use the invertible partition, which is composed of $M_I = 3$ regions, when iterating backwards. For map 1 the invertible partition is the same as the natural partition, $\mathcal{P}_I = \mathcal{P}_N = \{E_1, E_2, E_3\}$, whereas for map 2 the invertible partition is $\mathcal{P}_I = \{E_1 \cup E_2, E_3, E_4 \cup E_5\}$. In this second case, $E_1 = [-1, -(1+c)/2]$ and $E_2 = [-(1+c)/2, -c]$ are associated to $b[n] = 0$, whereas $E_4 = [c, (1+c)/2]$ and $E_5 = [(1+c)/2, 1]$ correspond to $b[n] = 1$. The inverse function, used for coding, is straightforward to obtain for the first map:

$$f(x) = \begin{cases} \frac{(1-c)y - (1+c)}{2}, & s = 1; \\ cy, & s = 2; \\ \frac{(1-c)y + (1+c)}{2}, & s = 3. \end{cases} \quad (11.46)$$

For the second map, the inverse function can also be easily obtained, but now we have to take into account the region to which y belongs in addition to the corresponding symbol:

$$x = f_s^{-1}(y) = \begin{cases} (y - 1)/2, & s = 1, \ y \in E_1 \cup E_2; \\ (y - (1 + 2c))/2, & s = 1, \ y \in E_4 \cup E_5; \\ (y + (1 + 2c))/2, & s = 3, \ y \in E_1 \cup E_2; \\ (y + 1)/2, & s = 3, \ y \in E_4 \cup E_5. \end{cases} \quad (11.47)$$

The good performance of the proposed chaotic communications system is shown in Figure 11.9 for map 2 (the performance of map 1, not shown, is very similar) and several values of the parameter c. The BER obtained

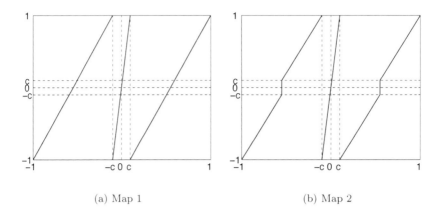

(a) Map 1   (b) Map 2

FIGURE 11.8
PWL maps used for the inverse symbolic chaotic encoding system.

follows the curve of the binary phase shift keying (BPSK) modulation, with approximately 1.5 dB loss for a 10^{-5} probability of error when $c = 0$ and hardly any loss in performance for values of $c > 0.5$.

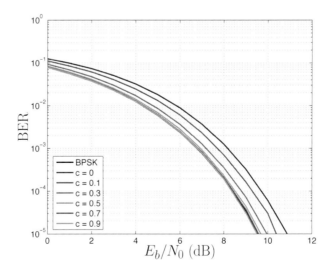

FIGURE 11.9
Bit error rate (BER) for the inverse symbolic chaotic encoding approach using map 2 with several values of c.

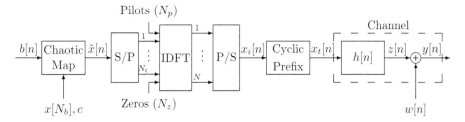

FIGURE 11.10
Block diagram for the proposed OFDM communications system with chaotic coding in the sub-carriers: transmitter and channel.

11.3.2   Combination with OFDM for multipath channels

The coding approach described in Section 11.3.1 provides very good results for the Gaussian channel. However, it suffers from a severe degradation in performance in the presence of multipath interference (unavoidable in wireless communications), just as most chaotic and conventional modulation schemes. In order to provide a certain degree of protection against the distortion introduced by the channel, in this section we propose to combine the chaotic coding with a modulation format robust against multipath fading: orthogonal frequency division multiplexing (OFDM).

The structure of the modulator is shown in Figure 11.10. The idea is simple: substitute the conventional coding used in each subcarrier of the OFDM system (BPSK, QPSK or M-QAM usually) by the chaotic coding described in Section 11.3.1. These coded symbols are then serial-to-parallel converted (S/P) to form blocks of $N_c = N_b$ symbols which, altogether with N_p pilots (used to estimate the channel) and N_z null subcarriers (used as guard intervals), serve to generate the OFDM symbol by means of an $N = N_c + N_p + N_z$ points inverse discrete Fourier transform (IDFT). A parallel-to-serial (P/S) conversion is performed next, a cyclic prefix of M samples is added to avoid intersymbol and intercarrier interference (ISI and ICI), and the signal is finally transmitted through the channel.

The receiver, shown in Figure 11.11, is simply the dual of the transmitter. First, the cyclic prefix is discarded, followed by an S/P conversion and an N point DFT, from which we discard the N_z guard zeros, and use the N_p pilots to update the estimate of the channel. Then, the N_c carriers that contain the useful information are equalized in frequency using the current estimate of the channel. A P/S conversion follows, and finally the N_c equalized samples are passed to the chaotic demodulator to estimate the transmitted information bits.

The good performance of the combined inverse symbolic chaotic encoding with OFDM is shown in Figure 11.12 for two different channels with perfect equalization. In Figure 11.12(a) a simple two-ray minimum phase channel,

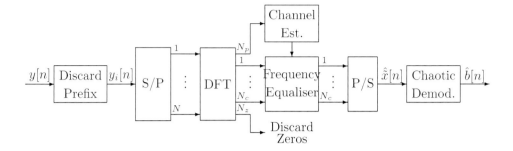

FIGURE 11.11
Block diagram for the proposed OFDM communications system with chaotic coding in the sub-carriers: receiver.

$h_1[n] = \delta[n] - 0.5\delta[n-1]$, is used, whereas in Figure 11.12(b), a more complex non-minimum phase channel, $h_2[n] = -0.3\delta[n-1] + 0.7\delta[n-2] + 0.4\delta[n-3] + 0.1\delta[n-4]$, is used. In both cases the performance of the proposed scheme is similar to the one for the AWGN channel.

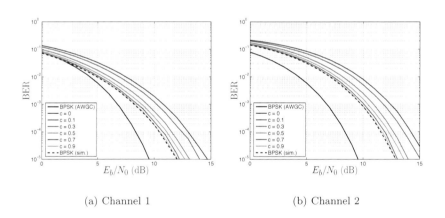

(a) Channel 1   (b) Channel 2

FIGURE 11.12
Bit error rate (BER) for the inverse symbolic chaotic encoding approach plus OFDM for two different channels using map 2 with several values of c.

11.4 Conclusions

In this chapter we have described optimal and suboptimal approaches for the estimation of chaotic sequences, showing how they can be applied to a chaotic communications scheme. First of all, the MLE has been derived for sequences obtained by iterating PWL maps. An efficient approximation to the MLE based on the VA, which provides almost the same performance as the MLE with a substantial reduction of its computational cost, is developed next. Finally, the VA-based MLE is applied to detect the transmitted bits in a chaotic communications approach based on symbolic encoding and backward iteration of PWL maps. The resulting scheme provides an increased degree of security with respect to conventional approaches such as BPSK with a similar performance. Furthermore, we show how it can be combined with OFDM to mitigate the effects of multipath fading typically appearing in wireless communications.

11.5 Acknowledgments

This work has been partly financed by the Spanish government through the CONSOLIDER-INGENIO 2010 Program (Project CSD2008-00010), as well as projects DEIPRO (TEC2009-14504-C02-01), COSIMA (TEC2010-19545-C04-03), and DISSECT (TEC2012-38058-C03-01).

11.6 Appendix: Cramer-Rao lower bound

The CRLB is a lower bound on the variance that can be attained by an unbiased estimator [11]. If an estimator attaining the CRLB is found, then we can ensure that it is the optimum estimator, i.e., the minimum variance unbiased estimator. Otherwise, it can be used as a reference to evaluate the quality of the estimators considered.

The CRLB for the estimation of $x[N]$ is given by

$$\text{Var}(\hat{x}[N]) \geq \left[-\text{E}\left(\frac{\partial^2 \ln p(\boldsymbol{y}; x[N], \boldsymbol{s})}{\partial x[N]^2} \right) \right]^{-1}, \tag{11.48}$$

where $p(\boldsymbol{y}; x[N], \boldsymbol{s})$ is the likelihood, as given by Equation (11.19), but expressed as a function of $x[N]$ and $\boldsymbol{s}$. After some algebra, it can be shown (see [14]) that the CRLB may be expressed as

$$\text{Var}(\hat{x}[N]) \geq \frac{\sigma^2}{1 + \sum_{n=1}^{N} \left(\dot{f}_{\boldsymbol{s}}^{-n}(x[N]) \right)^2}$$

$$= \frac{\sigma^2}{1 + \sum_{n=1}^{N} \prod_{k=0}^{n-1} \left(\dot{f}_{s[N-k-1]}^{-1}(x[N-k]) \right)^2}, \tag{11.49}$$

where $\dot{f}^{-n}$ is used to denote the derivative of f^{-n} $(1 \leq n \leq N)$ and the chain rule has been used to obtain the last expression. Finally, using the analytical expressions for the backward iteration of a generic PWL map given by Equations (11.25) and (11.26), the CRLB for this class of maps becomes

$$\text{Var}(\hat{x}[N]) \geq \frac{\sigma^2}{1 + \sum_{n=1}^{N} \left(B_{\boldsymbol{s}}^{1,n} \right)^2}$$

$$= \frac{\sigma^2}{1 + \sum_{n=1}^{N} \prod_{k=0}^{n-1} a_{s[N-k-1]}^{-2}}. \tag{11.50}$$

The CRLB can be obtained similarly for any other sample of the sequence, as shown in [14]. In particular, the CRLB for the initial condition, $x[0]$, is

$$\text{Var}(\hat{x}[0]) \geq \frac{\sigma^2}{1 + \sum_{n=1}^{N} \left(A_{\boldsymbol{s}}^{0,n} \right)^2}$$

$$= \frac{\sigma^2}{1 + \sum_{n=1}^{N} \prod_{k=0}^{n-1} a_{s[k]}^{2}}, \tag{11.51}$$

where $A_{\boldsymbol{s}}^{0,n}$ are the coefficients for the forward iteration of the PWL map, given by [14, 26]

$$A_{\boldsymbol{s}}^{0,n} = \prod_{k=0}^{n-1} a_{s[k]}. \tag{11.52}$$

Finally, note that a simpler closed-form expression for the CRLB cannot be developed in general for most PWL maps. However, for some particular cases, such as the BSM used in Section 11.2.4, simple analytical expressions can be provided, since the slope is identical in both regions of the natural partition (i.e., $a_1 = a_2 = 2$). In this case, the CRLB of the initial condition is [14]

$$\text{Var}(\hat{x}[0]) \geq \frac{3\sigma^2}{4^{N+1} - 1}, \tag{11.53}$$

whereas the CRLB of the final condition is given by

$$\text{Var}(\hat{x}[N]) \geq \frac{3 \cdot 4^N \sigma^2}{4^{N+1} - 1}. \tag{11.54}$$

These two limits show that $\mathrm{Var}(\hat{x}[0]) \to 0$ as $N \to \infty$, confirming that a consistent estimator for $x[0]$ can theoretically be found, whereas $\mathrm{Var}(\hat{x}[N]) \to 3\sigma^2/4$ as $N \to \infty$, showing the unconsistency in the estimation of $x[N]$. Unfortunately, several authors have proved that the MLE of $x[0]$ is inconsistent for low SNRs, thus not being able to attain the CRLB below a certain SNR threshold that depends on the chaotic map, its parameters, and even the initial condition [8, 12, 34]. Finally, note that the CRLB does not depend on the sample of the chaotic sequence chosen as a reference. Hence, all the equations in this appendix are valid regardless of whether $x[0]$, $x[N]$, or $x[n]$ ($1 \leq n \leq N-1$) is used as a reference point for the estimation.

Bibliography

[1] M. Ciftci and D. B. Williams. Optimal estimation and sequential channel equalization algorithms for chaotic communications systems. *EURASIP Journal on Applied Signal Processing*, 4:249–256, 2001.

[2] M. Ciftci and D. B. Williams. Optimal estimation for chaotic sequences using the Viterbi algorithm. In *IEEE Int. Conf. on Acoustics, Speech and Signal Processing (ICASSP)*, Salt Lake City, NV, 2001.

[3] P. Collet and J. P. Eckmann. *Iterated Maps on the Interval as Dynamical Systems*. Birkhauser, Boston, 1980.

[4] L. Cong, W. Xiaofu, and S. Songgeng. A general efficient method for chaotic signal estimation. *IEEE Transactions on Signal Processing*, 47(5):1424–1428, 1999.

[5] W. de Melo and S. van Strien. *One-Dimensional Dynamics*. Springer-Verlag, Berlin, 1993.

[6] R. L. Devaney. *An Introduction to Chaotic Dynamical Systems*. Addison Wesley, Redwood City, CA, 2nd edition, 1989.

[7] M. Hasler and Y. L. Maistrenko. An introduction to the synchronization of chaotic systems: Coupled skew tent maps. *IEEE Transactions on Circuits and Systems I*, 44(10):856–866, 1997.

[8] I. Hen and N. Merhav. On the threshold effect in the estimation of chaotic sequences. *IEEE Transactions on Information Theory*, 50(11):2894–2904, 2004.

[9] S. H. Isabelle and G. W. Wornell. *The Digital Signal Processing Handbook*, chapter Nonlinear maps. CRC Press & IEEE Press, Boca Raton, FL, 1998.

[10] S. H. Isabelle and G. W. Wornell. Statistical analysis and spectral estimation techniques for one-dimensional chaotic signals. *IEEE Transactions on Signal Processing*, 45(6):1495–1506, June 1997.

[11] S. M. Kay. *Fundamentals of Statistical Signal Processing: Estimation Theory*. Prentice-Hall, Upper Saddle River, NJ, 1993.

[12] S. M. Kay. Asymptotic maximum likelihood estimator performance for chaotic signals in noise. *IEEE Transactions on Signal Processing*, 43(4):1009–1012, April 1995.

[13] S. M. Kay and V. Nagesha. Methods for chaotic signal estimation. *IEEE Transactions on Signal Processing*, 43(8):2013–2016, August 1995.

[14] D. Luengo. *Estimación Óptima de Secuencias Caóticas con Aplicación en Comunicaciones*. PhD thesis, Red (TDR), 2006. Available at http://www.tesisenred.net/handle/10803/10663 (in Spanish).

[15] D. Luengo, C. Pantaléon, and I. Santamaría. Competitive chaotic AR(1) model estimation. In *Proc. XI IEEE Neural Networks for Signal Processing (NNSP) Workshop*, pages 83–92, 2001.

[16] D. Luengo, C. Pantaléon, and I. Santamaría. Bayesian estimation of discrete chaotic signals by MCMC. In *Proc. 11th European Signal Processing Conference (EUSIPCO)*, Toulouse, France, pages 333–336, 2002.

[17] D. Luengo and I. Santamaría. Asymptotically optimal maximumlikelihood estimation of a class of chaotic signals using the Viterbi algorithm. In *Proc. European Signal Processing Conference (EUSIPCO)*, Antalya (Turkey), 2005.

[18] D. Luengo and I. Santamaría. Secure communications using OFDM with chaotic modulation in the subcarriers. In *Proc. IEEE 61st Semiannual Vehicular Tech. Conf. (VTC2005-Spring)*, Stockholm (Sweden), 2005.

[19] G. M. Maggio and L. Reggiani. Applications of symbolic dynamics to UWB impulse radio. In *Proc. IEEE Int. Symp. on Circuits and Systems (ISCAS), volume III*, pages 153–156, Sydney (Australia), 2001.

[20] G. M. Maggio, N. Rulkov, and L. Reggiani. Pseudo-chaotic time hopping for UWB impulse radio. *IEEE Transactions on Circuits and Systems I*, 48(12), December 2001.

[21] N. Masuda, G. Jakimoski, K. Aihara, and L. Kocarev. Chaotic block ciphers: From theory to practical algorithms. *IEEE Transactions on Circuits and Systems I*, 53(6):1341–1352, June 2006.

[22] A. Mooney, J. G. Keating, and I. Pitas. A comparative study of chaotic and white noise signals in digital watermarking. *Chaos, Solitons & Fractals*, 35:913–921, 2008.

[23] C. Myers, S. Kay, and M. Richard. Signal separation for nonlinear dynamical systems. In *Proc. IEEE Int. Conf. Acoustics, Speech, and Signal Processing (ICASSP)*, pages 129–132, San Francisco, CA (USA), 1992.

[24] C. Myers, A. C. Singer, B. Shin, and E. Church. Modeling chaotic systems with hidden markov models. In *Proc. IEEE Int. Conf. Acoustics, Speech, and Signal Processing (ICASSP)*, pages 23–26, San Francisco, CA (USA), 1992.

[25] C. Pantaleón, D. Luengo, and I. Santamaría. Bayesian estimation of a class of chaotic signals. In *In Proc. IEEE Int. Conf. Acoustics, Speech and Signal Processing (ICASSP)*, Istanbul, pages 193–196, 2000.

[26] C. Pantaleón, D. Luengo, and I. Santamaría. Optimal estimation of chaotic signals generated by piecewise-linear maps. *IEEE Signal Processing Letters*, 7(8):235–237, 2000.

[27] C. Pantaléon, D. Luengo, and I. Santamaría. Chaotic AR(1) model estimation. In *Proc. IEEE Int. Conf. Acoustics, Speech and Signal Processing (ICASSP)*, pages 3477–3480, Salt Lake City, NV, 2001.

[28] C. Pantaléon, D. Luengo, and I. Santamaría. Estimation of a certain class of chaotic signals: An EM-based approach. In *Proc. IEEE Int. Conf. Acoustics, Speech, and Signal Processing (ICASSP)*, pages 1129–1132, Orlando, 2002.

[29] C. Pantaléon, L. Vielva, D. Luengo, and I. Santamaría. Bayesian estimation of chaotic signals generated by piecewise-linear maps. *Signal Processing*, 83:659–664, March 2003.

[30] H. C. Papadopoulos and G. W. Wornell. Optimal detection of a class of chaotic signals. In *Proc. IEEE Int. Conf. Acoustics, Speech, and Signal Processing (ICASSP)*, Minneapolis, pages 117–120, 1993.

[31] H. C. Papadopoulos and G. W. Wornell. Maximum-likelihood estimation of a class of chaotic signals. *IEEE Transactions on Information Theory*, 41(1):312–317, January 1995.

[32] J. G. Proakis. *Digital Communications*. McGraw-Hill, Singapore, 1995.

[33] M. D. Richard. Properties and discrimination of chaotic maps. In *Proc. IEEE Int. Conf. Acoustics, Speech, and Signal Processing*, Minneapolis, pages 141–144, 1993.

[34] I. Rosenhouse and A. J. Weiss. Consistent estimation of symmetric tent chaotic sequences with coded itineraries. *IEEE Transactions on Signal Processing*, 56(11):5580–5588, November 2008.

[35] T. Schimming and M. Hasler. Chaos communication in the presence of channel noise. *Journal of Signal Processing*, 4(1):21–28, 2000.

[36] T. Schimming and J. Schweizer. Chaos communication from a maximum likelihood perspective. In *Proc. Int. Conf. on Nonlinear Theory and Applications (NOLTA)*, pages 77–80, Crans Montana (Switzerland), September 1998.

[37] S. Smale. Differentiable Dynamical Systems I. Diffeomorphisms. *Bull. Am. Math. Soc.*, 73:747–817, 1967.

[38] H. L. Van Trees. *Detection, Estimation and Modulation Theory*. John Wiley & Sons, New York, 1968.

[39] S. Wang, P. C. Yip, and H. Leung. Estimating initial conditions of noisy chaotic signals generated by piece-wise linear markov maps using itineraries. *IEEE Transactions on Signal Processing*, 47(12):3289–3302, 1999.

12

Blind source separation in the context of deterministic signals

Diogo C. Soriano, Ricardo Suyama

Centro de Engenharia, Modelagem e Ciências Sociais Aplicadas (CECS)
Universidade Federal do ABC (UFABC)

Rafael A. Ando, Romis Attux

School of Electrical and Computer Engineering (FEEC)
University of Campinas (UNICAMP)

Leonardo T. Duarte

School of Applied Sciences (FCA)
University of Campinas (UNICAMP)

CONTENTS

This chapter presents some central issues concerning the problem of blind source separation (BSS) and extraction (BSE) in the context of deterministic and random sources. First of all, a brief introduction to these problems is provided followed by the classical strategy to solve them using independent component analysis (ICA). In addition to that, a new approach for source separation and extraction based on recurrence quantification analysis is introduced, which is illustrated by a representative set of simulations. As main applications, such strategies can be used in order to separate and extract sources in multiple-input multiple-output (MIMO) chaos-based communication systems or even for denoising chaotic time series in the multirecording context.

12.1 Introduction

The possibility of sending chaotic signals for communication purposes [9, 20] gives rise to some interesting theoretical discussions concerning the employment of classical signal processing techniques in such a context. Interestingly, the idea of exploiting the advantages of deterministic systems capable of engendering chaos for performing a communication task is closely connected to a more fundamental issue in signal processing, that of suitably employing the entire available a priori information (e.g., the full or partial knowledge of state equations, the underlying statistical properties of chaotic signals, etc.) in order to achieve efficient solutions to deal with a specific signal processing scenario.

This matter is clarified if the problem of BSS or BSE is considered in the context of chaotic signals. In essence, the BSS and BSE problems aim to recover a source or a set of sources based on different time series observations that consist in mixed versions of the original signals, without knowledge about the coefficients of the mixing process [12]. Despite the lack of explicit information, some reasonable assumptions about the sources to be recovered and about the structure of the mixing process can be made in order to build a separating system able to invert the mixing effect. For instance, there are many situations in which the sources are associated with independent events and the mixing model assumes a simple linear and instantaneous form. In this case, the original signals can be recovered by means of the classical ICA approach [12].

Obviously, different types of a priori information concerning the sources or the mixing model can be available for each separating scenario and to explore the specificities of each one has become an active research issue in modern signal processing. In fact, from the theoretical and practical standpoint, the BSS and BSE problems define common challenges found in geophysics, biomedical signal processing, audio processing, speech processing, and also in communication systems [6, 12, 21].

Therefore, given the relevance of BSS and BSE problems in different signal processing contexts, this chapter presents a brief introduction of such paradigms, their classical solutions based on independent component analysis, and some possible alternative criteria when deterministic signals are considered in order to enhance the separation/extraction efficiency. In particular, it is shown here that recurrence quantification analysis (RQA) — a signal processing technique commonly used in the analysis of deterministic time series [10, 15, 27, 29] — can be used to perform a more robust separation or extraction of chaotic sources in different mixing scenarios, considering mixtures of different deterministic signals with one another or even mixtures of deterministic and stochastic sources. More than that, this chapter shows that RQA-based measures can also perform the classical ICA solution (i.e., out of the deterministic scenario), which is done by the link between RQA and information theoretic-measures. Concerning the application of such techniques in the context of chaos-based communication systems, we can mention the possibility of denoising chaotic time series or simply separate different chaotic signals in MIMO communication systems [14,22], something that is illustrated here by a representative set of simulations.

The present chapter is divided as follows: Section 12.2 presents a brief introduction of the blind source separation/extraction problem, which is followed by ICA-based strategies to solve them in Section 12.3. Section 12.4 introduces the principle of complexity-based separation by means of RQA measures in the context of deterministic sources. Section 12.5 closes the chapter providing a discussion concerning applications and perspectives in the framework of chaos-based communication systems.

12.2 Blind source separation and blind source extraction

An intuitive way to introduce the BSS and BSE problems is by means of the cocktail party problem [6, 12]. Consider that you are in a party hearing several conversations at the same time, and, as a major task, you wish to separate or just pay attention at only one speaker. In fact, the human brain has the amazing property of performing this kind of separation/extraction task by means of high-level cognitive processing, which implies in intuitively exploiting the underlying information related to attributes like the timber or sparsity of a voice.

In practical terms, understanding the mechanisms underlying the cognitive processing performed by the brain and adapting them into efficient and reliable signal processing algorithms is quite a hard task. Fortunately, BSS and BSE can be described (in a more restrictive scenario) using classical signal processing tools, which, obviously, is more restrictive and requires a whole set of assumptions. Thus, within this artificial framework, the separation/extraction problem can be described in terms of several sensors that record time series that consist of mixed versions of the original sources. This mixing process obeys a known model with unknown coefficients, and the separation/extraction task relies on the estimation of a suitable transformation to be applied to the observations in order to invert the mixing effect and restore the original sources [6, 8, 12].

Formally, let us consider a set of N sources, whose samples define a vector $\mathbf{s}(k)$, and a set of M signals captured by the sensors that define the observations, given by a vector $\mathbf{x}(k)$. Generically, we can state that the observation vector obeys the description:

$$\mathbf{x}(k) = \mathbf{F}(\mathbf{s}(k), \dots, \mathbf{s}(k - L), \mathbf{n}(k), k) \tag{12.1}$$

$\mathbf{F}(.)$ being, in principle, a general function (e.g., linear, nonlinear, time variant, etc.), which depends on the original sources and a noise $\mathbf{n}(k)$ vector.

If the mapping defined by $\mathbf{F}$ is known, we can estimate the original sources by means of the application of its inverse $\mathbf{F}^{-1}$ (supposing that such an inverse exists) to the observation vector. On the other hand, in the blind scenario, the coefficients of the mixing process are not available and the estimation of $\mathbf{F}^{-1}$ can be rather complex. In fact, the problem as stated in this general form has a strong practical appeal, given the high number of models that fit this description. On the other hand, canonical solution for this general formulation is impractical, and, usually, simplified models are adopted for solving with success particular cases, some of which are described in the following.

A mixing system is said to be *linear* if the mapping $\mathbf{F}(.)$ obeys the superposition principle, i.e.,

$$\mathbf{F}(\alpha_1 \mathbf{s}_1(k) + \alpha_2 \mathbf{s}_2(k)) = \alpha_1 \mathbf{F}(\mathbf{s}_1(k)) + \alpha_2 \mathbf{F}(\mathbf{s}_2(k)) \tag{12.2}$$

for any constants α_1 and α_2, and signal vectors $\mathbf{s}_1$ and $\mathbf{s}_2$. If the mixing process does not obey this condition, it is said to be *nonlinear*.

In addition to that, the mixing system can also be classified as instantaneous or, alternatively, with memory. This classification is directly related to the parameter L in Equation (12.1), and, if $L > 0$, the observation reflects mixtures of current samples with past ones, and the system, if linear, is said to be *convolutive*. On the other hand, if $L = 0$, the system is said to be *instantaneous*.

Concerning the number of original sources (N) and sensors (M), when the latter is greater than the former ($M > N$), the mixing model is called *overdetermined*. On the other hand, when $M < N$, the system is said to be *underdetermined*. In particular, it is also important to distinguish the BSS problem from the BSE one. In the first case, it is wished to recover all the original sources, while in the latter just a subset of signals is restored.

Following these definitions, it can be stated that the most simple mixing model is given by a linear, instantaneous, and determined system (which implies $M = N$), allowing the description of the observation vector in terms of a linear transformation $\mathbf{A}_{M \times N}$ applied to the original sources, in the form

$$\mathbf{x}(k) = \mathbf{A}\mathbf{s}(k), \tag{12.3}$$

i.e., the observations are given by linear combinations of the source signals, which are defined by the mixing matrix $\mathbf{A}_{M \times N}$.

Finally, aiming at a clear exposition of the main concepts related to the

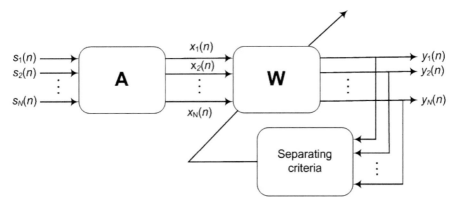

FIGURE 12.1
Blind source separation/extraction scheme. In this description, the N original sources that define the source vector $\mathbf{s}(n)$ are shown, as well as the M observations that lead to the observation vector $\mathbf{x}(n)$; the mixing matrix $\mathbf{A}$; the separating system $\mathbf{W}$ that is adapted following a specific separating/extracting criteria; the components of the output vector $\mathbf{y}$.

BSS and BSE problems, it is assumed here, for sake of conciseness, that the observations do not contain noise and that the number of sources equals the number of sensors, which defines a square mixing matrix $\mathbf{A}_{N \times N}$. These considerations lead to the separation/extraction scheme shown in Figure 12.1, which will be henceforth assumed to be valid in this work.

This approach defines a natural strategy not only for exposition purposes, but also for setting a first scenario for testing new ideas and solutions, as the one proposed here in the case of deterministic sources. Details concerning nonlinear mixing models and convolutive mixtures can be found in [6] and some works related to them outline the current state of art in source separation.

12.3 Solving the BSS/BSE problem based on independent component analysis

Assuming that the mixing model is given by linear, instantaneous, and invertible mapping, the original signals can be recovered by means of a separating system $\mathbf{W}$ such that

$$
\begin{aligned}
\mathbf{y}(k) &= \mathbf{W}\mathbf{x}(k) \\
&= \mathbf{W}\mathbf{A}\mathbf{s}(k) \\
&= \mathbf{s}(k),
\end{aligned} \tag{12.4}
$$

i.e., $\mathbf{W}^{-1} = \mathbf{A}$. However, despite the information available about the structure of the mixing processing, the elements of $\mathbf{A}$ are not known, which demands additional assumptions to accomplish the separation/extraction task.

A possible solution for determining $\mathbf{W}$ relies on exploiting the statistical information underlying the sources, for instance, assuming (if reasonable for the problem under study) statistical independence between them. In fact, this is the central consideration that gives support to the use of independent component analysis, and, despite being apparently restrictive, it can be observed in many practical applications.

Mathematically, a vector of random variables $\mathbf{s}$ has statistically independent elements if

$$p_s(\mathbf{s}) = p_{s1}(s_1)p_{s2}(s_2)\ldots p_{sN}(s_N), \qquad (12.5)$$

$s_1, s_2, \ldots, s_N$ being the vector components and $p_{si}(\cdot)$ the probability density function of the i-th component.

If the statistical independence condition is valid for the signals of interest and there is no more than one Gaussian source [5], the recovery task can be performed by applying a suitable transformation $\mathbf{W}$ to the observation vector that causes the elements of $\mathbf{y}(k)$ to be independent, hence estimating the signals of $\mathbf{s}(k)$. Additionally, as the coefficients of the mixing process are unknown, it is impossible to determine the power of output components $y_i(n)$ and also their relative order. These two limitations imply that the original signals are recovered up to scaling and ambiguity factors. Having these considerations in mind, the use of ICA in BSS and BSE can be formalized as:

Definition 1 *The* ICA *of a random vector* $\mathbf{x} = [x_1 \ x_2 \ \ldots \ x_M]^T$ *consists in the determination of a matrix* $\mathbf{W}$ *such that the elements of the random vector* $\mathbf{y} = \mathbf{W}\mathbf{x}$ *maximize a cost function that quantifies the degree of independence between the components of* $\mathbf{y}$.

From Definition (1), it can be clearly noted that source vector estimation is based on two main concepts: statistical independence and non-Gaussianity. Obviously, different criteria can be employed depending on the available *a priori* information about the sources. Usually, the knowledge of specific characteristics of the signals allows an increased efficiency and robustness in separation/extraction at the cost of a smaller degree of generality, something that will be illustrated here in the context of deterministic signals.

Before dealing with such a case, it is important to show how the separating system $\mathbf{W}$ can be generally adjusted using ICA [4]. The first step will involve a parameterization of $\mathbf{W}$, which can be achieved assuming that the mixing process assumes an orthogonal structure. Such assumption can be done without lack of generality, since a whitening procedure can always be applied to the observation vector as a pre-processing stage, producing orthogonal vectors as inputs to the separating system. Details about how to perform the whitening process can be found in [12, 21]. To illustrate such a situation, consider the case in which two sources are mixed by an $A_{2 \times 2}$ mixing matrix producing an observation vector that is preprocessed by means of a whitening stage, allowing the parameterization of the separating matrix also by means of an orthogonal matrix with a single variable θ in the form

$$\mathbf{W} = \begin{bmatrix} \cos\theta & \sin\theta \\ \text{-}\sin\theta & \cos\theta \end{bmatrix}. \tag{12.6}$$

ICA implies choosing a value of θ that maximizes a measure of independence between the output vector. The optimization problem underlying the choice of $\mathbf{W}$ can be solved by an exhaustive search for this particular case, which can be inappropriate for a higher number of sources. Alternatively, if it is wished to extract just one source, a separating vector $\mathbf{w}$ in the form shown in Equation (12.7) can be employed:

$$\mathbf{w} = \begin{bmatrix} \cos\theta & \sin\theta \end{bmatrix}^T. \tag{12.7}$$

In fact, there are ICA-based methods (like the popular FASTICA algorithm [12]) specially devoted to deal with this optimization problem in a time-efficient way. For those interested in performing blind source separation/extraction in the context of a high number of sources, we strongly recommend [4, 21] for classical and computational-intelligence based approaches for adjusting the parameters of $\mathbf{W}$.

12.3.1 ICA-based measures

There are different practical ways of directly or indirectly seeking for statistical independence in order to adapt $\mathbf{W}$ [6, 12, 21]. A common approach relies on the maximization of the non-Gaussianity of the components of the output vector. This strategy follows directly from the central limit theorem [19], which states, in simple terms, that the sum of independent random variables tends to generate components with a probability density function closer to Gaussian variables, a tendency that is illustrated in Figure 12.2.

In addition to that, there are measures based on higher order statistics that can be used to define a non-Gaussianity cost based function (fit_k). A commonly used option is the kurtosis, defined as in

$$\text{fit}_k = K(\alpha) = E\{\alpha^4\} - 3(E\{\alpha^2\})^2, \tag{12.8}$$

$E\{.\}$ being the statistical expectation operator [7]. Thus, a possible approach for performing ICA of a random vector is to determine $\mathbf{W}$ such that the absolute value of the kurtosis of the output components y_i are maximum, which leads to

$$\max_{\mathbf{W}} \mid \text{fit}_k(y_i) \mid. \tag{12.9}$$

Another possible way to seek for independence in the output vector can be defined by an information-theoretic approach. In general, this strategy aims at

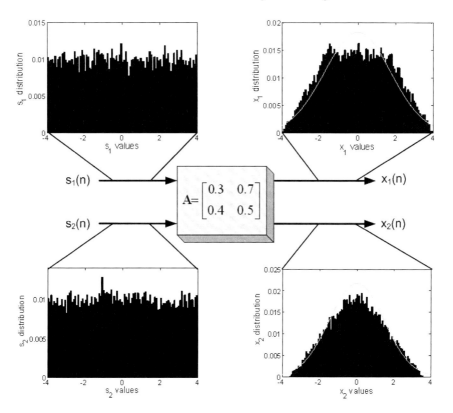

FIGURE 12.2
Two independent uniform distributed sources mixed by a linear transformation giving rise to random variables closer to Gaussian distributed observations.

the minimization of the mutual information between the $\mathbf{y_i}$ components [7]. More specifically, the mutual information between two random variables α and β can be defined as

$$I(\alpha, \beta) = H(\alpha) - H(\alpha \mid \beta), \tag{12.10}$$

$H(\alpha)$ being the Shannon entropy of the random variable α and $H(\alpha \mid \beta)$ the conditional entropy of α given β. Thus, the mutual information can be interpreted in terms of the information gained about α by knowing β. This provides a quantity that is null for independent events (i.e., the conditional uncertainty $H(\alpha \mid \beta)$ equals the uncertainty about α) and positive otherwise. In this case, independency is implicity achieved when the mutual information cost function (fit$_{\mathrm{MI}}$) between the y_i components is minimized, i.e.,

$$\min_{\mathbf{W}}\{\mathrm{fit}_{\mathrm{MI}} = \sum_{i=1}^{N} H(y_i) - H(\mathbf{y})\}. \qquad (12.11)$$

In practical terms, the mutual information can be obtained by means of the estimation of the probability density functions of the random variables followed by the evaluation of Shannon entropy for the marginal and joint distributions associated with each component y_i [7]. Moreover, there are also possible alternative ways of evaluating independence (e.g., using the principle of information maximization INFOMAX [2] and a maximum likelihood formulation [3]), but, for our purposes, the criteria exposed in Equations (12.9) and (12.11) are enough to illustrate the potential applicability of ICA for separating deterministic signals.

Before introducing the whole separation scenario, which includes the preprocessing techniques and the parameterization of the separating systems to be adapted following the exposed ICA-based measures, the next subsection presents an alternative approach for performing separation/extraction when it is known that the original sources to be recovered have a deterministic structure.

12.3.2 Chaotic signals and the generation of recurrence plots

In formal terms, a chaotic signal is defined as a continuous-valued signal with finite and positive entropy rate and infinite redundancy rate [13], but, for our purposes, a chaotic signal should be simply understood as one generated by a chaotic system, which means that its properties are defined by the dynamical system that generates it and the behavior of its trajectories in the phase space. In order to capture this behavior, the Takens theorem [1] can be applied for reconstructing the underlying attractor from a single observed signal, defining a state vector $\mathbf{x}(k)$ such that

$$\mathbf{x}(k) = \left[\begin{array}{cccc} x(k) & x(k-\tau) & \cdots & x(k-(d_e-1)\tau) \end{array} \right], \qquad (12.12)$$

where d_{e} represents the embedding dimension and τ represents the delay between samples. Even though this trajectory may not be exactly the same generated by the system, it will be topologically equivalent thereto [1] providing the same invariant measures.

Thus, after reconstruction, it is possible to characterize the attractor with the aid of its revisited states, which can be done with a recurrence plot [17, 27, 28]. This approach allows a great versatility, since it can provide means of quantifying different invariant measures (e.g., correlation dimension, Kolmogorov-Sinai entropy, etc. [11, 17]) and also some classical information-theoretic ones, as the mutual information itself. This particular capability of quantifying both classical nonlinear dynamic measures and also information-theoretic ones confers to the RQA the possibility of defining a general approach for being employed in mixing scenarios of different natures (i.e., for deterministic and random sources).

More specifically, the recurrence plot is created by determining a binary distance matrix that indicates which states in the reconstructed attractor are closer (in terms of a threshold distance ϵ) to one another. This can be performed by generating the reconstructed state $\mathbf{x}(k)$ and building an $N \times N$ matrix, where the element (i, j) will be a black dot (or a bit 1 from the binary alphabet $(0, 1)$) whenever $\mathbf{x}(i)$ is sufficiently close to $\mathbf{x}(j)$, i.e., whenever $\|\mathbf{x}(i) - \mathbf{x}(j)\| < \epsilon$. Formally, the recurrence matrix $\mathbf{R}_{i,j}$ can be defined as [17]

$$\mathbf{R}_{i,j} = \Theta(\varepsilon - \|\mathbf{x}(i) - \mathbf{x}(j)\|), \tag{12.13}$$

$\Theta(.)$ being the Heaviside function given by

$$\Theta(x) = \begin{cases} 0, & \text{if } x < 0 \\ 1, & \text{otherwise} \end{cases}. \tag{12.14}$$

The characterization and applicability of recurrence plots become clear by comparing maps obtained from signals of different natures. In Figure 12.3, maps generated from a periodic signal (a), a chaotic signal (b), a random signal (c), and a mixture of chaotic and random signals (d) are presented. The patterns clearly differ in their structure and their regularity, and, more than simply capturing the correlation characteristics, offer means of accessing the small-scale structure of the time series. This quantitative analysis established in terms of different statistical measures of the obtained recurrence patterns is the core of recurrence quantification analysis and the underlying contrast functions used here to perform the source separation and extraction. For details in the construction of recurrence plots see [15, 16].

The applicability of RQA is evident in Figure 12.3 when it is noticed that chaotic time series tend to generate shorter diagonals than those associated with periodic signals, but longer than those associated with a random process. This suggests that statistics based on the structure of the diagonals in a recurrence plot can be used to define contrast functions to separate deterministic from random signals. These contrasts can be built from classical recurrence measures [17, 27, 28] like the percentage of determinism (fit_d), the entropy of the diagonals (fit_e), and the longest diagonal (fit_l) found in the map. Formally, if $P(\varepsilon, l)$ is defined as the frequency distribution of line lengths l for a map with resolution ε, the score function associated with the deterministic content in a window of diagonal lengths a to b can be mathematically described as

$$\text{fit}_\text{d} = \sum_{l=a}^{l=b} lP(l) \left/ \sum_{l=1}^{l=N} lP(l) \right. \tag{12.15}$$

The score function related to the longest diagonal line (excluding the main diagonal) found in the map can be defined as

$$\text{fit}_\text{l} = \max(\{l_i\}_{i=1}^{N_l}), \tag{12.16}$$

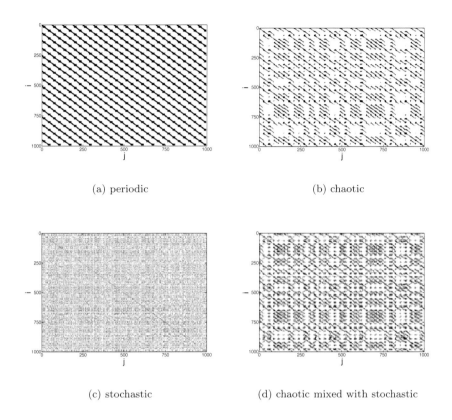

(a) periodic

(b) chaotic

(c) stochastic

(d) chaotic mixed with stochastic

FIGURE 12.3
Panels (a), (b), (c), and (d) show, respectively, the recurrence plots ($N = 1000$ samples, $d_e = 3, \tau = 3, \varepsilon = 0.5$) from a periodic oscillation ($sin(10t)$), a chaotic Lorenz time series, a Gaussian random source (zero mean and unitary variance), and a mixture of random and chaotic sources.

where N_l is the total number of diagonal lines (excluding the main one). Finally, the score function based on Shannon entropy associated with the probability distribution $p(l) = P(l)/N_l$ of finding a diagonal of length l is defined by

$$\text{fit}_e = - \sum_{l=l\,\text{min}}^{N_l} p(l) \ln(p(l)) \qquad (12.17)$$

and is related to a measure of complexity of the recurrence plot (e.g., uncorrelated noise has a small fit_e, reflecting low complexity, while periodic signals — as the one presented in Figure 12.3(a) — tend to exhibit more uniform distributed diagonals, which is associated with a higher fit_e and higher complexity). The threshold $l_{\min}$ excludes the diagonal lines which are formed by tangential motion of phase space trajectories [17].

In this context, analogously to what has been defined for ICA, it is also possible to define a separating/extracting criterion that attempts to maximize the deterministic character of the output components y_i. This optimization problem can be mathematically described as

$$\max_{\mathbf{W}} \mid \mathrm{fit}_x(y_i) \mid, \tag{12.18}$$

$\mathrm{fit}_x(y_i)$ being any RQA-based cost function: $\mathrm{fit}_d(y_i)$, $\mathrm{fit}_e(y_i)$, or $\mathrm{fit}_l(y_i)$. The maximization of $\mathrm{fit}_d(y_i)$ in the context of an interval of diagonals (a, b) that excludes points in short segments (an intrinsic characteristic of white random signals) will necessarily lead to the most deterministic component (with highest deterministic content in (a, b)), while the maximization of $\mathrm{fit}_l(y_i)$ will recover the component associated with the most regular source (as illustrated in Figure 12.3). Finally, regular and chaotic sources tend to generate more uniformly distributed diagonals, which engender higher entropy values, justifying the maximization of $\mathrm{fit}_e(y_i)$ for the search of deterministic sources.

This approach allows the identification of the deterministic components at each output, being, therefore, an attractive way for performing the blind extraction of deterministic sources, something that eliminates the ambiguity factor presented in ICA. Obviously, in the scenario where just stochastic sources are considered, the extraction performance will depend on the correlation characteristics of each source, the extraction with RQA-based measures being unreliable for white signals.

12.4   Results

12.4.1   Extracting chaotic sources from mixtures with stochastic signals

Let us consider at first the scenario in which a chaotic signal obtained from the Lorenz model - preprocessed to have zero mean and unit variance - is mixed with a stochastic Gaussian source with zero mean and power 2 dB below the deterministic source. Assume also that these sources are mixed by an orthogonal mixing matrix $\mathbf{A}$, in the form

$$\mathbf{A} = \begin{bmatrix} \sin\theta & -\cos\theta \\ \cos\theta & \mathrm{sen}\theta \end{bmatrix}, \tag{12.19}$$

with $\theta = \pi/6$, which tacitly considers that a preprocessing whitening stage has already being applied to the observations. This situation is illustrated in Figure 12.4, where the time series of the original sources and the observation

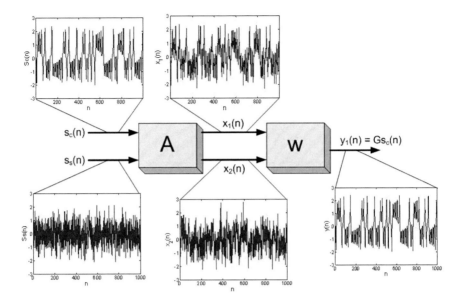

FIGURE 12.4
Extraction scenario defined by mixtures of chaotic $(s_c(n))$ and stochastic $(s_s(n))$ sources and an instantaneous linear mixing model $\mathbf{A}$ leading to the observations $x_1(n)$ and $x_2(n)$. In this case, the extracting structure $\mathbf{W}$ is designed to invert the mixing process and just recover the deterministic source up to a scaling factor.

are available. Within this framework, the goal is to build a separating vector as defined in Equation (12.7), in order to find the values θ that recover the deterministic source.

To accomplish this task, an exhaustive search in this parameter has been performed with two criteria: mutual independence and determinism maximization. Note that the computation of the mutual information necessary requires operation in a source separation scenario, since all the components of the output vector are needed to evaluate this measure. Specially in this case, the separation matrix defined in Equation (12.6) was used, while all the other cost functions (fit$_k$, fit$_d$, fit$_e$, and fit$_l$) were evaluated just for the component y_1 obtained after the application of the separating vector in Equation (12.7) to the observation vector. The evaluation of the cost functions in terms of θ gives rise to Figure 12.5, in which it is clear that both the maximization of the independence (maximization of kurtosis of y_1 and the minimization of mutual information between the output components) and of the complexity criteria evaluated by the exposed RQA-based measures leads to the perfect solution that inverts the mixing process ($\theta = \pi/6$, as shown by the first vertical line in the upper panel).

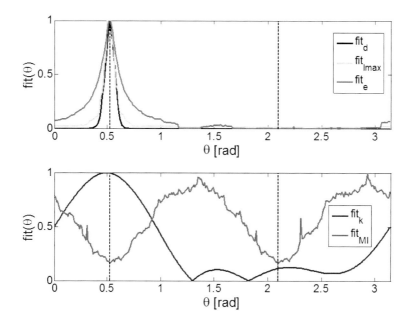

FIGURE 12.5 (SEE COLOR INSERT)

Exhaustive search in θ parameter for the maximization of statistical independence and deterministic characteristics of the output component y_1. All the cost functions were normalized to the unity. The recurrence parameters $d_e = 3$, $\tau = 6$, ϵ were adjusted to produce a recurrence rate - the density of points in the recurrence plot - of 1%, $N = 2000$ samples.

With the results shown in Figure 12.5, it is clear that the RQA-based measures have the desired requirements to be considered a suitable contrast function in order to perform the extraction of the deterministic signal. More than that, while the RQA-based measures seem to precisely indicate the solution that inverts the mixing process (given by $\theta = \pi/6$ - the first vertical line in the panels), the ICA-based estimators seem to be less exact (something that is supported here by the imprecision in the kurtosis peak and also by the fluctuations observed in mutual information cost function), which is probably associated with the requirement of more samples to provide reliable estimates. This is a typical instance of how more specific information (i.e., knowledge about the deterministic nature of the desired source) can improve the extraction robustness. In fact, despite the fact that mutual information had achieved the same performance obtained for the RQA cost functions in this extracting scenario (a mean squared error — MSE — of 9×10^{-4} between the original and recovered source), an extensive analysis increasing the noise power and reducing the number of samples [23, 25] reveals a much more critical effect for the ICA approach.

It is also interesting to observe that, for the exposed separating scenario, the classical ICA permutation ambiguity is not present when recurrence-based measures are employed, since the separation criterion is founded on information that specifies the source to be recovered. On the other hand, the kurtosis of the y_2 component or even the mutual information as previously defined can also provide the solution for restoring the stochastic source (note that fit$_{MI}$ points to a second minimum at $\theta = 2\pi/3$, i.e., to the solution that would perfectly recover the stochastic source), which, in this term, can be interpreted as a more general framework, i.e., not restricted to the extraction of deterministic signals.

As a major application aspect, it can be noticed that this extraction scenario is intrinsically associated with the denoising problem established in the multi-recording context. Although the requirement for several observations of a process would sound a little bit restrictive, there are several practical instances where they can be available (e.g., in electroencephalography or electrocardiography), which strongly motivate the development of a deterministic component analysis based on RQA measures.

12.4.2 Separating deterministic sources

An interesting perspective concerning chaos-based communications systems is the possibility of separating different deterministic signals using both ICA- and RQA-based measures. This separation scenario is illustrated in Figure 12.6, where two chaotic time series (one obtained from the Lorenz system and the other from the Rössler model) are mixed with a full-rank matrix $\mathbf{A}$, exactly as stated in the previous subsection (i.e., taking $\theta = \pi/6$ in Equation (12.19)).

In this case, the separating system is given by Equation (12.6), and, once again, the solution in θ that restores the original sources can be found by an exhaustive search aiming to maximize independence and the deterministic character of the output vector. The evaluation of all previously defined score functions is shown in Figure 12.7, where it can be observed that two main deterministic components are present. In fact, the first component is associated with the recovery of the Lorenz signal in the y_1 component when $\theta = \pi/6$, while the second one refers to the recovery of the Rössler time series when $\theta = 2\pi/3$. It can also be noted that the separation can be performed by means of ICA-based measures, since the kurtosis peak and the mutual information minimum provide the same solutions obtained in the RQA context.

Figure 12.7 also clarifies how the RQA score functions can achieve different performances depending on the nature of the original sources. For instance, when just deterministic sources are considered in the mixing model, fit$_e$ naturally tends to point to mixtures more organized, i.e., with diagonals resembling pure deterministic signals, which clearly introduces an offset that

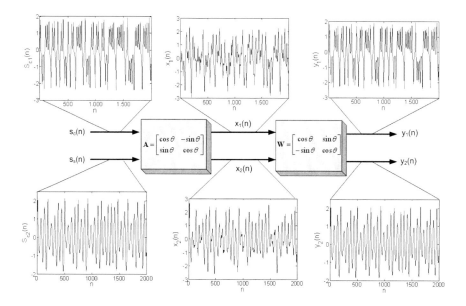

FIGURE 12.6
Separation scenario defined by mixtures of a Lorenz $(s_{c1}(n))$ and Rössler time
series $(s_{c2}(n))$ considering an instantaneous linear mixing model $\mathbf{A}$ leading to
the observations $x_1(n)$ and $x_2(n)$.

does not stand in the previous mixing scenario (i.e., where deterministic and
stochastic sources were mixed). This kind of offset induced by the deterministic
nature of the observations is also observed for fit_{lmax} measure, which, despite
such behavior, has a more precise peak definition (at least on $\theta = 2\pi/3$) and
provides a better estimation of the perfect solution that recovers the Rössler
signal (note that at this point fit_e exhibit a double peak, which does not fulfill
the formal requirement for defining a contrast function).

Moreover, the definition of the diagonal length interval according to which
the percentage of points of the recurrence plot is calculated, taken for computing fit_d, is crucial to determine the profile of this score function. For instance,
Figure 12.8 shows the the deterministic content — i.e., the percentage of points
in diagonals from length a to b — for different intervals: it is clear that the
main peak of fit_d is shifted according to the chosen interval. This illustrates
that it is possible to separate deterministic signals if the recurrence criterion
is adopted, although additional specification is needed with respect to the
recurrence characteristics, since both sources are deterministic.

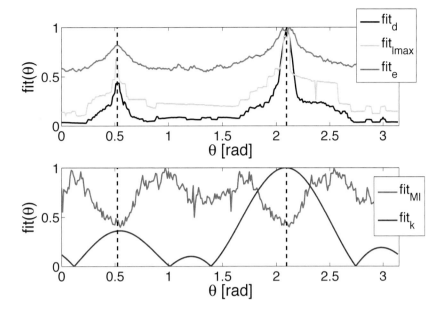

FIGURE 12.7 (SEE COLOR INSERT)
Exhaustive search in θ parameter that enhances the statistical independence or deterministic characteristics of the output component y_1. All the cost functions were normalized to the unity. Recurrence parameters: $d_e = 3$, $\tau = 6$, $\epsilon = 0.1$, $N = 2000$ samples. The perfect solutions that restore the Lorenz and Rössler signals in the y_1 component are given by, respectively, $\theta = \pi/6$ and $\theta = 2\pi/3$.

12.4.3   Performing ICA by means of RQA-based measures

More than simply quantifying the deterministic "fingerprint" of a time series, the RQA can alternatively define useful complexity and information-theoretic measures. As shown in [17], the first step to achieve the mutual information definition based on recurrence depends on the calculation of the second Rényi entropy for the signal in the form [17]

$$H_2 = -\ln\left(\frac{1}{N^2}\sum_{i,j=1}^{N}\mathbf{R}_{i,j}\right) = -\ln(\mathrm{RR}(\varepsilon)), \qquad (12.20)$$

$RR(\varepsilon)$ being the recurrence rate, i.e., the density of points in the recurrence plot. For our purposes, it is extremely convenient not to consider the points in the main diagonal, since these points mandatorily appear in the recurrence plots of different signals and may introduce an undesirable bias in the mutual information estimator as presented further. This implies changing Equation

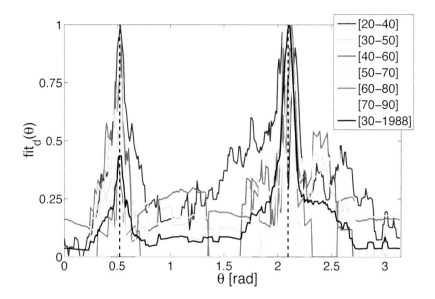

FIGURE 12.8 (SEE COLOR INSERT)

Normalized $\mathrm{fit_d}$ as a function of θ for different diagonal length intervals ($[a\ b]$ — legend) for the output component y_1. Recurrence parameters: $d_e = 3$, $\tau = 6$, $\epsilon = 0.1$, $N = 2000$ samples. The perfect solutions that restore the Lorenz and Rössler signals in the y_1 component are given by, respectively, $\theta = \pi/6$ and $\theta = 2\pi/3$.

(12.20) to

$$\widehat{H}_2 = -\ln\left(\frac{1}{N(N-1)}\sum_{i,j=1 i\neq j}^{N}\mathbf{R}_{i,j}\right). \qquad (12.21)$$

Hence, a recurrence-based mutual information estimator between $y_1(n)$ and $y_2(n)$ can be defined following the same definition presented in Equation (12.10)

$$\widehat{I}_2^{y_1,y_2} = \widehat{H}_2^{y_1} + \widehat{H}_2^{y_2} - \widehat{H}_2^{y_1,y_2}, \qquad (12.22)$$

where $\widehat{H}_2^{y_1,y_2}$ is the estimated joint second-order Rényi entropy of $y_1(n)$ and $y_2(n)$, given by

$$\widehat{H}_2^{y_1,y_2} = -\ln\left(\frac{1}{N(N-1)}\sum_{i,j=1 i\neq j}^{N}\mathbf{JR}_{i,j}^{y_1,y_2}\right), \qquad (12.23)$$

$\mathbf{JR}_{i,j}^{y_1,y_2}$ being the joint recurrence matrix, i.e.,

$$\mathbf{JR}_{i,j}^{y_1,y_2}(\varepsilon^{y_1},\varepsilon^{y_2}) = \Theta(\varepsilon_{y_1} - \|y_1(i) - y_1(j)\|)\Theta(\varepsilon_{y_2} - \|y_2(i) - y_2(j)\|). \quad (12.24)$$

The possibility of defining a mutual information measure from RQA opens the perspective of not only performing the blind source separation and extraction in the context of deterministic signals, but also of performing classical ICA, which implies separating also purely random signals. In fact, this information-theoretical approach defines an important link between these two most relevant signal analysis strategies.

To illustrate this point with the aid of a representative set of simulations, we have mixed two uniform distributed sources and adapted the separating system following the independence criteria based on the minimization of the mutual information of the output vector components. Mutual information estimation was performed both by means of the classical histogram and recurrence approaches as defined, respectively, in Equation (12.10), which defines the fit_{MI} — and Equation (12.22) — that leads to fit_{I2}. The employed mixing matrix was identical to that one presented in the previous examples.

In can be clearly observed in Figure 12.9 that both mutual information estimators were able to find the ideal solutions that recover the original sources, attesting that it is possible to perform ICA by means of RQA. Concerning the obtained performance, it can be observed that fit_{I2} exhibit a little shift from the exact solutions (traced lines), but with a lower offset (i.e., a measure close to zero, which is in accordance with the ideal values considering independent uniform distributions).

It is also interesting to observe that, in the separation of deterministic sources, the different mutual information estimators can achieve distinct performances. For instance, if we consider the separation exclusively of deterministic source (as in the previous mixing scenario), e.g., the Lorenz time series and a periodic source (e.g., $\sin(5t)$), it can be noticed that mutual information evaluated by fit_{I2} has a better performance than fit_{MI} or even the kurtosis measure (Figure 12.10). In fact, the quality of the sources restored using ICA-based measures in this case is significantly worse than the one obtained with recurrence.

Finally, with this set of simulations, we have shown the separation/extraction of chaotic signals in different mixing scenarios, which include mixtures of deterministic and stochastic signals, only of deterministic signals, or even only random sources using ICA and RQA measures.

12.5   Discussions and conclusions

In the present chapter, we have presented some central issues concerning the problem of blind source separation and extraction in the context of deterministic and random sources. After formulating the BSE and BSS problems, the classical approach to solving them using independent component analysis was

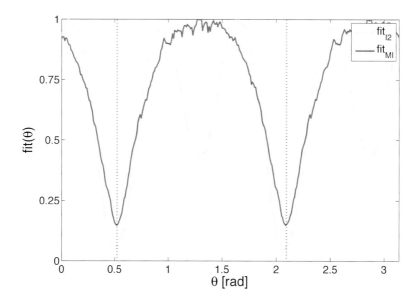

FIGURE 12.9 (SEE COLOR INSERT)
Estimated mutual information of the output vector for different values of θ using classical histogram — (fit$_{MI}$) and recurrence (fit$_{I2mod}$) — based approaches. All the cost functions were normalized to have maximum value equal to one. The recurrence parameters are: $d_e = 1$, $\epsilon = 0.4$, $N = 6000$ samples. The solutions that perfectly restore the original sources are given by, respectively, $\theta = \pi/6$ and $\theta = 2\pi/3$.

discussed. This classical paradigm was then contrasted with a new strategy for performing separation/extraction based on recurrence quantification analysis.

We have shown here that classical RQA can establish a solid criterion for recovering not only deterministic sources mixed with stochastic ones, but also mixed chaotic signals. When only random signals are considered, recurrence quantification is still capable of performing the separation/extraction, since it also offers means of defining information theoretic measures (e.g., mutual information), which allows performing classical ICA. Moreover, this separation/extraction approach seems to be in accordance with an idea previously introduced by Pajunen [18], which attempts to minimize the complexity of the components of the output vector in order to restore the original sources. Implicitly, the maximization of the recurrence-based measures as performed here implies seeking for more "regular" sources in the output vector, which means less (algorithmic) complex components as proposed in [18].

To illustrate the practical relevance of ICA and RQA approaches in the context of the exposed problem, we can mention the applicability of source separation to different scientific areas, in which a deterministic model can

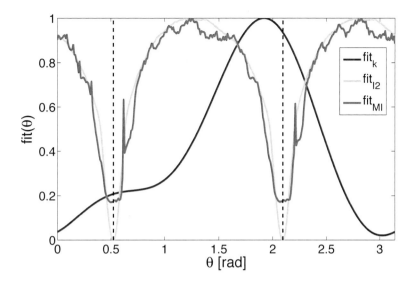

FIGURE 12.10 (SEE COLOR INSERT)
Normalized fit_k, fit_{I2}, and fit_{MI} for the y_1 component as a function of θ. Recurrence parameters: $d_e = 3$, $\tau = 6$, $\epsilon = 0.1$, $N = 2000$ samples. Original sources: Lorenz time series and $\sin(5t)$. The solutions that perfectly restore the original sources are given by, respectively, $\theta = \pi/6$ and $\theta = 2\pi/3$.

play a key role in describing physical phenomena. In particular, a first natural application refers to the denoising of chaotic time series in the context of a multirecording environment, which is intrinsically related to the problem of sending information in noisy environments. In fact, the use of a denoising stage can significantly enhance the robustness of synchronization in master-slave systems [24], which is an essential issue in chaos-based communication systems. Moreover, source separation can define one of the most promising approaches for source separation for MIMO chaos-based communication systems, as classically performed in current communication paradigms [26].

We feel that RQA provides a distinct way of employing *a priori* information concerning the deterministic character of source models, which can be of particular interest in different practical and theoretical problems in dynamical systems. Therefore, the development and potential application of this deterministic component analysis deserve careful attention.

Bibliography

[1] H. D. I. Abarbanel. *Analysis of Observed Chaotic Data*. Springer Verlag, New York, 1996.

[2] A. J. Bell and T. J. Sejnowski. An information-maximization approach to blind separation and blind deconvolution. *Neural Computation*, 7:1129–1159, 1995.

[3] J.-F. Cardoso. Infomax and maximum likehood of blind source separation. *IEEE Signal Processing Letters*, 4:112–114, 1997.

[4] J.-F. Cardoso. High-order contrasts for independent component analysis. *Neural Computation*, 11:157–192, 1999.

[5] P. Comon. Independent component analysis, a new concept? *Signal Processing*, 36:287–314, 1994.

[6] P. Comon and C. Jutten. *Handbook of Blind Source Separation - Independent Component Analysis and Applications*. Elsevier, New York, 2010.

[7] T. M. Cover and J. A. Thomas. *Elements of Information Theory*. Wiley-Interscience, Hoboken, NJ, 2nd edition, 2005.

[8] S. A. Cruces-Alvarez, A. Cichock, and S.-I. Amari. From blind signal extraction to blind instantaneous signal separation: Criteria, algorithms, and stability. *IEEE Transactions on Neural Networks*, 15:859–873, 2004.

[9] K. M. Cuomo, A. V. Oppenheim, and S. H. Strogatz. Synchronization of Lorenz-based chaotic circuits with applications to communications. *IEEE Transactions on Circuits and Systems II*, 40:626–633, October 1993.

[10] J-P. Eckmann, S. O. Kamphorst, and D. Ruelle. Recurrence plots of dynamical systems. *Europhys. Lett.*, 4:973–977, 1987.

[11] P. Faure and H. Korn. A new method to estimate the Kolmogorov entropy from recurrence plots: Its application to neuronal signals. *Physica D: Nonlinear Phenomena*, 122:265–279, 1998.

[12] A. Hyvärinen, J. Karhunen, and E. Oja. *Independent Component Analysis*. Adaptive and Learning Systems for Signal Processing, Communications, and Control. Wiley Interscience, New York, 2001.

[13] G. Kubin. What is a chaotic signal ? In I. Pitas, editor, *Proc. IEEE Workshop on Nonlinear Signal and Image Processing*, pages 141–144, Greece, 1995.

[14] F. C. M. Lau and C. K. Tse. *Chaos-Based Digital Communication Systems: Operating Principles, Analysis Methods and Performance Evaluation.* Springer, Berlin, 2011.

[15] N. Marwan. A historical review of recurrence plots. *The European Physics Journal*, 164:2–12, 2008.

[16] N. Marwan. How to avoid potential pitfalls in recurrence plot based data analysis. *International Journal of Bifurcation and Chaos*, 21:1003–1017, 2011.

[17] N. Marwan, M. C. Romano, M. Thiel, and J. Kurths. Recurrence plots for the analysis of complex systems. *Physics Reports*, 438:237–329, 2007.

[18] P. Pajunen. Blind source separation using algorithmic information theory. *Neurocomputing*, 22:35–48, 1998.

[19] A. Papoulis and S. U. Pillai. *Probability, Random Variables and Stochastic Processes.* McGraw-Hill, New York, 4th edition, 2002.

[20] L. M. Pecora and T. L. Carroll. Synchronization in chaotic systems. *Phys. Rev. Lett.*, 64:821–824, February 1990.

[21] J. M. T. Romano, R. Attux, C. C. Cavalcante, and R. Suyama. *Unsupervised Signal Processing: Channel Equalization and Source Separation.* CRC Press, Boca Raton, FL, 2010.

[22] A. Sibille, C. Oestges, and A. Zanella, editors. *MIMO: From Theory to Implementation.* Academic Press, New York, 2010.

[23] D. C. Soriano, R. Attux, J. M. T. Romano, M. B. Loiola, and R. Suyama. Denoising chaotic time series using an evolutionary state estimation approach. In *Computational Intelligence in Control and Automation (CICA), 2011 IEEE Symposium*, pgs. 116–122, Paris, April 2011.

[24] D. C. Soriano, E. Z. Nadalin, R. Suyama, J. M. T. Romano, and R. Attux. Chaotic convergence of the decision-directed blind equalization algorithm. *Communications in Nonlinear Science and Numerical Simulation*, 17:5097–5109, 2012.

[25] D. C. Soriano, R. Suyama, and R. Attux. Blind extraction of chaotic sources from white gaussian noise based on a measure of determinism. In T. Adali, C. Jutten, J. M. T. Romano, and A. K. Barros, editors, *Lecture Notes in Computer Science*, vol. 5441, pgs. 122–129, Berlin, 2009.

[26] R. Suyama. *Proposal of blind source separation methods for convolutive and nonlinear mixtures.* PhD thesis, School of Electrical and Computer Engineering - UNICAMP, 2007.

[27] L. L. Trulla, A. Giuliani, J. P. Zbilut, and C. L. Webber. Recurrence quantification analysis of the logistic equation with transients. *Physics Letters A*, 223:255–260, 1996.

[28] J. P. Zbilut, A. Giuliani, and C. L. Webber. Recurrence quantification analysis as an empirical test to distinguish relatively short deterministic versus random number series. *Physics Letters A*, 267:174–178, 2000.

[29] J. P. Zbilut and C. L. Webber. Recurrence quantification analysis: Introduction and historical context. *International Journal of Bifurcation and Chaos*, 17:3477–3481, 2007.

13

Denoising chaotic time series using an evolutionary state estimation approach

Diogo C. Soriano, Murilo B. Loiola, and Ricardo Suyama

Centro de Engenharia, Modelagem e Ciências Sociais Aplicadas (CECS)
Universidade Federal do ABC (UFABC)

Marcio Eisencraft

Escola Politécnica, Universidade de São Paulo

Vanessa B. Olivatto, João Marcos T. Romano, and Romis Attux

School of Electrical and Computer Engineering (FEEC)
University of Campinas (UNICAMP)

CONTENTS

13.1 Introduction

Chaotic signals are those produced by nonlinear state mappings defined by a set of differential or difference equations that exhibit some particular characteristics such as strong dependence to initial conditions, aperiodicity, and broadband spectrum. These aspects can easily lead to confusion between chaotic signals and random processes, and, from a practical point of view, render the filtering procedure of experimental time series based on the classical Fourier approach not suitable, given the similarity between the spectral characteristics of noise and of the signal of interest [1].

In fact, denoising chaotic time series — or extracting chaotic time series from a mixture with a noise component — is a challenging open problem [1,13], the performance and limits of any algorithm to accomplish this task being strongly dependent on the amount of available a priori information. When the structure of the underlying dynamics and the free parameters are known, but not its initial condition, the denoising problem turns to be analogous to the shadowing problem [11,15], that is, that of finding the pure deterministic orbit that shadows the noisy observation. Reference [11] has shown that this problem can be solved by finding the deterministic orbit that minimizes the Euclidean distance from the noisy observation vector. This approach is strongly dependent on the initial noise level and requires local linear approximations for the dynamics, which can lead to numerical drawbacks associated to the presence of expansive directions of the chaotic dynamics. In fact, such numerical instabilities are a natural consequence of the presence of positive eigenvalues in the Jacobian matrices associated with the dynamics and numerical corrections that usually require a manifold decomposition technique, something that can be laborious without ensuring stable filtering algorithms. This scenario is more complex when the free parameters are also unknown. In this case, there are also works that try to approximate the noise-free trajectory by adjusting local polynomial maps for the attractor reconstructed using the Takens theorem [17–19], which also usually employs estimates of the Jacobian matrix of the underlying nonlinear dynamics and is strongly dependent on the quality of the performed reconstruction.

Apart from the possible benefits that such techniques may bring to the analysis of chaotic time series – for instance, denoising can be used as a preprocessing stage for estimating other important measures like the attractor dimension and the Lyapunov exponents – they may also be of great interest in a particular problem involving nonlinear dynamical systems: that of a communication system based on chaotic synchronization [3,4].

In communication systems based on chaotic synchronization of two systems (master-slave configuration), it is well known that when additive noise is present on the link between master and slave, the sensitive dependence on initial conditions that characterizes chaotic signals amplifies this error and synchronization is no longer obtained [8,10]. This problem can have a significant practical impact: in chaos-based communication systems, for instance, it would lead to a decrease in the quality and/or in the rate of information exchange [14].

Bearing that in mind, in this chapter we present an alternative to decrease the master-slave synchronization error when there is additive white Gaussian noise (AWGN) between master and slave: to use a denoising technique based on an evolutionary approach proposed by [21] to increase the signal-to-noise ratio (SNR) of the signal that arrives on the slave.

The proposed method assumes that the structure of the dynamics is known, but not the initial conditions and free parameters. It is shown here that it is possible, to a certain extent, to use an evolutionary algorithm in order to find the initial conditions and free parameters that lead to a pure deterministic chaotic signal that minimizes the deviation from the observed noisy vector, thus avoiding the need for local linear approximations to the deterministic trajectory, and, as a consequence, the numerical drawbacks present in the classical filtering paradigms for chaotic time series [1].

Evolutionary algorithms have been successfully employed in different contexts related to chaotic dynamics, including control, attractor reconstruction, synchronization, cryptography [24], and system identification [12]. In our case, we take advantage of the flexibility of the global search potential of an algorithm of this class, an artificial immune system [5,6], to find a trajectory that is as close as possible to the noiseless transmitted signal.

13.2 The state estimation problem

When the time series being analyzed is produced by a system whose underlying behavior can be characterized as low-dimensional chaos (as in the case of chaotic synchronization), denoising should be understood as reconstructing the trajectory of the dynamical system from noisy observations, which is, essentially, a state estimation problem.

In this sense, let us first consider the problem of extracting chaotic time series from an observation with additive noise under the premise that the dynamics is known, but not the free parameters and initial conditions, i.e., the system model $F(\mathbf{x}, \mathbf{p})$ underlying the observed signal is known, but not the parameter vector $\mathbf{p}$ and the initial condition $\mathbf{x}_0$. In this case, the available knowledge amounts to the observed noisy chaotic signal $z(n)$ and the underlying dynamical structure. For instance, in the Hénon map, we assume that the chaotic signal is generated according to the following state equations:

$$x_1(n+1) = x_2(n) + 1 - \alpha x_1^2(n)$$
$$x_2(n+1) = \beta x_1(n), \tag{13.1}$$

where the initial conditions $\mathbf{x}_0 = [x_1(1)\ x_2(1)]$ and the free parameter vector $\mathbf{p} = [\alpha\ \beta]^T$ are unknown and the observed signal is given by

$$z(n) = x_1(n) + \xi(n), \tag{13.2}$$

$\xi(n)$ being a Gaussian random variable with zero mean and $x_1(n)$ the first state variable of the Hénon model. In our simulation, the variance of the random variable $\xi(n)$ is used to control the SNR. It is well known that the system given in Equation (13.1) can lead to chaos for a large range of values of α and β. The idea in this work would be, in this case, to estimate the noise-free chaotic signal $x(n)$ exclusively from the a priori information regarding the dynamical structure as provided by Equation (13.1). Obviously, all the denoising methods employed here are quite general and can be applied to other discrete-time or even continuous-time nonlinear dynamical systems.

In the simulations performed here, we have also used the Logistic map — another classical chaotic well-known discrete-time dynamical system — as defined in Equation (13.3):

$$x_1(n+1) = \mu x_1(n)(1 - x_1(n)), \tag{13.3}$$

where once again the initial condition $\mathbf{x}_0 = x_1(1)$ and parameter vector $\mathbf{p} = \mu$ are unknown and a noisy observation $z(n)$ as defined in Equation (13.2) is available.

13.2.1   Benchmark I: the extended Kalman filter

The Kalman filter [16] is a widely known state-space estimator that provides recursive minimum mean-squared error estimates for linear systems embedded in Gaussian noise. However, the Kalman filter cannot be directly applied to nonlinear state-space models. Hence, to obtain an approximate solution that allows us to extend the Kalman filtering framework to nonlinear state-space models, we can use the extended Kalman filter (EKF) [16]. The EKF, in fact, provides recursive estimates of the state variables by applying the standard KF to a linearized version of the nonlinear system. This is accomplished by linearizing the nonlinear functions of the state-space model about the current estimate of the state vector [16]. In other words, the basic idea of the EKF is to linearize the nonlinear state-space model at each time instant around the most recent state estimate. Once a linear model is obtained, the classical Kalman filter is applied.

13.2.2   Benchmark II: the Wiener filtering

The Wiener filter is arguably the cornerstone of the entire optimal filter theory [16], and it is based on the idea of obtaining an estimate that is "as close as possible" to the noise-free signal $x(n)$ in the mean-squared error sense. Mathematically, the parameters of the filter are obtained as the solution to the following optimization problem:

$$\min_{\mathbf{w}} J_{Wiener} = E\left\{[x(n) - \hat{z}(n)]^2\right\}, \tag{13.4}$$

where $\mathbf{w}$ denotes the filter parameter vector. If we consider a linear filtering structure such that its output is given by

$$\begin{aligned}
\hat{z}(n) &= w_0 x(n) + w_1 x(n-1) + \cdots + w_{L-1} x(n-L+1) \\
&= \mathbf{w}^T \mathbf{x}(n)
\end{aligned} \tag{13.5}$$

with

$$\mathbf{w} = \begin{bmatrix} w_0 & w_1 & \cdots & w_{L-1} \end{bmatrix}^T$$
$$\mathbf{x}(n) = \begin{bmatrix} x(n) & x(n-1) & \cdots & x(n-L+1) \end{bmatrix}^T, \tag{13.6}$$

then it can be shown that the optimal parameter vector is given by

$$\mathbf{w} = \mathbf{R}^{-1}\mathbf{p}, \tag{13.7}$$

where $\mathbf{R} = E\left[x(n)x^H(n)\right]$ is the autocorrelation matrix, and $\mathbf{p} = E\left[x_1(n)z(n)\right]$ is the cross-correlation vector.

13.3   Denoising of chaotic time series using an evolutionary approach

In order to reduce the noise in the observed signal under the premise that the system model is known, one could try to estimate the true parameters and initial conditions of the system, which would allow a perfect reconstruction of the deterministic orbit, thus eliminating the noise. However, it is well known that any attempt to estimate the initial condition would lead to a deterministic orbit that inevitably exponentially diverges from the observed one, given the finite precision of the estimate of the initial state and the chaotic nature of the system.

Therefore, in [21], the authors proposed a piecewise estimation, i.e., for each segment of N_s samples of the observed signal, a set of parameters and initial conditions is obtained and an orbit close to the true one is reconstructed. Mathematically, the task consists of the following minimization problem

$$\arg\min_{\mathbf{x}_0, \mathbf{p}}\{f_{\text{score}}(\mathbf{x}_0, \mathbf{p}) = [\frac{1}{N_s}\sum_{n=1}^{N_s}(s'(n) - z(n))^2]\}, \tag{13.8}$$

where $\mathbf{x}_0$ and $\mathbf{p}$ are, for the Hénon map, $\mathbf{x}_0 = [x_1(0)\ x_2(0)]^T$ and $\mathbf{p} = [\alpha\ \beta]^T$). In simple terms, the idea is to minimize the score function $f_{\text{score}}(\cdot)$ given by the mean-squared error between a candidate chaotic time series solution $s'(n)$ (dependent on the optimization parameters) and the observed vector $z(n)$. The candidate solution $s'(n)$ is obtained by solving (iterating or integrating) $\mathbf{F}(\mathbf{x}, \mathbf{p})$ given the parameter vector $\mathbf{p}$ and initial condition $\mathbf{x}_0$.

As the cost function presented in Equation (13.8) depends on the nonlinear state equations, which are strongly sensitive to the dynamics parameters and to the initial conditions, classical optimization methods (e.g., gradient-based methods) are usually not suitable to deal with the underlying multimodality, which justifies the use of alternative optimization strategies. In this sense, we have opted for employing a search heuristic that uses the immune system as an inspiration, an optimization approach that has a significant global search potential [6]. It is important to remark that other optimization metaheuristics such as particle swarm or even differential evolution could also be employed, but, from an analytical standpoint, we feel that the best path to assess the performance of the proposal is by means of a comparison with filtering techniques that are classical in signal processing and state estimation theory.

13.3.1   Optimization by artificial immune system

Artificial immune systems (AIS) correspond to a bio-inspired computational strategy that uses concepts derived from the study of immune systems of superior organisms [5,6]. It has several applications to engineering problems, and is particularly useful for optimization, presenting a good performance in terms of global convergence rate even in problems with significant multimodality.

Among the various AIS employed for solving optimization tasks, we decided to employ, in this work, a version of the CLONALG algorithm [7] adapted to operate with real coding. This choice was motivated by two features of this tool: (1) a relatively simple modus operandi and (2) an interesting balance between local and global search mechanisms.

The modus operandi of the CLONALG is based on two conceptual pillars: clonal selection and affinity maturation [6]. The clonal selection principle establishes that, when an organism is invaded by antigens (e.g., virus or bacteria), specific cells of the immune system recognize the exogenous element and are selected to proliferate, which gives rise to a cloning process with rates proportional to the affinity - defined by some measure of recognition - of these cells to the antigens. In the affinity maturation process, the individuals produced in the current generation can exhibit mutations with rates inversely proportional to their affinity with the antigens, and the mutated generation can eventually present individuals with higher affinity [6,7].

To effectively convert these ideas into an efficient optimization algorithm, it is necessary to make some considerations. First, each candidate solution to the optimization problem corresponds to an individual, that is, a vector of real values that represents, in simple terms, the structure of an immune cell. The quality of an individual (called here *fitness* measure) is defined by the cost function (with the caveat that it is necessary to convert minimization into maximization), providing means of quantifying the antibody-antigen affinity [6,7].

Finally, there is also a periodic insertion of new randomly generated individuals to replace individuals with poor fitness in order to perform a better exploration of the search space. These steps can be summarized in the CLONALG [7] pseudo-code shown in Algorithm 1.

It is important to note that the CLONALG algorithm performs the maximization of a fitness measure (defined as J_{FIT}) and not the direct minimization of the cost function presented in Equation (13.8). Hence, we have used the following relation between the cost function and the fitness measure to be

---

Algorithm 1 Pseudo-code for CLONALG algorithm

1. Randomly initialize the population (N);
2. Determine the fitness of each individual: J_{FIT};
3. While the maximum number of generation is not attained, do

   (a) Create N_c clones for each individual;
   (b) Keep the original individual, and apply a mutation process for each clone as described in Equation (13.10):

$$c' = c + \epsilon Y(0, 1) \tag{13.10}$$
$$\epsilon = \frac{1}{\eta} exp(-J_{FIT}),$$

   where c' and c represent the clones modified by mutation and the original one, respectively. $Y(0, 1)$ represents Gaussian random variable with zero mean and unitary variance and η represents a control parameter to establish the applied mutation;

   (c) Evaluate the fitness of each individual of the population and keep in the population only the best solution of each group given by the individual and its derived mutated clones;
   (d) At each t generations, eliminate the m elements with lowest fitness and substitute them by randomly generated individuals;

4. Return to Step 3.

---

maximized

$$J_{FIT} = \frac{1}{f_{score}(\mathbf{x}_0, \mathbf{p})}. \tag{13.9}$$

13.4 Simulation results I

In our simulations, after a number of preliminary tests, we decided to use $N = 50$ individuals and $N_c = 20$ clones. An adaptive value of η - which represents the mutation rate - was used. η was initially defined as 1 and, after half of the total number of generations (which was set in 4000), it linearly increased until a final value of 1000 was reached. It is also valuable to remark that the algorithm parameters (as N, N_c, η, number of generations) define its performance and some numerical experimentation is recommended before setting these parameters (for more details, see [21]). In this case, some "thumb-rules" or heuristic settings can be applied to perform a reasonable search, which consists of one of the major issues in bio-inspired optimization [6].

Figure 13.1 shows the mean square error (MSE) obtained for different SNRs and for different time series sizes for the Hénon map. The original chaotic time series corrupted by noise was generated using $\mathbf{p} = [1.4 \ 0.3]^T$, $\mathbf{x}_0 = [0.2 \ 0.2]^T$ in Equation (13.1), and iterating the map 100 times. It can be noted that the proposed evolutionary state estimation approach leads to bad denoising results for larger windows of samples (curve 1 — Figure 13.1), which is a consequence of a limited shadowing capacity that any candidate solution would exhibit, since the system operates in chaos. The performance of the estimation process is also degraded for low SNR values, given that the deterministic character of the observation tends to be masked by the influence of noise. However, if a windowing process is employed with a suitable number of samples per window ($N_{s/w}$), e.g., 20 in curve 2, the proposed method can achieve a much better denoising performance, except when the SNR is too low (e.g., 4 dB). On the other hand, curve 3 in Figure 13.1 shows that a windowing with just a few samples ($N_{s/w} = 10$) does not lead to such a satisfactory MSE. In fact, when an inadvertently shorter $N_{s/w}$ is used, noise is incorporated as an uncertainty in the initial conditions and dynamics parameters, which leads to a bad estimation of the shadowing trajectory.

From Figure 13.1 it is possible to conclude that the time series length has a twofold effect on the denoising proposal. On the one hand, the $N_{s/w}$ cannot be too large given the natural limited shadowing capacity of any candidate solution imposed by the chaotic phenomenon. On the other hand, the employed $N_{s/w}$ has to be large enough to guarantee that the estimated time series captures the deterministic features of the observation vector, minimizing the effect of the stochastic fluctuations on the estimated parameters.

The significant denoising potential of the proposal when a suitable number of $N_{s/w}$ are employed is illustrated in Figure 13.2. The upper panel of this figure shows the estimated signal for an observation vector with an SNR of 5.7 dB and $N_{s/w} = 20$, while the lower panel shows a typical bad purely deterministic solution found when $N_{s/w} = 100$. In the latter case, it is clear that the estimated noise-free trajectory only approximates the observation and the original chaotic signals for specific time intervals, which leads to a bad overall denoising performance, something that does not stand for the simulation scenario of Figure 13.2(a).

13.4.1 Testing and comparing the performance of the proposal

In practical terms, it is necessary to define the number of samples per window to be employed accessing only the observation vector and the obtained denoising results, which characterizes an unsupervised scenario. In this case, an adequate $N_{s/w}$ can be achieved based on the convergence of the estimated parameters of the dynamics and on the observational MSE (MSEO), defined

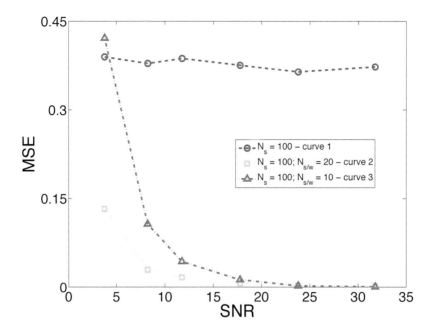

FIGURE 13.1
Mean square error (MSE) obtained for the proposed method considering different SNRs and samples per window ($N_{s/w}$) for first state variable of Hénon map. N_s is the total number of samples.

here as the MSE between the estimated deterministic orbit and the observed one. Thus, if a small value for $N_{s/w}$ is employed, fluctuations in the estimates of the dynamical parameters should be observed due to the influence of noise, which implies an increase of the standard deviation of the estimates. On the other hand, a large number of samples per window lead to bad solutions that diverge from the original one, and also from the observation, which implies an increase in the observational error.

In this sense, if we consider an observation vector with 1000 samples obtained from the Hénon map, the heuristics for finding a suitable number of samples per window can be supported by the results shown in Figure 13.3. In this case, Figure 13.3(a) and (b) show the mean $\pm$ the standard deviation (error bar) of the estimated dynamical parameters for the Hénon map considering different values of $N_{s/w}$, and an SNR of 11.7 dB. It can be clearly observed that, for $N_{s/w} = 10$ samples, the obtained dynamics parameters averages diverge from their ideal values and a high standard deviation is also obtained (when compared, for instance, with the results obtained for $N_{s/w} = 25$), a fact that follows from the influence of noise in the estimation process. For

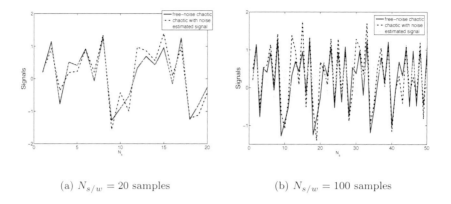

(a) $N_{s/w} = 20$ samples   (b) $N_{s/w} = 100$ samples

FIGURE 13.2
Panel (a) and panel (b) show the original chaotic time series, the chaotic signal with noise, and the estimated noise-free chaotic signal for $N_{s/w} = 20$ and $N_{s/w} = 100$ samples, respectively, considering an SNR of 5.7 dB for the first state variable of Hénon map. Panel (b) only shows the first 50 samples to achieve a better illustration.

a progressive increase in $N_{s/w}$, the estimates get close to the original values, reducing the MSE (Figure 13.3(c)) and the standard deviation until an optimum is attained, which happens between 20 and 25 samples. After this ($N_{s/w} > 30$), the MSE starts to increase again, which is also accompanied by a smooth increase in the MSEO (in Figure 13.3(d)). This increase in MSEO is associated with bad solutions of the optimization problem, and defines the limits of the shadowing capability of the estimation achieved by the proposed method. Thus, the search for convergence of the estimated parameters and for a low MSEO in Figure 13.3(d) leads to $N_{s/w} = 25$, which is in agreement with the minimal MSE and supports the proposed heuristics towards an efficient windowing for denoising the chaotic time series.

It can be noted in Figure 13.3(c) that, for an adequate $N_{s/w}$, the MSE obtained with the proposal approximates the MSE provided by the Wiener filter in the supervised scenario, that is, is close to a significant benchmark performance for linear filtering structures [16]. Moreover, the proposed method has provided a better filtering performance than the extended Kalman filter considering $N_{s/w} > 45$, without requiring information about the dynamics parameters. It is important to stress here that the benchmark methods have the advantage of not requiring a windowing process, but, from the comparison of the proposal with these two benchmark methods, and considering the information required by each of them, it is reasonable to affirm that the proposed

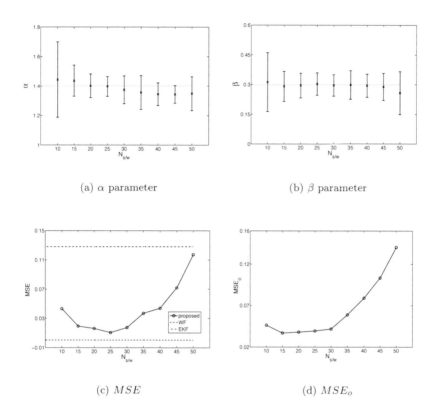

(a) α parameter

(b) β parameter

(c) MSE

(d) MSE_o

FIGURE 13.3
Panels (a) and (b) show, respectively, the mean $\pm$ standard deviation (error bar) for parameters α and β for different samples per window $(N_{s/w})$. Panel (c) shows the obtained MSE for the proposed method, for the Wiener filter (WF), and for the extended Kalman filter (EKF). Panel (d) shows the observational mean square error (MSEO). The SNR of the observation signal is approximately 11.7 dB and 1000 samples were considered. The traced lines in panels (a) and (b) denote the original dynamics parameters values ($\alpha = 1.4$ and $\beta = 0.3$).

method has a good denoising potential when the heuristics established here are employed.

In order to provide a better insight on the performance of the proposal, Figure 13.4 shows the estimated time series obtained by the different state estimation approaches studied here. In this context, Figure 13.4(a) shows the time series obtained using WF, while 13.4(b) shows the time series estimated via the proposed method and 13.4(c) shows the time series obtained by means

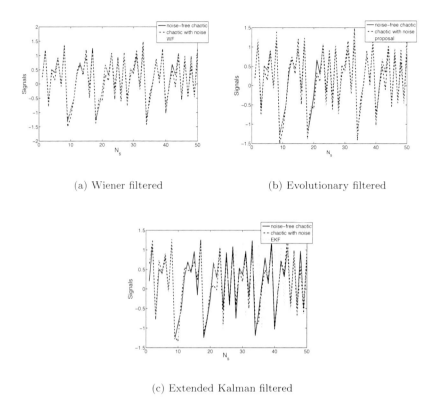

(a) Wiener filtered   (b) Evolutionary filtered

(c) Extended Kalman filtered

FIGURE 13.4

Panel (a) shows the original noise-free chaotic time series, the corrupted observation with an SNR of 11.7 dB, and the Wiener filtered signal in the supervised
scenario (WF). Panels (b) and (c) are analogous to the first one, but bring
the estimated signal by the proposed method considering $N_{s/w} = 25$ and the
estimated signal by the EKF approach, respectively.

of the EKF approach. It is clear from Figure 13.4 that the proposal exhibits
a performance close to that of the WF, even working on an unsupervised
scenario.

   To show that the proposed method is not restricted to low-noise situations,
Figure 13.5 brings the results of a simulation scenario similar to that shown
in Figure 13.3, but with an SNR of 5.6 dB. In this case, it can be clearly
observed that the increase in the noise power has led to an increase of the
standard deviation of the estimated parameters for low values of $N_{s/w}$. In
addition to that, the increase in the MSEO follows a pattern similar to that
of the previous scenario, smoothly increasing for $N_{s/w} \geq 30$, which suggests

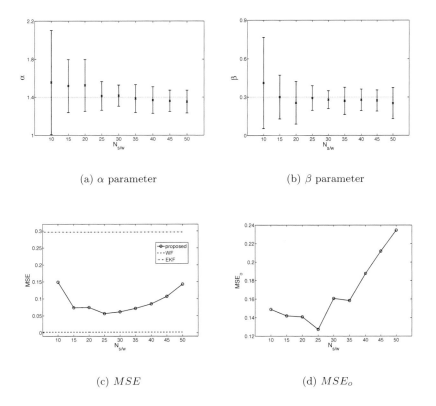

(a) α parameter

(b) β parameter

(c) MSE

(d) MSE_o

FIGURE 13.5
Panels (a) and (b): mean $\pm$ standard deviation for parameters α and β for different samples per window $(N_{s/w})$. Panel (c): obtained MSE for the proposed method, for the Wiener filter (WF), and for the extended Kalman filter (EKF). Panel (d): observational mean square error (MSEO). The SNR of the observation signal is approximately 5.6 dB and 1000 samples were considered.

that the suitable value of $N_{s/w}$ is not shifted. In fact, the minimal MSE is also achieved for $N_{s/w} = 25$, something that illustrates the robustness of the heuristic to noisy environments.

The obtained time series using the different state estimation approaches are shown in Figure 13.6. Again, it can be noted that the proposed method has a performance that is much closer to that of the Wiener approach than the EKF.

The proposed heuristic to estimate a suitable window length and, as consequence, to ensure a high-quality denoising of the chaotic time series is also supported by the results found when dealing with the Logistic map. In this

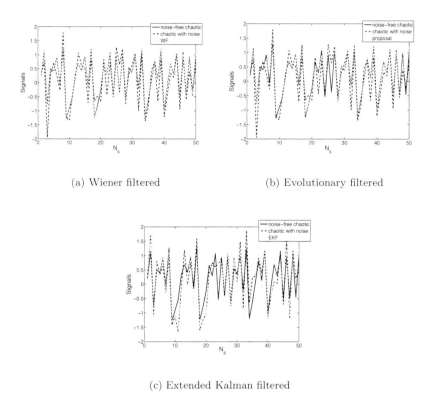

(a) Wiener filtered   (b) Evolutionary filtered

(c) Extended Kalman filtered

FIGURE 13.6
Panel (a) shows the original noise-free chaotic time series, the corrupted observation with an SNR of 5.6 dB and the Wiener filtered signal in the supervised
scenario (WF). Panels (b) and (c) are analogous to the first one, but bring
the estimated signal by the proposed method considering $N_{s/w} = 25$ and the
estimated signal by the EKF approach, respectively.

case, the observation vector is generated by adding noise to the chaotic series
obtained when $\mu = 3.75$ and $x(1) = 0.2$ in order to obtain a $z(n)$ with an SNR
of 10.7 dB. Figure 13.7(a) shows the mean $\pm$ standard deviation of the parameter μ, while Figure 13.7(b) shows the MSE obtained by the proposal and
by the Wiener and extended Kalman approaches. The observational MSE is
shown in Figure 13.7(c). As previously observed for the Hénon map, the MSE
error decreases as the estimation parameter converges with the progressive increase in $N_{s/w}$. After that, the optimal performance of the proposed method
is attained and the MSE increases again, which is also accompanied by an
increase in the observational error.

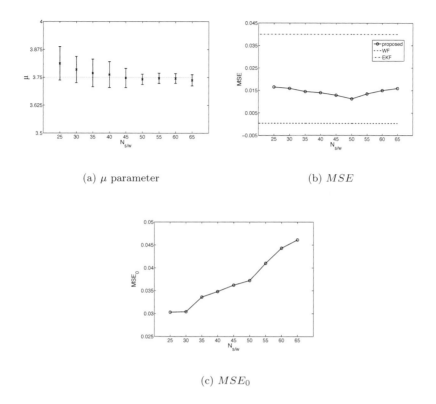

(a) μ parameter   (b) MSE

(c) MSE_0

FIGURE 13.7
Panel (a) shows the mean $\pm$ standard deviation (error bar) for dynamics parameter μ for different number of samples per window $(N_{s/w})$. Panel (b) shows the obtained MSE for the proposed method, Wiener filter (WF), and extended Kalman filter (EKF). Panel (c) shows the observational mean square error (MSEO). The SNR of the observation signal is approximately 10.7 dB and 1000 samples were considered. The traced lines in the first panel denote the original dynamics parameters value $(\mu = 3.75)$.

Finally, Figure 13.8 shows the obtained time series for the different state estimation approaches. In this case, a $N_{s/w} = 50$ was used for the proposed method, which illustrates an interesting shadowing capacity, i.e., despite the chaotic behavior, it is possible to achieve a trajectory that remains close to the (desired) chaotic noise-free signal for a large number of samples.

These results attest that the proposed method has a performance close to that of a supervised linear approach (WF), but in a blind context, in which it performs better than a classical unsupervised state estimation method (EKF).

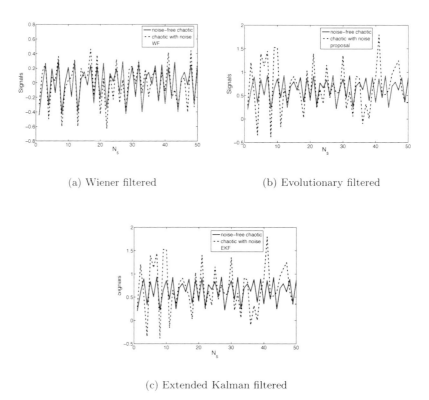

(a) Wiener filtered   (b) Evolutionary filtered

(c) Extended Kalman filtered

FIGURE 13.8
Panel (a) shows the original noise-free logistic chaotic time series, the corrupted observation with an SNR of 10.7 dB, and the Wiener filtered signal in the supervised scenario (WF). Panels (b) and (c) are analogous to the first panel, but bring the estimated signal by the proposed method considering $N_{s/w} = 50$ and the estimated signal by EKF approach, respectively.

The tests employed two of the most emblematic discrete-time dynamical systems used to study chaos theory, and indicate a significant potential for general applications.

13.5 Evolutionary denoising in discrete-time chaotic synchronization

In their work, Pecora and Carrol use the conditional Lyapunov exponents to check the asymptotic stability of the slave system and hence the possibility of synchronism [20]. A slightly different approach is given by Wu and Chua [23], proposing that the master and slave equations be written in such a way that the dynamics of the synchronization error is simple enough to allow direct verification of its convergence to zero. Both works, however, consider only the continuous-time case. The adaptation to discrete-time systems, proposed by [9], is succinctly described in the following.

Consider two discrete-time systems defined by

$$\mathbf{x}(n+1) = A\mathbf{x}(n) + \mathbf{b} + \mathbf{f}(\mathbf{x}(n)) \qquad (13.11)$$

$$\mathbf{y}(n+1) = A\mathbf{y}(n) + \mathbf{b} + \mathbf{f}(\mathbf{x}(n)), \qquad (13.12)$$

where $\{\mathbf{x}(n), \mathbf{y}(n)\} \subset \mathbb{R}^K$, $\mathbf{x}(n) = [x_1(n), \ldots, x_K(n)]^T$, $\mathbf{y}(n) = [y_1(n), \ldots, y_K(n)]^T$, and $n \in \mathbb{N}$. The real-valued matrix $A_{K \times K}$ and the vector $\mathbf{b}_{K \times 1}$ are constants. The function $\mathbf{f}(\cdot)$, $\mathbb{R}^K \to \mathbb{R}^K$ is nonlinear.

The system described by Equation (13.11) is autonomous and is called *master*. The one described by Equation (13.12) depends on $\mathbf{x}(n)$ and is called *slave*.

The dynamics of the synchronization error between the two systems $\mathbf{e}(n) = \mathbf{y}(n) - \mathbf{x}(n)$, in this case, is given by

$$\mathbf{e}(n+1) = A\mathbf{e}(n). \qquad (13.13)$$

They are said to be *completely synchronized* if $\mathbf{e}(n) \to \mathbf{0}$ as n grows. Consequently, master and slave synchronize completely if the eigenvalues λ_i of A satisfy [2]

$$|\lambda_i| < 1, \ 1 \le i \le K. \qquad (13.14)$$

Hence, if a system can be written as in Equation (13.11) with $\mathbf{A}$ satisfying the condition defined in Equation (13.14), it is easy to set up a slave system that synchronizes with it.

In our numerical examples we employ, in the role of chaos generator, the two-dimensional Hénon map in Equation (13.1). In this case, the master can be written in the form of Equation (13.11) with A as

$$A = \begin{bmatrix} 0 & 1 \\ \beta & 0 \end{bmatrix}, \qquad (13.15)$$

$\mathbf{b} = [1; 0]$, and $\mathbf{f}(\mathbf{x}(n)) = \mathbf{f}(x_1(n)) = [-\alpha x_1^2(n); 0]^T$. Thus, only the scalar signal $x_1(n)$ must be transmitted from master to slave.

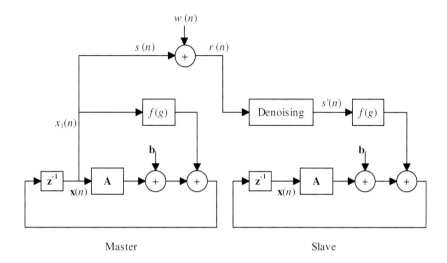

FIGURE 13.9
Block diagram of the synchronization set.

The eigenvalues of A are $\lambda_1 = -\lambda_2 = \sqrt{\beta}$. So there is master-slave synchronization whenever $|\beta| < 1$.

Now, we consider that the signal that arrives at the slave is not $x_1(n)$ but $r(n)$, given by

$$r(n) = x_1(n) + \xi(n), \tag{13.16}$$

where $\xi(n)$ is a zero-mean AWGN process with power σ_w^2. In this case, the receiver is described by

$$y_1(n+1) = 1 - \alpha r^2(n) + y_2(n) \tag{13.17}$$
$$y_2(n+1) = \beta y_1(n). \tag{13.18}$$

This setup is represented in block diagram form in Figure 13.9.

Now, the synchronization error is not given by Equation (13.13) anymore, and complete synchronization is no longer attainable. Figure 13.10 shows $x_1(n)$, $y_1(n)$, and $|e(n)|^2$ for the noiseless case ($\sigma_w^2 = 0$) and for an SNR $\gamma = 15$ dB. The SNR is defined here as

$$\gamma(dB) = 10 \log \left(\frac{\overline{s^2(n)}}{\sigma_w^2} \right), \tag{13.19}$$

where $\overline{s^2(n)}$ is the mean-squared value of $s(n)$, the transmitted signal.

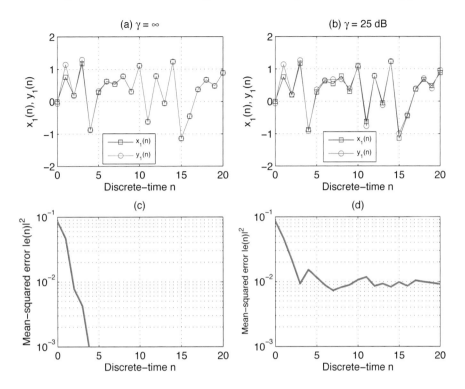

FIGURE 13.10
Chaotic synchronization: master and slave states $x_1(n)$, $y_1(n)$, and mean-squared error $|\mathbf{e}(n)|^2$ for $\gamma = +\infty$ ((a) and (c)) and $\gamma = 15$ dB ((b) and (d)).

As Figure 13.10 shows, for a noiseless channel ($\gamma = \infty$), complete synchronization is quickly attained. However, in this example, for a relatively high SNR of 15 dB, which means that the noise power is approximately 3% of the power of $s(n)$, $|\mathbf{e}(n)|^2 \approx 10^{-1}$, which is approximately 17% of the mean value of $|\mathbf{y}(n)|^2$. Thus, a small additive noise in the transmitted signal can result in a significant higher synchronization error due to the feedback and intrinsic nonlinearity of the involved systems.

This nonlinear effect of channel noise in the synchronization error is certainly one of the factors responsible for the relatively poor performance of digital communication systems based on chaos coherent compared to their conventional counterparts in terms of bit error rate in AWGN channels [22]. That is why denoising methods may have a positive impact in communication systems based on chaotic synchronization.

13.6   Simulation results II

In order to assess the possible gain of employing a denoising technique in the synchronization of two dynamical systems, we considered the scenario depicted in Figure 13.9. For an observation vector with 200 samples obtained from the Hénon map, we have applied the denoising methodology described in Section 13.3, considering an estimation window of $N_s = 10$ samples. Once a denoised version $s'(n)$ of the received signal is obtained, it is used as the input of the slave system, as indicated in Figure 13.9. The synchronization error is then evaluated, and is defined by

$$\text{SyncError} = \frac{1}{180} \sum_{n=20}^{200} ||\mathbf{x}(n) - \mathbf{y}(n)||^2, \qquad (13.20)$$

where $\mathbf{x}(n)$ and $\mathbf{y}(n)$ denote, respectively, the state vectors of the master system and slave system. The summation starts with $n = 20$ in order to discard the transient in the synchronization process.

As mentioned earlier, the denoising procedure employs a piecewise estimation of the received signal, thus obtaining a set of parameters and initial conditions for each estimation window. Even though it would not be possible to perfectly estimate the parameters and the initial conditions due to the chaotic nature of the system and the noise, the obtained parameters fluctuate around the true ones used to generate the transmitted signal – $\alpha = 1.4$ and $\beta = 0.3$, as indicated in Figure 13.11, which depicts the evolution of the estimated parameters along the various segments of the signal. As a matter of fact, this observation was used to increase the convergence speed of the algorithm. Instead of randomly initializing all parameters for each estimation window, the estimated parameter values for one data window were used as the initial values for the parameters in the subsequent estimation window.

Figure 13.12 shows the synchronization error for different values of γ, and it becomes clear that the denoising procedure is effective for a wide range of noise levels ($\gamma > 10$ dB). In other words, it means that the CLONALG algorithm was able to obtain a set of parameters that allowed a very reasonable reconstruction of the original trajectory, thus enhancing the synchronization performance of the slave system. For this range of SNR, it can be noted that the denoising method exhibits a 3-dB gain with respect to the non-processed data, i.e., it is possible to achieve the same synchronization error with an SNR 3 dB lower, which represents a significant improvement.

For low γ values, it was not possible to obtain the same performance level achieved for higher SNR. This limitation is partially explained by the fact that the CLONALG parameters were kept the same for all SNR values - in a practical situation, if we know in advance that the SNR is not so high, these parameters could be adjusted and a better synchronization error would be obtained.

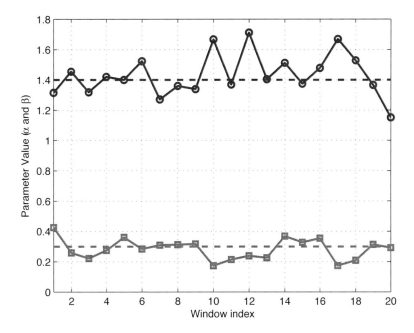

FIGURE 13.11
Evolution of the system parameters – α and β – among the estimation windows.

13.7 Conclusions

The presented approach shows the flexibility and capacity of employing an evolutionary strategy to find a reliable estimate of the chaotic signal for an SNR higher than 5 dB. The method was compared here with the Wiener and extended Kalman filters, having a performance close to that of the former, but using much less information. The proposal can be considered structurally simple, as it does not require any manipulation of the linearized version of the state equations nor attractor reconstruction, although a heuristic method has to employed to find a suitable number of samples per window that lead to a successful denoising task.

When applied to the specific problem of chaotic synchronization, the method, for a wide range of SNR values, was able to increase the channel SNR, thus reducing the synchronization error of the slave system. These encouraging results indicate many possibilities to be explored in future works, which include the evaluation of the methodology for different dynamical sys-

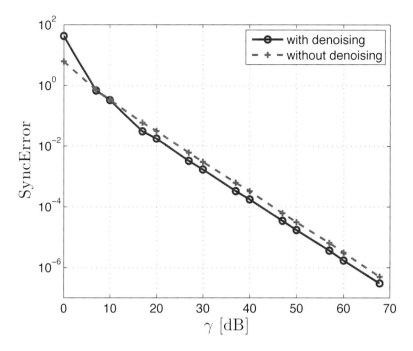

FIGURE 13.12
Synchronization MSE for different values of γ.

tems, and its application in chaos-based communication systems with the aim of understanding how the denoising procedure would affect the bit error rate of the system.

Bibliography

[1] H. D. I. Abarbanel. *Analysis of Observed Chaotic Data*. Springer Verlag, New York, 1996.

[2] R. P. Agarwal. *Difference equations and inequalities*, volume 155 of *Monographs and Textbooks in Pure and Applied Mathematics. Theory, methods, and applications*. Marcel Dekker Inc., New York, 1992.

[3] K. M. Cuomo and A. V. Oppenheim. Circuit implementation of synchronized chaos with applications to communications. *Phys. Rev. Lett.*, 71(1):65–68, July 1993.

[4] K. M. Cuomo, A. V. Oppenheim, and S. H. Strogatz. Synchronization of Lorenz-based chaotic circuits with applications to communications. *IEEE Transactions on Circuits and Systems II*, 40:626–633, October 1993.

[5] L. N. de Castro. *Fundamentals of Natural Computing*. Chapman & Hall, Boca Raton, FL, 2006.

[6] L. N. de Castro and J. Timmis. *Artificial Immune Systems: A New Computational Intelligence Approach*. Springer Verlag, London, 2002.

[7] L. N. de Castro and F. J. Von Zuben. Learning and optimization using the clonal selection principle. *IEEE Transactions on Evolutionary Computation*, 6:239–251, 2002.

[8] M. Eisencraft and A. M. Batista. Discrete-time chaotic systems synchronization performance under additive noise. *Signal Processing*, 91(8):2127–2131, 2011.

[9] M. Eisencraft, R. D. Fanganiello, and L. Baccalá. Synchronization of discrete-time chaotic systems in bandlimited channels. *Mathematical Problems in Engineering*, 2009:1–12, 2009.

[10] M. Eisencraft, R. D. Fanganiello, and L. H. A. Monteiro. Chaotic synchronization in discrete-time systems connected by bandlimited channels. *IEEE Communications Letters*, 15(6):671–673, June 2011.

[11] J. D. Farmer and J. J. Sidorowich. Optimal shadowing and noise reduction. *Physica D: Nonlinear Phenomena*, 47:373–392, 1991.

[12] F. Gao, J.-J. Lee, Z. Li, H. Tong, and X. Lü. Parameter estimation for chaotic system with initial random noises by particle swarm optimization. *Chaos, Solitons & Fractals*, 42:1286–1291, 2009.

[13] P. Grassberger, R. Hegger, H. Kantz, C. Schaffrath, and T. Schreiber. On noise reduction methods for chaotic data. *Chaos: An Interdisciplinary Journal of Nonlinear Science*, 3(2):127–141, 1993.

[14] J. Grzybowski, M. Eisencraft, and E. Macau. Chaos-based communication systems: current trends and challenges. In S. Banerjee, M. Mitra, and L. Rondoni, editors, *Applications of Chaos and Nonlinear Dynamics in Engineering — Vol. 1*, volume 71 of *Understanding Complex Systems*, pages 203–230. Springer Berlin/Heidelberg, 2011.

[15] S. M. Hammel. A noise reduction method for chaotic systems. *Physics Letters A*, 148(8–9):421–428, 1990.

[16] S. Haykin. *Adaptive Filter Theory*. Prentice-Hall, Upper Saddle River, NJ, 4th edition, 2002.

[17] E. J. Kostelich and T. Schreiber. Noise reduction in chaotic time series data: a survey of common methods. *Phys. Rev. E*, 48:1752, 1993.

[18] E. J. Kostelich and J. A. Yorke. Noise reduction in dinamical systems. *Phys. Rev. A*, 38(3):1649–1652, 1988.

[19] E. J. Kostelich and J. A. Yorke. Noise reduction: finding the simplest dynamical system consistent with the data. *Physica D: Nonlinear Phenomena*, 41:183–196, 1990.

[20] L. M. Pecora and T. L. Carroll. Synchronization in chaotic systems. *Phys. Rev. Lett.*, 64:821–824, February 1990.

[21] D. C. Soriano, R. Attux, J. M. T. Romano, M. B. Loiola, and R. Suyama. Denoising chaotic time series using an evolutionary state estimation approach. In *Computational Intelligence in Control and Automation (CICA), 2011 IEEE Symposium*, pages 116–122, Paris, April 2011.

[22] C. Williams. Chaotic communications over radio channels. *IEEE Transactions on Circuit and Systems I*, 48(12):1394–1404, December 2001.

[23] C. W. Wu and L. O. Chua. A simple way to synchronize chaotic systems with applications to secure communication systems. *International Journal of Bifurcation and Chaos*, 3(6):1619–1627, December 1993.

[24] I. Zelinka, S. Celinkovsky, H. Richter, and G. Chen. *Evolutionary Algorithms and Chaotic Systems*. Springer, Berlin, 2010.

14

Isochronal synchronization and digital chaos-based communication

José M. V. Grzybowski and Takashi Yoneyama

Technological Institute of Aeronautics (ITA)

Elbert E. N. Macau

Laboratory for Applied Mathematics and Computing (LAC)
National Institute for Space Research (INPE)

CONTENTS

Isochronal synchronization (IS) is amongst the most intriguing collective behaviors observed in coupled chaotic oscillators and networks. The oscillators' dynamics behave identically in time, despite time-delays in the coupling signals among them. Such a phenomenon has attracted the attention of researchers due to its potential applications in chaos-based communication schemes [26, 27]. In this context, a major advantage of isochronal synchronization is the possibility of handling simultaneous bidirectional transmission requiring as little as one chaotic oscillator at each end of the communication line [26]. Moreover, recent studies found that increased security in the negotiation of secret keys through public channels can be attained [16, 24, 27]. Thus, chaos-based communication using IS may provide increased efficiency and security as compared to other chaos-based schemes, yet maintaining a simple conceptual framework.

For illustration, a simplified scheme of bidirectional communication is shown in Figure 14.1; in contrast, unidirectional communication is pictured in Figure 14.2. Both bidirectional and unidirectional coupling schemes can participate in the communication process using IS, although unidirectional coupling is only necessary when a third oscillator is used to synchronize the other two. Regardless of the coupling configuration, the implementation of simultaneous bidirectional communication requires stable isochronal synchro-

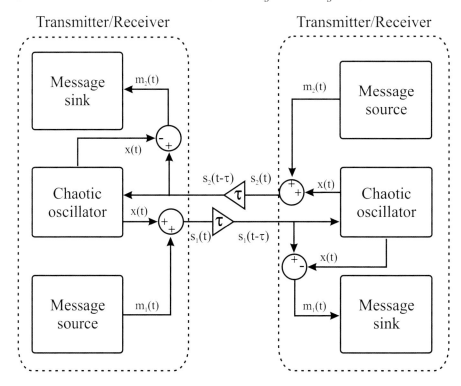

FIGURE 14.1

Full-duplex communication framework using chaotic systems: stable isochronal synchronization can be exploited in the implementation of simultaneous bidirectional communication, as shown in [26, 27].

nization. In a broader sense, this is a requirement of any coherent chaos-based communication schemes, and this fact reveals the first challenge towards the implementation of such communication schemes: the understanding of the conditions leading to synchronization. In the particular case of simultaneous bidirectional communication, the understanding of the conditions leading to isochronal synchronization is needed.

From several studies in the literature, it seems established that two chaotic systems do not display stable isochronal synchrony when coupled exclusively through mutual delayed coupling. In such case, the presence of any slight asymmetries or parameter mismatches will induce the emergence of unstable behavior and synchronization is not sustained. As a consequence, setups featuring only mutual delayed coupling are not adequate for simultaneous bidirectional communication purposes using isochronal synchronization. Considering this scenario, other forms of coupling were studied in recent years, especially in the context of lasers and electronic circuits. Numerical simulations and experimental setups revealed that stable isochronal synchronization

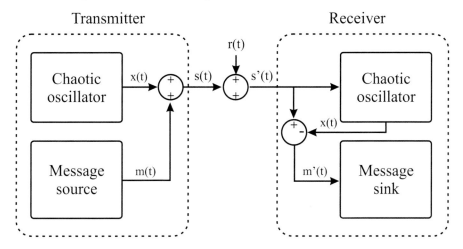

FIGURE 14.2
Chaos-based communication scheme based on unidirectional coupling, as proposed originally: the message $m(t)$ is added to the chaotic signal $x(t)$, resulting in the chaotic waveform $s(t)$ in the transmitter's end. The message extraction at the receiver's end is performed by subtraction, taking advantage of the chaos-pass filtering effect. In this case, time-delays due to propagation of the signal do not have consequences upon the stability of synchronization.

is achievable in a variety of coupling schemes and even for different chaotic oscillators, once adequate conditions are met [6, 17, 23–27]. The objective of this chapter is to present these ideas and explore their uses in chaos-based communication schemes.

14.1 Preliminaries

Since the first publications on chaos synchronization [8, 21], much effort has been deployed in the development of devices and techniques that can support and sustain chaos-based communication [4, 5]. Such schemes would mainly exploit the chaos-pass filtering effect (CPFE), a mechanism belonging to the realm of synchronized chaotic systems that performs nonlinear filtering of perturbations and distortions affecting a chaotic signal. The CPFE conserves the chaotic part of a chaotic waveform (chaotic waveform = chaotic signal + message + noise) and damps the components not belonging to the chaotic dynamics (such as message signals and noise). As explored for communication, the CPFE allows message signals to be separated from their chaotic carriers in the receiver's end, such that communication is possible.

As a signal travels the distance between transmitter and receiver, it will almost surely be impaired by attenuation distortion, delay distortion, and noise. Attenuation distortion is the decrease in the amplitude of the signal as it travels through the channel. Delay distortion, on its turn, is the effect of transmission media in which not all frequency components travel at the same speed. As a result, different frequencies arrive at different times, causing distortion of the transmitted signal at the receivers' end. Finally, noise is an inherent condition of physical systems and it consists of parasitic or unmodeled dynamics of any kinds, which add energy to (or subtract it from) the system [1]. In the context of communications, it is common to consider thermal noise, cross talks, and others.

Concerning chaos communication based upon IS, yet another element must be considered: time delay due to the signal propagation through the transmission media. When two systems are coupled together by means of a transmission line, the transmitted signals have to travel through the line to the other end, which generates delays. Although propagation delay has no further consequences when communication is unidirectional, mutual delayed coupling is closely related to several synchronization stability issues.

In this context, a fundamental consequence may arise in the context of IS which is not observed for other forms of synchronization: the error system is a delay differential equation (DDE), whose behavior can greatly surpass that of ordinary differential equations (ODEs) in complexity. Recall that oscillations and instability may arise even in the simplest of the DDE solutions, due to the (rather common) occurrence of Hopf bifurcations, i.e., the loss of stability of a fixed point into a pair of complex conjugate eigenvalues. In many cases, the occurrence of such bifurcations can be imputed to the increase in the time-delay magnitude. An example is shown in Figure 14.3, as a fixed point solution turns into a limit-cycle solution for the Nicholson blowflies equation [7] as delay magnitude increases. Thus, it is a fact that coupling delays and stability issues are closely related and this has direct consequence on the implementation of chaos-based communication schemes. Therefore, such stability issues must be properly handled before any form of communication is implemented.

Towards the study and numerical implementation of simultaneous bidirectional communication schemes, an assumption regarding time delays is generally made in the literature: they are considered to have a specific source, namely, the propagation of communication signals through the physical media. This is not true in general for digital communication systems but, apart from further discussions regarding the validity of such simplifying assumption, it certainly isolates the main relevant elements of the problem and allows the construction of simple models for theoretical, numerical, and experimental studies. As such, time delays are generally assumed to be given by $\tau = distance/speed$, i.e., the distance that the signal must travel divided by propagation speed. From this definition, time delays can be constant or time-varying, and time-varying delays arise from time-varying distance or speed, or both.

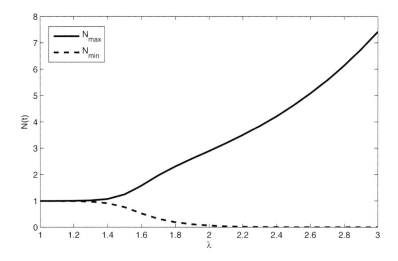

FIGURE 14.3
Bifurcation diagram for the Nicholson blowflies equation [7] for the parameter $\lambda = r\tau$, where $r > 0$: increase in the magnitude of time-delay τ causes the fixed-point solution to become an oscillatory limit-cycle solution.

As mentioned beforehand, the influence of propagation delay (or, as more commonly called, coupling delay) is fundamentally related to the unidirectional or bidirectional nature of the communication scheme. In unidirectional schemes, in the form transmitter/receiver, the effect of transmission delay is limited to delay in the information reception. For example, suppose that a bit is sent by the transmitter system x at the instant t and that the channel is subject to delay τ due to its intrinsic physical characteristics. In this case, the bit is received at the receiver system y at the instant $t + \tau$, and the principle behind decoding the bit is achronal synchronization between transmitter and receiver systems. In other words, as the equality $x(t) = y(t + \tau)$ holds, the receiver follows the transmitter with time lag τ to recover the transmitted bit by means of the CPFE of the synchronized oscillators [26]. In some codification schemes, the simple subtraction $x(t) - y(t + \tau)$ will reveal the encoded message. On the other hand, in bidirectional schemes, the effect of transmission delay can be critical, since it acts upon synchronization stability.

First, as the coupling setup is symmetric in respect to time delays, isochronal synchronization may arise, depending also on other conditions related to the coupling setup. For asymmetric delays, other forms of synchronization may take place, such as leader-laggard synchronization (LLS) [2]. Although the symmetry condition in general does not affect synchronization quality, it induces one or another form of synchronization. Regardless of what form of synchronization is established, its use for communication relies on the understanding of stability conditions.

Second, as stability issues are overcome, the conception of simultaneous bidirectional communication schemes relies on the fact that $x(t) = y(t)$ when IS establishes, despite the delay time introduced by the channel [25, 26]. In this case, bits are encoded simultaneously at both ends of the transmission line. Due to delay, bits that are transmitted at time t by x and by y are received at $t + \tau$ at both ends. The mechanics of communication relies on the identification of the bits transmitted at the other end by evaluating bursts in the synchronization error [25, 26]. Following this approach, IS allows information to be simultaneously injected and retrieved properly at the cost of one oscillator at each end of the communication line. As communication is performed, perturbations are introduced in the synchronous dynamics of x and y and the stability of IS guarantees that synchronous state will reestablish after a perturbation is injected, such that the subsequent bit can be encoded and transmitted.

As simultaneous transmission occurs, there are two possibilities of bit encoding at a given moment [26]: (i) either both systems are encoding the same bit (e.g., 0 or 1) or (ii) each system is encoding a different bit (e.g., system x is encoding '1,' while system y is encoding '0,' or the other way around). That being considered, two different situations may arise. In the first case, as systems are coding the same bit, the synchronization is maintained since both systems are subject to the same perturbation (either '0' or '1,' at both ends) [26]. In the second case, as the systems encode different bits (e.g., one encodes '0' and the other encodes '1'), the message injection causes a perturbation in the synchronous state. At the one end, the perturbation is interpreted as the coding (at the other end) of a bit that is different from the one sent. Following this procedure, the bit streams can be retrieved at both ends by monitoring synchronization errors and performing XOR operations [26].

Figure 14.4 illustrates the outcome of simultaneous data transmission: m_x and m_y are bit streams transmitted by systems x and y, respectively. Note that the dynamics of the synchronization error allows the identification of spikes that correspond to each system encoding a different bit. It follows that the received bit streams can be recovered after a simple XOR operation, as illustrated in Figure 14.4. As a result, the messages are transmitted and recovered in both ends of the transmission line.

To properly evaluate the possibilities of implementation of digital chaos-based communication based on IS, it is necessary to understand the principles behind the construction of coupling setups that enable the emergence of stable isochronal synchronization. The next section reviews recent developments in the understanding of isochronal synchronization and its use for simultaneous bidirectional communication.

14.2 Isochronal synchronization and chaos-based communication

Isochronal synchronization is estimated to bear simultaneous bidirectional communication at bit rates higher than 10 Gbps [15], while it is also demonstrated to allow the negotiation of secret keys through a public communication channel [16]. These and other technical predicates have attracted the attention of researchers for its application in chaos-based communication schemes. At the current stage, the efforts are mainly devoted to understanding the mechanisms behind the emergence and maintenance of synchronous state in the face of small perturbations, mismatches, or asymmetries.

In the face of unstable isochronal synchronization in coupling configurations where only mutual delayed coupling is present [2,12,15], some techniques have been proposed to provide and enhance stable synchronous behavior. As

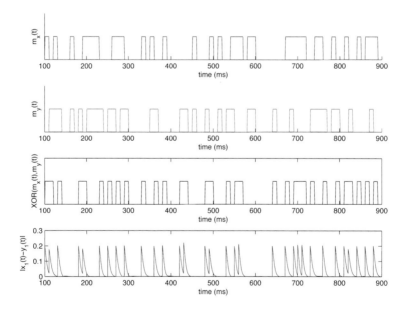

FIGURE 14.4
Bidirectional communication based on isochronal chaos synchronization allows systems to transmit and to receive digital messages simultaneously: (a) m_x, binary message sent by user x; (b) m_y, binary message sent by user y; (c) XOR operation between binary messages m_x and m_y and dynamics of the synchronization error.

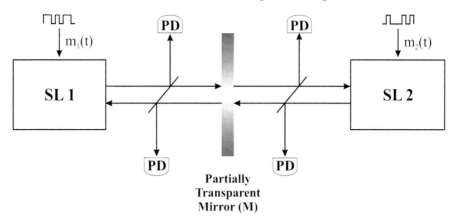

FIGURE 14.5
Mutually coupled semiconductor lasers with optical self-feedback due to effect of a partially transparent mirror: for this configuration, isochronal synchronization can be obtained for arbitrary distances between the lasers [24].
Photo-diodes (PD) detect the outputs of SL1 and SL2; binary messages are
simultaneously injected in SL1 and SL2 at rates of 1 Gpbs and recovered at
the other end.

a backbone for communication schemes, synchronization stability issues were
addressed using different approaches and techniques. These techniques mainly
consider the use of delayed [24] and direct [9] self-feedback or the inclusion of
a third oscillator to induce the injection-locking effect [27].

In this context, Fischer et al. [24] studied simultaneous bidirectional transmission using coupled semiconductor lasers (SLs) operating in chaotic regime.
To avoid LLS and induce stable IS, the mutual coupling of the SLs was combined with optical delayed self-feedback. To obtain such coupling configuration, a partially transparent mirror was positioned in the communication channel between the lasers, as shown in Figure 14.5. The coupling signal in this
scheme is a combination of the delayed outputs of both lasers, where delays
are caused by propagation time and delay magnitude is larger than the time
scale of the relaxation oscillations. It is remarked that synchronization establishes for arbitrary distances between the lasers [24]. From the communication
viewpoint, this is highly desirable, since it means that the phenomenology of
isochronal synchronization is not a barrier for long-distance communication.
Thus, although other limiting technical factors for long-distance communication certainly exist, IS may establish under this approach as symmetry
conditions are met, thus distance not being a restriction.

Also interesting from the results in [24] is that the position of the mirror
does not affect quality of synchronization; rather, it induces different forms
of synchronization to take over, such as IS or LLS. While isochronal synchronization is achievable for symmetric positioning of the mirror, it is not for its

asymmetric positioning. In the latter case, LLS is achievable, being the time shift of the systems' dynamics given by the difference between delay times [24]. Despite synchronization quality asymmetric operation may engender, LLS is recognized to make the decryption process more complex [15] and, as a consequence, less desirable from the communication viewpoint. Thus, considering the current development of the theory and the technique, the symmetry condition seems to be essential for the practical implementation of simultaneous bidirectional communication, since it allows the maintenence of simple encoding/decoding processes.

Regarding the performance of simultaneous bidirectional chaos-based communication schemes, numerical studies estimated the maximum coding and transmission rates in delay-coupled semiconductor lasers [24], based on the oscillator's resynchronization time. After a synchronization burst occurs and before the next bit from stream can be encoded, resynchronization must establish. For semiconductor lasers, it was found that a maximum bit rate of around 3 Gbps is feasible, as resynchronization takes about 0.3 ns [24]. More recently, using a different coupling setup, [15] argued that such rate can be made higher than 10 Gbps.

On their turn, security aspects are evaluated on the basis of the realization that encryption keys can be negotiated in the process. Fischer et al. [24] argue that, although the outputs of SL1 and SL2 are readily available for an eventual eavesdropper, the transmitted bits would only be accessible as the systems encode different bits. Thus, a key could be obtained by gathering the N first bits that coincide, a situation in which an eavesdropper would have no evidence of whether the transmitted bit is 0 or 1, due to the absence of difference between the outputs of SL1 and SL2. This problem was further explored by Kanter et al. [16] and, as a result, the authors confirmed the unfeasibility of efficient software attacks to this technique.

Still in the realm of lasers, the Ikeda ring oscillator (IRO), a simplified representation of a ring laser, was tested by Zhou and Roy [27] in a scheme for simultaneous bidirectional communication based on IS using a different coupling configuration. To overcome the unstable operation observed for mutual delay-coupling of two IROs, say, IRO1 and IRO2, the authors propose the use of a third oscillator, IRO3, to drive the other two unidirectionally, as shown in Figure 14.6. As a result of the injection-locking effect, stable isochronal synchronization can be obtained between IRO1 and IRO2 due to the drive signal from IRO3 and communication can be established through the mutual coupling established between IRO1 and IRO2. The authors indicate that synchronization is maintained as the messages $m_1(t - \tau_{12})$ and $m_2(t - \tau_{21})$ are injected into the transmitting ring cavity, which are delayed by the propagation times τ_{12} and τ_{21}, respectively. According to the study, this procedure guarantees that the disturbance caused by the message injection will be symmetric and synchronization will be preserved. Another important factor to be considered is the relation among the coupling strengths. While κ_{31} and κ_{32} have to be chosen sufficiently large, since unidirectional drive induces isochronal

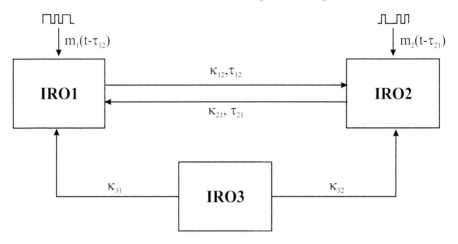

FIGURE 14.6
Ikeda ring oscillators with round-trip time τ_R where IRO1 and IRO2 are mutually coupled: isochronal synchronization between IRO1 and IRO2 is obtained by the unidirectional drive performed by IRO3 in both systems simultaneously. For isochronal synchronization to emerge, the delays from IRO3 to IRO1 and IRO2 must be the same, that is, $\tau_{31} = \tau_{32}$.

synchronization between IRO1 and IRO2, due to the injection-locking effect, the values of κ_{12} and κ_{21} should be chosen small enough not to disturb the synchronous state, yet large enough to allow communication between IRO1 and IRO2.

For this scheme, maximum encoding rates were obtained as 5 bits per round-trip of the IRO. As the round-trip time for the IRO is defined as L/c, where L is the length of the cavity and c is the velocity of light in the media (taken as 2.3×10^8 in optical fibers [22]), this would yield encoding rates of the order of Gbps. Moreover, according to the authors [27], security aspects are enhanced due to the presence of a third laser, since the reconstruction of the decoding waveform could not be performed by using the unidirectional drive signal and the mutual coupling signal alone. Further, as trademark of simultaneous bidirectional communication schemes based on IS, public keys can be negotiated 'on the fly.'

As an inconvenience, there is the need for the third oscillator, which makes the scheme more resource demanding than two-oscillator schemes. Besides demanding a third oscillator, it would also demand a channel and bandwidth for the transmission of the synchronization signal from IRO3 to IRO1 and IRO2 during operation. Furthermore, a direct consequence of the presence of IRO3 is that IS between IRO1 and IRO2 depends on the symmetry of delays from the signals coming from IRO3, i.e., it is required that $\tau_{31} = \tau_{32}$. To have this requirement met, either the third laser has to be equally distant from the other two or else artificial mechanisms have to be used to adjust delay magnitudes, such as electronic buffering of the signals. Either way, the setup becomes more demanding and complicated.

Following a similar scheme, Jiang et al. [15] studied isochronal synchronization in the context of mutually coupled SLs subject to identical unidirectional injection. The results agree with previous ones in the sense that the presence of a sufficiently strong unidirectional coupling is needed for stable IS. In this sense, the strength of the unidirectional injection from SL3 was shown to be related to that of the mutual coupling: high-quality stable isochronal synchronization was achieved as unidirectional injection from the third laser is sufficiently large as compared to mutual coupling between SL1 and SL2 [15]. This fact suggests that the coupling setups can be chosen from a large set of values, once they obey certain symmetry and proportionality conditions, in opposition to self-feedback schemes which seem to offer smaller parameter ranges for stable synchronization, as shown from the results in [9].

In coupling configurations using three lasers and the combination of unidirectional and mutual couplings, robustness against parameter mismatches was reported, indicating that the setup may handle inherent mismatches arising in practical applications. Particularly, from the scheme presented by Jiang et al. [15], stable isochronal synchronization between SL1 and SL2 is argued to tolerate up to tens of percentage of parameter mismatch, which is a sufficient amount to bear inherent detuning observed in experimental setups [6]. Similar results were obtained for other chaotic systems, such as those studied in [6,27], and this fact suggests that, to some extent, the practical detuning in the oscillators' parameters is counteracted by inherent characteristics of the phenomenology of isochronal synchronization, regardless of the chaotic oscillator under consideration. Nevertheless, while the presence of a certain amount of parameter mismatches does not affect stability, it does affect synchronization quality, which is observed to degrade as the magnitude of parameter mismatches increases.

The realization that the phenomenon of isochronal synchronization is revested with a certain amount of robustness is quite important as practical applications are aimed: on the one hand, mismatches are fully expected to be present in practical realizations while, on the other, they are generally limited to a combination of inherent differences of physical components and detuning. As such, it seems established that mismatches, inaccuracies, and uncertainty observed in practice can be naturally overcome once maintained, restricted to acceptable levels. As further evidence of this fact, robustness was also observed in the context of IROs driven by a third oscillator [27] and in that of SLs with optoelectronic feedback [6]. This suggests that, up to a considerable extent, robustness of IS does not depend directly on the dynamics of the chaotic oscillators involved.

More recent developments using three lasers show that isochronal synchronization is not only attainable under injection-locking effects. In fact, considering coupling setups featuring three lasers, all of them can relax into isochronal synchronization, i.e., $x_1(t) = x_2(t) = x_3(t)$, as subject to bidirectional chain coupling. Consider the coupling scheme pictured in Figure 14.7, based on the work of Illing et al. [13]. Note that the chain is coupled bidirectionally and all couplings are time-delayed. Considering this setup, a main requirement for IS to establish among the three oscillators is that the middle laser must be subject to self-feedback. Interestingly enough, for the coupling configuration presented in Figure 14.7, isochronal synchronization is not observed between the center and outer oscillators in the absence of self-feedback, i.e., $\gamma_{22} = 0$, despite the symmetric nature of the coupling configuration. Such phenomenon had been studied earlier by Landsman and Schwartz [18], who explained that a common solution exists for the outer subsystems due to internal symmetry. In this case, the Lyapunov exponents calculated with respect to directions which are transversal to the synchronization manifold $s(t)$ are all negative, indicating that synchronization of the outer subsystems would emerge, i.e., $x_1(t) = x_3(t) = s(t)$. This coupling configuration is shown in Figure 14.8, while the isochronal synchronization of the dynamics of outer lasers is pictured in Figure 14.9, and while the unsynchronized behavior of the inner one of the outer lasers is shown in Figure 14.10.

As further contribution, the authors [13] also showed that an analysis of such simple coupling configuration can lead to more general and profound comprehension of isochronal synchronization in networks of delay-coupled oscillators in general. As such, in a sense, simple configurations featuring three delay-coupled lasers can unveil mechanisms leading to synchronization in large

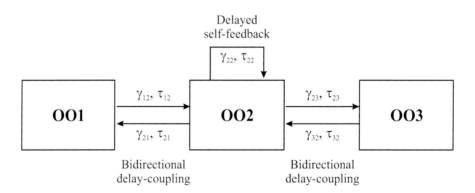

FIGURE 14.7
Chain of optoelectronic oscillators (OO) bidirectionally coupled with time-delays τ_{12}, τ_{21}, τ_{23}, τ_{32}, and self-feedback strength τ_{22}, with respective coupling strengths γ_{12}, γ_{21}, γ_{23}, γ_{32}, and γ_{22}: for zero self-coupling ($\gamma_{22} = 0$), balanced bidirectional coupling ($\gamma_{21} = \gamma_{23}$, $\gamma_{12} = \gamma_{32}$), and equal coupling delays ($\tau_{12} = \tau_{21} = \tau_{23} = \tau_{32}$), the conditions are created for the emergence of isochronal synchronization between OO1 and OO3; as the self-feedback strength is raised above a threshold, isochronal synchronization among the three oscillators is possible [13].

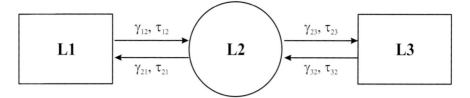

FIGURE 14.8
Chain coupling of three lasers, as proposed in [18]: the outer lasers are coupled to the central laser by means of bidirectional coupling. For isochronal synchronization, coupling delays have to be symmetric, i.e., $\tau_{12} = \tau_{21} = \tau_{23} = \tau_{32}$. As sufficient coupling strength is provided, the dynamics of the identical lasers SL1 and SL3 can relax into isochronal synchrony. Within this framework, the central oscillator (SL2) can be different from the other two [25].

networks. This is an important direction of research, since the understanding of stability of isochronal synchronization in networks may allow the extension of current chaos-based communication techniques to such a realm [10, 11].

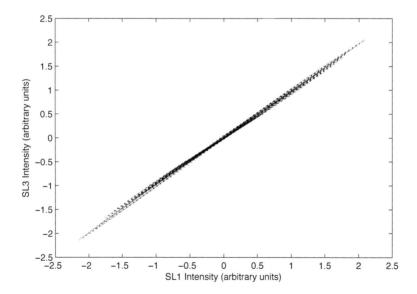

FIGURE 14.9
Isochronal synchronization between outer lasers SL1 and SL3 in a three-laser chain: even in the absence of direct coupling between the oscillators, they can be driven to isochronal synchronization, by means of the interplay between the central and outer lasers.

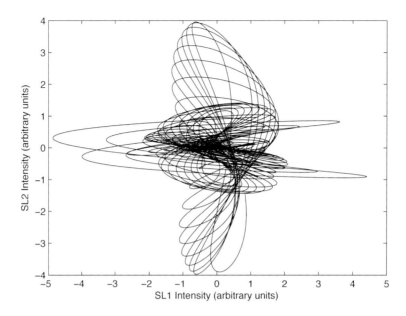

FIGURE 14.10
Unsynchronized dynamical behavior of SL1 and SL2 in a three-laser chain:
the central laser in the chain does not relax into isochronal synchronization
as it is not subject to self-coupling, i.e., $x_1(t) \neq x_2(t)$ as $\gamma_{22} = 0$.

As observed in the results of aforementioned works, isochronal synchronization is highly dependent on symmetry of the coupling, as those in Figures
14.5, 14.6, 14.7, and 14.8. On the one hand, its emergence requires that time-
delays and the sum of coupling inputs are nearly identical. This may become
a drawback in practice: while the coupling strength is a matter of design, time
delays do not depend only upon design choices. This somewhat restricts the
direct application of isochronal synchronization for communication in many
forms of coupling setups featuring asymmetric characteristics, including most
networks arising in real situations. In some cases, time delays can be elec-
tronically compensated and balanced by buffering signals, in order to adjust
time-delays for the design of communication systems for specific purposes, as
in the case when spatial asymmetry is present. However, it is undeniable that
the stringent requirement for symmetry adds an extra load of complexity to
the practical realization of communication.

Regarding two mutually coupled systems subject to direct self-feedback,
according to the setup presented in Figure 14.11, stable isochronal synchronization can be obtained and, beyond, it can be proven stable by means of
analytical results [9, 20]. Note that direct self-feedback makes the practical
implementation of the coupling setup easier, since direct self-feedback is eas-

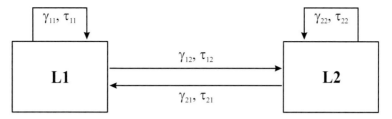

FIGURE 14.11
Two mutually delay-coupled systems featuring direct self-feedback: although simpler than most of the coupling setups studied in the literature, this coupling setup may severely restrict the magnitude of coupling delays.

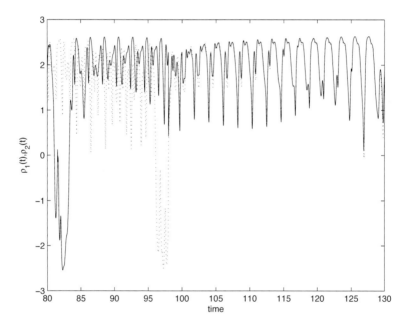

FIGURE 14.12
Isochronal synchronization between two mutually coupled Lang-Kobayashi equations with direct self-feedback: coupling delay is taken as $\tau_{12} = \tau_{21} = 10^{-2}$.

ier to implement than delayed self-feedback. In practice, direct self-feedback offers a means for stable IS upon a coupling configuration that can be considered the simplest one. Note that the condition on the symmetry of delay magnitudes is easily respected in this case, and isochronal synchronization can emerge on the basis of the addition of mutual delayed coupling. Figures 14.12

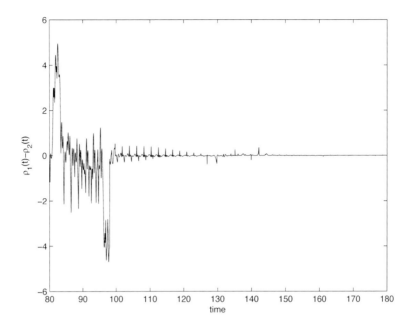

FIGURE 14.13
Isochronal synchronization error for mutually delay-coupled Lang-Kobayashi
equations subject to self-coupling: the error stabilizes at the origin of the error
system, indicating that synchronous behavior was established.

and 14.13 show the dynamics of Lang-Kobayashi equations and the dynamics
of the synchronization, respectively, as systems get in synchrony by mutual
delayed feedback and direct self-feedback. On its turn, Figure 14.14 presents
the classic signature of synchronization, as state variables of the two systems
are plotted one against the other.

Note that, for coupling setups considering direct self-feedback, it is required that the systems feature chaotic dynamics regardless of the existence
or not of delayed self-feedback. Under this framework, the requirement that
time delay of self-couplings are adjusted for symmetric operation can be relaxed, since this condition ceases to exist in this case; on the other hand,
this framework seems to impose an upper bound on the allowable value of
delay in the mutual coupling such that the chaotic behavior of the coupled
systems and synchronization stability are preserved. This is undesirable under
the viewpoint of communication, since upper bounded coupling delays mean
that communication cannot be established if the distance between oscillators
is large. The root of the problem is that isochronal synchronization becomes
unfeasible for large delays under this framework. In addition, the parameter

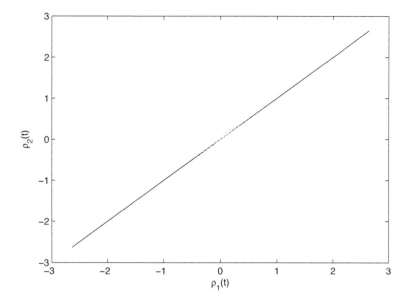

FIGURE 14.14
Identity of trajectories of the mutually delay-coupled Lang-Kobayashi equations subject to direct self-feedback.

ranges for stability seemingly become narrower under direct feedback setups and, more importantly, time-delays become more influential upon synchronization stability [9]. Mainly due to these reasons, direct self-feedback does not seem to offer a consistent framework for the implementation of communication based upon IS.

One step ahead of numerical simulations, Wagemakers et al. [26] conceived the first experimental demonstration of chaos-based simultaneous bidirectional communication. Using Mackey-Glass electronic circuits as a basis, the authors implemented full-duplex transmission of messages between delay-coupled chaotic systems, using delayed mutual coupling and delayed self-feedback. After dealing with intrinsic noise of the systems and evaluating the amplitude of the message to be added in the chaotic dynamics, the transmission of binary sequences at the encoding rate of 80 bps was performed at both ends of the communication line, simultaneously. Numerical simulations are shown in Figures 14.15–14.18 that illustrate the functioning of this scheme. While not particularly attractive for its bit rate performance, such implementation promoted experimental evidence that IS can indeed sustain communication, as supposed in computational simulations using laser equations [24, 27].

Needless to say, while they provide simple implementation, chaotic elec-

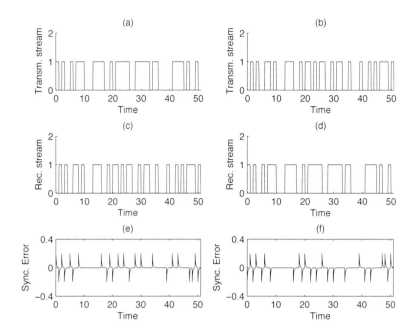

FIGURE 14.15
Numerical realization of bidirectional simultaneous transmission of binary messages, using the methodology proposed in [26]: (a) binary message transmitted; (b) received by system 1; (c) synchronization error, as measured in system 1; (d) binary message transmitted; (e) received by system 2; and (f) synchronization error, as measured in system 2.

tronic circuits are not particularly suited for applications in communication due to their low encoding rate capacity. In this context, lasers have superior encoding performance while they also benefit from low noise levels that are a desirable feature of optical systems in general. Thus, due to their operation at time scales of the order of nano- to picoseconds (10^{-9} to 10^{-12} s), lasers can be subject to much higher injection rates while they also benefit from low noise level of optical fiber as compared to that of copper-based systems [22]. As such, the next step towards the development of simultaneous bidirectional transmission using isochronal synchronization seems to be its experimental implementation using lasers, such that numerical and theoretical results using laser equations could be confirmed.

In a broad context, beyond the realm of lasers or electronic circuits, isochronal synchronization was analytically shown to be achievable in pairs of delay-coupled chaotic oscillators in general, indicating that its occurrence, a physical phenomenon, is not restricted to a particular chaotic oscillator [9].

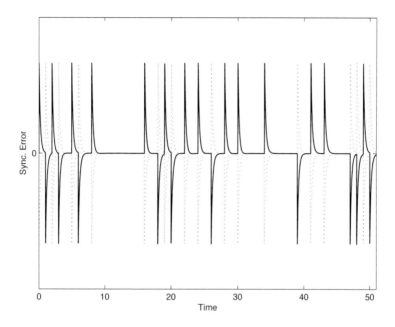

FIGURE 14.16
Anti-synchronization between the error variables from each of the systems against time: the error dynamics in both ends of the communication channel remain synchronized as random binary streams are input simultaneously in the dynamics of both chaotic oscillators.

Further, the phenomenon was shown to emerge even in network configurations, in prospective studies aiming its application for communication in flying formation of satellites [10, 11]. Seemingly, the symmetry of oscillator couplings, both regarding its strength and delay magnitude, seems to be a stringent requirement in the context of networks as well. Moreover, as a synchronous isochronal solution does not exist for every network topology, particularly as one considers coupling delays and direct self-feedback, as shown in [20], it seems that the development of network communication schemes based on isochronal synchronization is summarily restricted to specific topologies that are not commonly found in real networks. These conditions seem to be mainly related to network symmetry conditions [10, 11, 20].

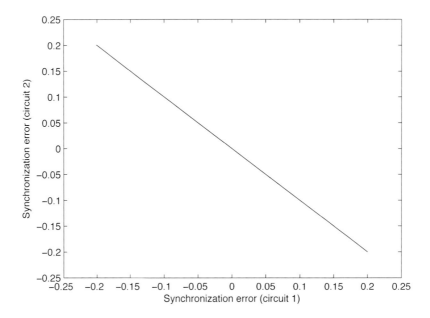

FIGURE 14.17
Anti-synchronization between error variable measures in each of the systems, against each other: persistent synchronous behavior shows that simultaneous bidirectional communication can be exploited in mutually delay-coupled chaotic oscillators, as experimentally demonstrated in [26].

14.3   Further possibilities and perspectives

Although mutually delay-coupled chaotic systems are sensitive to the magnitude and symmetry of time delay, parameter mismatches and noise, successful realizations of chaos-based communication in simple numerical and experimental setups reinforce the possibility of fully exploiting chaotic dynamics to transmit information in such real environments [6, 24, 26, 27]. Considering the source coding efficiency and data compression rates allowed by chaotic dynamics [3], chaos-based simultaneous bidirectional communication based on isochronal synchronization tends to be of increasing research interest. It is expected that experiments considering optical simultaneous bidirectional communication will appear in the following years, maybe even in the form of a field experiment under real-world conditions.

Beyond coupling configurations featuring two chaotic oscillators, recent developments have enhanced the comprehension of IS in networks of chaotic

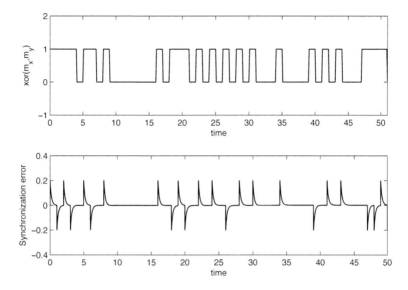

FIGURE 14.18
XOR between the binary streams $m_x(t)$ and $m_y(t)$ and synchronization error $e_x(t)$: spikes in the synchronization error from zero indicate the encoding of different bits in each end of the communication channel.

oscillators, such that chaos-based communication schemes might soon be designed to deal with network communication issues, covering hot topics such as that of communication and coordination of maneuvers in Flying Formation of Satellites [10, 11, 19]. Towards this end, it can be speculated that isochronal synchronization can be explored for network communication using current simultaneous bidirectional communication schemes as a basis.

Besides, multiple access techniques could be incorporated into current chaos-based communication schemes using isochronal synchronization. As verified by Itoh and Chua [14], frequency division multiplexing (FDM), time division multiplexing (TDM), and amplitude division multiplexing (ADM) can be advantageously explored in the context of multi-user chaos-based communication. In this scenario, the unique features of isochronal chaos synchronization could be combined to successful techniques from conventional communication, such that chaos-based communication schemes definitely establish as options for upcoming applications.

Bibliography

[1] K. J. Astrom. *Introduction to Stochastic Control Theory*. Dover, New York, 2006.

[2] J. M. F. Avila and J. R. Rios Leite. Time delays in synchronization of chaotic coupled lasers with feedback. *Optics Letters*, 17:21442–21451, 2009.

[3] M. S. Baptista, E. E. N. Macau, C. Grebogi, Y. C. Lai, and E. Rosa Jr. Integrated chaotic communication scheme. *Phys. Rev. E*, 62:4835–4845, 2000.

[4] K. M. Cuomo and A. V. Oppenheim. Circuit implementation of synchronized chaos with applications to communications. *Phys. Rev. Lett.*, 71(1):65–68, July 1993.

[5] K. M. Cuomo, A. V. Oppenheim, and S. H. Strogatz. Synchronization of Lorenz-based chaotic circuits with applications to communications. *IEEE Transactions on Circuits and Systems II*, 40:626–633, October 1993.

[6] T. Deng, G.-Q. Xia, L.-P. Cao, J.-G. Chen, X.-D. Lin, and Z.-M. Wu. Bidirectional chaos synchronization and communication in semiconductor lasers with optoelectronic feedback. *Optics Communications*, 282(11):2243–2249, June 2009.

[7] T. Erneux. *Applied Delay Differential Equations*. Springer, New York, 2009.

[8] H. Fujisaka and T. Yamada. Stability theory of synchronized motion in coupled-oscillator systems. *Progress of Theoretical Physics*, 69:32–47, 1983.

[9] J. M. V. Grzybowski, E. E. N. Macau, and T. Yoneyama. Isochronal synchronization of time delay and delay-coupled chaotic systems. *J. Phys. A: Math. Theor.*, 44:175103, 2011.

[10] J. M. V. Grzybowski, M. Rafikov, and E. E. N. Macau. Chaotic communication in a satellite formation flying - the synchronization issue in a scenario with transmission delay. *Acta Astronautica*, 14:2793–2806, 2009.

[11] J. M. V. Grzybowski, M. Rafikov, and E. E. N. Macau. Synchronization analysis for chaotic communication on a satellite formation flying. *Acta Astronautica*, 67:881–891, 2010.

[12] T. Heil, I. Fischer, W. Elsässer, J. Mulet, and C. R. Mirasso. Chaos synchronization and spontaneous symmetry-breaking in symmetrically delay-coupled semiconductor lasers. *Phys. Rev. Lett.*, 86:795–798, January 2001.

[13] L. Illing, C. D. Panda, and L. Shareshian. Isochronal chaos synchronization of delay-coupled optoelectronic oscillators. *Phys. Rev. E*, 84:016213, 2011.

[14] M. Itoh and L. O. Chua. Multiplexing techniques via chaos. In *Proc. IEEE Int. Symp. Circuit Syst.*, June 9-12, Hong Kong, pages 905–908, 1997.

[15] N. Jiang, W. Pan, L. Yan, B. Luo, W. Zhang, S. Xiang, L. Yang, and D. Zheng. Chaos synchronization and communication in mutually coupled semiconductor lasers driven by a third laser. *Journal of Lightwave Technology*, 28:1978–1985, 2010.

[16] I. Kanter, E. Kopelowitz, and W. Kinzel. Public channel cryptography: chaos synchronization and Hilbert's tenth problem. *Phys. Rev. Lett.*, 101:084102, 2008.

[17] E. Klein, N. Gross, M. Rosenbluh, W. Kinzel, L. Khaykovich, and I. Kanter. Stable isochronal synchronization of mutually coupled chaotic lasers. *Phys. Rev. E*, 73:066214, June 2006.

[18] A. Landsman and I. B. Schwartz. Complete chaotic synchronization in mutually coupled time-delay systems. *Phys. Rev. E*, 75:026201, 2007.

[19] C. M. P. Marinho, E. E. N. Macau, and T. Yoneyama. Chaos over chaos: a new approach for satellite communication. *Acta Astronautica*, 57:230–238, 2005.

[20] T. Oguchi, H. Nijmeijer, and T. Yamamoto. Synchronization in networks of chaotic systems with delay-coupling. *Chaos*, 18:037108, 2008.

[21] L. M. Pecora and T. L. Carroll. Synchronization in chaotic systems. *Phys. Rev. Lett.*, 64:821–824, February 1990.

[22] K. V. Prasad. *Principles of Digital Communication Systems and Computer Networks*. Charles River Media, Hingham, MA, 2004.

[23] I. B. Schwartz and L. B. Shaw. Isochronal synchronization of delay-coupled systems. *Phys. Rev. E*, 73:046207, 2006.

[24] R. Vicente, C. R. Mirasso, and I. Fischer. Simultaneous bidirectional message transmission in a chaos-based communication scheme. *Optics Letters*, 32(4):403–405, February 2007.

[25] A. Wagemakers, J. M. Buldu, and M. A. F. Sanjuan. Isochronous synchronization in mutually coupled chaotic circuits. *Chaos*, 17:023128, 2007.

[26] A. Wagemakers, J. M. Buldu, and M. A. F. Sanjuan. Experimental demonstration of bidirectional chaotic communication by means of isochronal synchronization. *Europhys. Lett.*, 81:40005, 2008.

[27] B. B. Zhou and R. Roy. Isochronal synchrony and bidirectional communication with delay-coupled nonlinear oscillators. *Phys. Rev. E*, 75(2):026205, February 2007.

15

Cryptography based on chaotic and unsynchronized elements of a network

Romeu M. Szmoski, Fabiano A. S. Ferrari, and Sandro E. de S. Pinto
Department of Physics
Universidade Estadual de Ponta Grossa

Ricardo L. Viana
Department of Physics
Universidade Federal do Paraná

Murilo S. Baptista
Institute for Complex Systems and Mathematical Biology, SUPA
University of Aberdeen

CONTENTS

15.1 Introduction

Communication is the action of transmitting and eventually receiving another message in response, as in an exchange of e-mails, for example. In a technical approach, one may say that communication is the process that involves the transmission and reception of messages between a source and an addressee receiver, in which information transmitted through physical resources

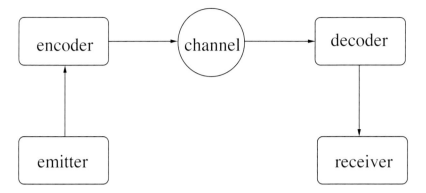

FIGURE 15.1
Schematic diagram of a communication system.

or equipment, and technological devices are coded at source and decrypted at the destination using agreed systems of signs or sound symbols, written or iconographic. Schematically (Figure 15.1), there are five elements in any communication system, namely, an emitter, an encoder, a transmission channel, a decoder, and a receiver [24]. The encoder handles the message, so it passes through the channel, and the decoder carries out the reverse process. In order to be practical and safe, the construction of a communication system may require other components [23].

The message that we are talking about consists of a number of data, often called primary data, transmitted from a source to a receiving station. This is not about imparting knowledge, but simply data. These are usually signals that can take many forms and often translated by numerical terms so that one can accurately measure the amount of information transmitted. When the message is composed of binary digits, each unit of information is called bit (short for binary digit).

The information may be considered independent of all semantic content, and in this case is defined statistically. This is the field of information theory. It is characteristic of the study of information in which, given a signal or an information unit, there must be some indeterminacy in relation to the next signals. Indeed, an "information" that offers no indeterminacy is not, properly speaking, information. The amount of information provided by a signal is a function of its probability. It is considered that the amount in question is equal to the negative base-2 logarithm of the probability of the signal. Therefore, the information in the sense indicated here does not refer to what "you say" but what "you could say."

A major concern of communication is the security in data transmission. In many cases, the speed and reliability for transmitting a message with low probability of errors are not enough, but also the transmission has to be carried out in an extremely secure way. In 1949, C.E. Shannon took a decisive step toward showing that if the length of the key is not an inconvenience, a message can be securely sent [24]. In a communication system, traditional cryptography functions in the software level rather than in the high speed physical level. Besides, security requires the use of ergodic and mixing transformations also in the software level. In the early 1990s an attempt was made to show how to use chaos synchronization to create a secure communications systems [9,19]. This kind of synchronization has enabled one to create a fast cryptographic system that operates in the physical (hardware) level of the communication system. Besides, chaotic systems are naturally ergodic and mixing, which provides security. This initial idea was shown to be insecure [21]; however, the complexity of nonlinear systems enables other approaches. On account of this, different communication systems based on chaos synchronization have been proposed and investigated [14, 25] . In general, such systems assume chaotic dynamics for both the decoder and receiver. Thus, the encoder codifies a message using some property of the chaotic signal and, after being transmitted, the message is decoded by the receiver, which is synchronized to the emitter [13].

An eavesdropper trying to determine the parameters of a cryptographic system will always make an error. If the dynamics of the system are chaotic, a small error grows exponentially that makes it difficult to decrypt the intercepted message [2,3,5,15–17], if the decoding relies on a receiver that is exactly identical to an emitter. This is one of the properties that made chaos-based cryptosystems popular.

Synchronization offers a communication system where information is transmitted and received in real time and that operates in a physical level. However, the need for synchronization reduces the parameter range of the transmitter and receiver within which the system can be considered secure. For example, in [10] it was shown that receiver and emitter do not need to match for the retrieval of information.

A different approach to secure communication using coupled chaotic systems was presented by Hung and Hu [11]. In their method, a binary message is codified considering the coupling direction between chaotic maps on a ring. The receiver decodes the message, determining the transfer entropy [22] between succeeding maps. Indeed, for any interacting system, the transfer entropy can be used for determining which variables influence the dynamics of each other. The novel feature of this method is that it does not require synchronization for transmitting data; it only requires the determination of the transfer entropy. One inconvenience is that, as this quantity is statistically defined, many observations from the dynamics of the maps are necessary in order to decode the message. The observation interval needed by the receiver to determine the transfer entropy between the maps was taken as the relaxation time [11]. Therefore, the transmission of information through this mechanism requires operating times that are multiples of the relaxation time.

In this work, we propose an improvement over the previous method of Hung and Hu consisting of a communication system that uses both chaos synchronization as well as transfer entropy. While Hung and Hu's method uses only one network of coupled maps, our model contains two networks, the emitter and the receiver, the latter being a replica of the former. The transmission of a message through the system is accomplished by two stages that we call pre- and post-synchronization. Here, synchronization occurs just between the elements with the same label, in the emitting and receiving networks. Importantly, the elements in each network are out of synchronization. The use of desynchronized networks allows one to explore the good properties of synchronization, namely, fast transmission of information, but without losing the parameter range for which the system is secure [29]. Most important, since all nodes in the emitter network are desynchronous, the encoded signal generated by this network is composed of variables that are not correlated. In our method, it is strictly necessary that receiver is identical to emitter in order to decode the message. Since the transmitter is composed of a large network of desynchronous elements, it is very difficult to determine the parameters of this network by analyzing the transmitted signal. An eavesdropper cannot reproduce equally the emitter. Even if an intruder manages to find all values of the state variables from the emitter and receiver, if one does not know precisely the coupling function and intensity, one is not capable of decoding the message correctly. Assuming the coupling prescription to be secret keys of the communication system, if the key is altered after each communication, then security would be increased further.

In the stage of preasynchronization, the networks start interacting directly with each other until every node of the emitting network becomes identical to every other node with the same label of the receiving network. Once the emitter and receiver networks are synchronized, the second stage starts transmitting and decoding the message. During this stage, the interaction between networks is deactivated and the message is encoded in a binary signal through an on-off device. The advantage of this procedure consists of transmitting N bits of information for each unit of the relaxation time. In other words, the communication device transmits N bits of information using a bit stream, which makes data transmission fast.

The proposed method relies on the fact that a long scalar quantity, composed of the trajectory of a uni-dimensional system (s) coupled in a master-slave fashion with an N-dimensional emitter network by a connecting matrix representing the binary message to be transmitted, carries information about these couplings and, therefore, the message. The message can only be decoded by a person who has complete knowledge of the emitter network. Imagine the system s as a node in a large dynamical network. With the exception of the node s, assume that the information about all other nodes is either known or can be precisely measured. Our method works because it is possible to determine which are the nodes in this dynamical network that are connected to s by measuring the flow of information created by a connection. Hence, nodes coupled to s influence its dynamical behavior. Furthermore, the receiver uses the transfer entropy to decode the message.

The fundamentals behind the success of our cryptographic method share similarities with one possible way in which the information is believed to be processed and transmitted in the brain. Reservoir Computing (RC) [12] is a machine-learning paradigm employed to retrieve from a dynamical network, the reservoir (the âbrainâ), information of an external perturbation driving it, input. The assumption behind RC is that the information about the input is spread out all over the network, and reliable retrieval of it can be accomplished by making a weighted average from the trajectories of some selected nodes of the reservoir, the output. Discovering which nodes should be selected is a remarkable task that can be resolved by a learning process whose purpose is to find an approximate match between input and output. In summary, one hopes to find the connecting topology between the reservoir and output (for a given random connecting topology between the input and the reservoir), such that the output matches the input. It has been recently demonstrated that RC can be performed by a reservoir composed of a single dynamical node operating as if it were a complex system [4]. In order to make the analogy between RC and the proposed cryptographic method, imagine the system s functioning as the reservoir, the emitter network being responsible to produce the input, the connecting topology between the input and s given by the message, and the output generated by the receiver network, which in our method equals the input in the post-synchronization stage. Similarly, to RC, which considers that the input perturbs any random set of nodes in the reservoir, the proposed cryptographic method works regardless of how the connecting topology is between the input and s, the message. Any random message can be decoded by one who has confidential information. The main assumption for the success of RC lies behind the belief that the reservoir stores information about the perturbation experienced by it. In this analogy we make, it is appealing to state that the reservoir s also manages to store information about the particular way it is being perturbed. In contrast to RC, that aims at discovering the connecting topology between the reservoir and output, in the proposed cryptographic method it is the connecting topology between input and the reservoir, the message that needs to be revealed. Finally, in RC input and output only roughly match, while in our method, they need to match perfectly; otherwise, the message cannot be decoded. Therefore, this analogy allows one to interpret the encoding system s as a sort of reservoir that has memory about the message, i.e., the way it is being perturbed.

This work is organized as follows: in Section 15.2, we describe the communication device we propose based on both chaos synchronization and transfer entropy analysis. Section 15.3 describes the pre-synchronization stage, computing the synchronization time of a lattice of coupled piecewise linear chaotic maps. In Section 15.4 we describe the post-synchronization stage, determining the relaxation time of the system. Section 15.5 examines an example of data transmission through this mechanism. We devoted Section 15.6 to improve security and efficiency of the method. The last section is devoted to our Conclusions.

15.2   Description of the communication mechanism

Consider the emitter (E) and receiver (R) networks as two identical coupled map lattices composed of N sites each [8, 20, 27]. The number of sites N is defined according to the amount of bits in a message that will be transmitted. If we assume the message as a binary sequence $m = \{m^{(1)}m^{(2)} \cdots , m^{(N')}\}$ of length N', with $m^{(i)}$ equal to 0 or 1, N is equal to the number of elements belonging to this sequence, that is, $N = N'$. The state of the E and R networks, at each discrete time n, is defined by the vectors $\mathbf{e}_n = (e_n^{(1)}, e_n^{(2)}, \cdots , e_n^{(N)})^T$ and $\mathbf{r}_n = (r_n^{(1)}, r_n^{(2)}, \cdots , r_n^{(N)})^T$, respectively, with $e_n^{(i)}$ and $r_n^{(i)}$ corresponding to a state variable $z \in \Omega$ whose time evolution is governed by a chaotic map $f : \Omega \mapsto \Omega$, with $\Omega \subset \mathbb{R}$.

There have been investigations of chaos synchronization between replicas of coupled map networks using continuous maps [1, 6, 26]. Based on such previous investigations we can restrict our analysis to continuous maps f over the set Ω. In particular, we focus on the piecewise-linear tent map $f(z) = 1 - 2|z - 0.5|$ [18].

Besides the individual dynamics, the sites in each network are submitted to a coupling prescription. This intra-network coupling is arbitrary, but, for transmitting and sending a message correctly both E and R networks must be taken as identical, sharing the same coupling prescription and parameters. So, our communication system uses a symmetric private key. Here we consider a Laplacian-local intra-network coupling with periodic boundary conditions and random initial conditions:

$$F(z^{(i)}) = (1 - \varepsilon)f(z_n^{(i)}) + \frac{\varepsilon}{2}\left[f(z_n^{(i-1)}) + f(z_n^{(i+1)})\right], \qquad (15.1)$$

where $z^{(i)} = z^{(N \pm i)}$ represents the state variable of the i-th site $(i = 1, ..., N)$ and $\varepsilon \in [0, 1]$ stands for the strength of intra-network coupling in each network.

The transmission data process between E and R networks is composed of two stages: the pre- and post-synchronization. When we refer to synchronization we mean the process in which $\mathbf{e}_n = \mathbf{r}_n$ for all n. The components of each state vector are necessarily not equal. If all maps of a network mutually synchronize, then the dynamics of the network, given by Equation (15.1), reduce to the dynamics of an uncoupled map. This is an undesirable feature for the point of view of the secure communication, since an intruder could determine the network state from knowing the state of only one site, that would endanger the security of the transmission. Thus, in order to avoid mutual synchronization in each network, we will assume that intra-network coupling intensity is sufficiently weak, which increases the dimension of the emitter.

15.3 Pre-synchronization stage

A way of synchronizing the state vectors of the E and R networks is to assume an interaction between them. This inter-network coupling may be unidirectional or bidirectional. In the first case, also known as master-slave coupling, one network influences the dynamics of the other but is not influenced by the latter; while, in the second case, both networks influence and are influenced by each other. We will take here the master-slave coupling, E as the master and R as the slave networks, such that the dynamics of the system are described by

$$\begin{aligned} e_{n+1}^{(i)} &= F(e_n^{(i)}) \\ r_{n+1}^{(i)} &= (1-\gamma)F(r_n^{(i)}) + \gamma F(e_n^{(i)}), \end{aligned} \tag{15.2}$$

in which γ is the strength of the inter-network coupling. Figure 15.2 illustrates the system described by Equations (15.1) and (15.10). Each site is represented by a disc and the lines indicate the connections among them. Note that the E-sites (black balls) interact with their neighbors inside the E-network, and R-sites (gray balls) interact with their neighbors inside the R-network (intra-network connections are bidirectional). However, the sites of E influence a corresponding site of R and its nearest neighbors, since the inter-network connections are unidirectional.

The coupling remains active while the networks do not synchronize with each other. When this occurs the network E stops sending information about its state variables. Since the synchronization of the E and R networks marks the end of the first stage of the process, we have to identify when it occurs. We use as a synchronization diagnostic the synchronization error average, given by

$$w_n = \frac{1}{N} \sum_{i=1}^{N} |e_n^{(i)} - r_n^{(i)}|. \tag{15.3}$$

If the state vectors of E and R networks are synchronous we have $w_n = 0$, otherwise $w_n > 0$. The synchronization time τ_s is the time it takes for this average error to be less than 10^{-14} over a time window of 1000 consecutive iterations of the system.

In order to make the $\mathbf{e}_n$ and $\mathbf{r}_n$ vectors identical we need to choose the γ and ε parameters in such a way that they allow synchronization. In the particular case in which $\varepsilon = 0$, synchronization is just observed when $\gamma > \gamma_c = 1/2$. For an inter-network coupling strength of $\gamma = 1/2 \pm \delta$, we investigated the synchronization time τ_s between the networks as a function of the intra-network coupling strength ε. Figure 15.3 shows the average synchronization time as a function of the intra-network coupling strength for 100 different randomly chosen initial conditions, $N = 21$ and for different values of δ. We

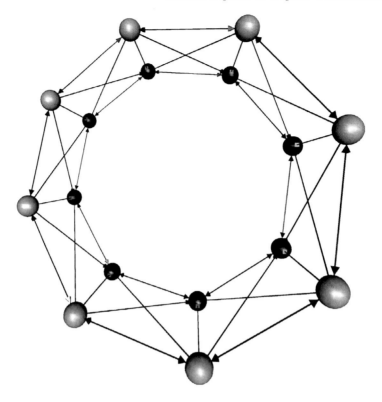

FIGURE 15.2
Schematic diagram of the connections between corresponding sites of two networks (E: emitter, as in black balls; R: receiver), with Laplacian-local intra-
network coupling and a master-slave inter-network coupling.

see that, for $\gamma = 0.5$ ($\delta = 0$) and $\varepsilon < 0.3$, the average synchronization time
scales with ε as a power-law

$$< \tau_s >= C\varepsilon^{-\alpha}, \tag{15.4}$$

in which C and α are obtained by the best fit of the points. In the case
presented we have the following values: $\alpha = 1.018$ and $C = 32.23$. Considering
networks of different sizes, we found that $\alpha \approx 1.0$ while C depends on the value
of w from which the networks are considered synchronized, more specifically,
$C = \ln w^{-1}$. Even though this relation, in general, is not a power-law for
any value of ε, it can be used nevertheless to estimate the synchronization
time between the networks. In the following we will consider just the case for
$\varepsilon \leq 0.1$ and, thus, Equation (15.4) allows us to estimate the time it takes for
the coupled networks to mutually synchronize.

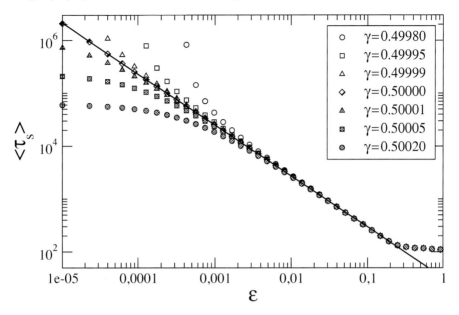

FIGURE 15.3
Average synchronization time as a function of the intra-network coupling parameter ε for different values of the inter-network coupling parameter γ. The solid line is a least squares fit with slope -1.0.

15.4 Post-synchronization stage

In the previous section, we showed how to synchronize the networks E and R by controlling the inter-network coupling parameter γ. Here, we will describe how to convey a particular message m between E and R in a safe and efficient way. A necessary condition is that both networks remain synchronized during all the process, otherwise the receiver will not be able to read the message correctly.

Remember that we consider the networks to be synchronized whenever the synchronization error w_n is less than a tolerance fixed at 10^{-14}. In order to keep the network synchronized we truncate the state variables of both E and R such that we get rid of differences $o(w)$ less than 10^{-14}. Then we turn off the inter-network coupling, since it is no longer necessary for keeping the networks synchronized. Indeed, E and R being identical networks, if $\mathbf{e}_n = \mathbf{r}_n$ at a given instant $n = \tau_s$ then they remain synchronized for all $n > \tau_s$.

After the inter-network coupling is switched off, the second stage of the communication process begins, in which we transmit the desired message. To compare our method with traditional cryptography, we briefly introduce

TABLE 15.1
XOR true table.

Message	0	0	1	1
Keystream	0	1	0	1
XOR	0	1	1	0

the famous Vernam cipher [28], a symmetrical key cipher where plain text digits are combined with a keystream. This combination produces a ciphertext using the operation XOR, symbolized by $\oplus$, whose true table is presented in Table (15.1). The operation is reciprocal: one uses an identical keystream both to encipher plain text to ciphertext and to decipher ciphertext to yield the original plain text.

According to Shannon, for a cipher to be considered secure it must satisfy the following conditions: (i) the keystream must have at least the same length of the message, (ii) it is changed at every communication, and (iii) binary symbols of keystream must be randomly decorrelated, then cipher is proven to be secure [23].

In our method, the XOR transformation is a sophisticated nonlinear transformation. Similarly to the Vernam cipher, the size of the emitter network is equal to the length of the message. The emitter network has a role similar to the keystream in the Vernam cipher method. Finally, the requirement that a keystream must have decorrelated symbols is analogously reproduced in our method by having decorrelated nodes in the emitter network.

We introduce a discrete-time dynamical system $S : \Omega \mapsto \Omega$ that is responsible for encoding the message in the signal consisting of an orbit of the system S. The map S defines the value of the signal s_n in each time instant associating the message characters to the state variables of the emitter network through an on-off device. If the i-th element of the message m, denoted as $m^{(i)}$, has a binary value of 1, then $e_n^{(i)}$ influences the dynamics of s_n (mode-on), whereas if $m^{(i)} = 0$, then $e_n^{(i)}$ does not influence the signal dynamics (mode-off). Moreover, the dynamics of the signal is given by the following map

$$s_{n+1} = S(s_n) = (1 - \beta)s_n + \frac{\beta}{\eta} \sum_{i=1}^{N} e_n^{(i)} m^{(i)}, \tag{15.5}$$

in which $\beta \equiv (N - 1)/N$ and $\eta = \sum_{i=1}^{N} m^{(i)}$ is a normalization factor. Given

TABLE 15.2
Cipher and decipher operations.

Plain Text	$\oplus$	Keystream	=	ciphertext
ciphertext	$\oplus$	keystream	=	Plain text

a randomly chosen initial condition s_0 the iteration of the map S yields a chaotic orbit $\{s_n\}_{n=0}$. The map S itself is not chaotic, but since it is driven by $e_n^{(i)}$, that itself being chaotic, the orbit $\{s_n\}_{n=0}$ results as chaotic as well.

Notice that the magnitude of the signal is not a feature of the message, but rather a feature of the dynamics of each element of the emitter network (E). Consequently, if the system initiates from different initial states, the same message will generally result in different signals. On the other hand, it may happen that two different messages result in the same signal for a limited and usually short period of time, however the signals will eventually diverge with time.

For recovering the message contained in the signal $\{s_n\}_{n=0}$ the receiver network R first verifies which state variables $e_n^{(i)} = r_n^{(i)}$ influence and which do not influence the s_{n+1}. Then the receiver network associates the symbol 1 to the former case and 0 to the latter case, observing the indexes of the sites in the network. The transfer entropy is the dynamical tool that allows the receiver network R to accomplish such verification. The transfer entropy $T_{r_n^{(i)} \to s_{n+1}}$ vanishes if and only if the dynamics of s_{n+1} does not depend on the dynamics of $r_n^{(i)}$, so if the transfer entropy is nonzero there is a statistical coherence between these signals. It is defined as

$$T_{r_n^{(i)} \to s_{n+1}} = \sum p(s_{n+1}, s_n, r_n^{(i)}) \log \frac{p(s_{n+1}|s_n, r_n^{(i)})}{p(s_{n+1}|s_n)}, \qquad (15.6)$$

where $p(\ , \ , \ )$ and $p(\ | \ )$ mean the join and conditional probabilities, respectively. These probabilities may be calculated using a box counting algorithm or a kernel estimator [22]. In this work we use the former procedure by considering the following coarse-grained variables:

$$\tilde{x} = \begin{cases} 0, & \text{if } 0 < x \leq 1/2, \\ 1, & \text{if } 1/2 < x < 1, \end{cases} \qquad (15.7)$$

in such a way that, instead of using the variables $r_n^{(i)}$, s_n, and s_{n+1} we use the binary variables $\tilde{r}_n^{(i)}$, $\tilde{s}_n$, and $\tilde{s}_{n+1}$.

Since there are only eight possible binary states, the summation in Equation (15.6) has only eight terms. Besides, instead of working with the signal s_n itself, it suffices to consider the binary variables $\tilde{s}_n$ without risk of turning the decoding process unsafe. In the binary form the signal can be transmitted through any public channel. In the following we explain how the determination of the transfer entropy of each variable from the receiver network to the signal enables the receiver to decode the message.

When the receiver network computes the transfer entropy to recover the message encoded in the signal, it hardly will achieve a vanishing contribution due to the fluctuations. Moreover the receiver network sites coupled with the signal yield a transfer entropy value much higher than those uncoupled sites. A high-pass filter is used to enable the receiver network to correctly recover

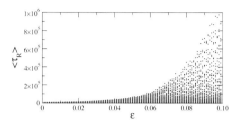

FIGURE 15.4
Average relaxation time as a function of intra-network coupling strength ε.

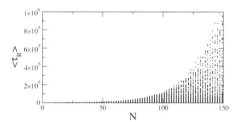

FIGURE 15.5
Average relaxation time as a function of message size N.

the message. Let m be the message sent by the emitter network and m' be the message recovered by the receiver network. The elements of the received message are given by

$$m'^{(i)} = \Theta(T_{r_n^{(i)} \to s_{n+1}} - \sigma(T)), \qquad (15.8)$$

where $\Theta(x)$ is the Heaviside unit step function and $\sigma(T)$ is the transfer entropy standard deviation of all network sites of the R network. Hence the received message is a string of 0's and 1's if the transfer entropy of the corresponding network site is less or greater than one standard deviation of T.

The decoding process performed by the receiver network is not instantaneous, but requires a finite time interval during which a sufficiently large amount of binary signal is sent, in order to correctly decode the message. We define the relaxation time τ_R the time it takes for R to correctly decode the message, i.e., such that $m' = m$ [11]. We have observed that the value of τ_R depends on the message, for a given set of parameters. Hence we work with the average relaxation time $< \tau_R >$, where the average is taken with respect to a number N_m of randomly chosen messages.

In numerical simulations, we consider a number $N_m = 100$ of different messages m consisting of randomly chosen strings of bits. The whole set of average relaxation times is plotted as a function of the intra-network coupling

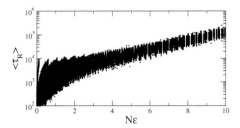

FIGURE 15.6
Average relaxation time as a function of $N\varepsilon$, for $N_m = 100$ different messages, each of them with $N_0 = 100$ initial conditions.

strength ε (Figure 15.4) and the network size N (Figure 15.5), where we verified that $< \tau_R >$ grows with both parameters. So, it is suggestive to analyze the τ_R-dependency with respect to the product $N\varepsilon$ (Figure 15.6), which shows a growth whose upper bound is an exponential curve

$$< \tau_R >= K e^{\kappa N \varepsilon}, \tag{15.9}$$

where $K = 4.9 \times 10^3$ and $\kappa = 1/3$ were obtained by fitting the maximum relaxation time points, presented in Figure 15.7.

During the transmission of the message it may well happen that some amount of noise corrupts the transmitted signal. It is thus important to verify if the receiver network remains able to decode correctly the message in the presence of external noise, within a time of the order of relaxation time. We consider that the signal s_n is subjected to a Gaussian noise of zero mean and variance σ. The bit error ratio (BER) is the fraction of erroneously transmitted bits with respect to the total number of bits in the message [7]. This ratio was computed as an average over $N_m = 500$ randomly chosen messages of fixed length $N = 51$. In Figure 15.8 we plot the BER as a function of σ and the intra-network coupling strength ε. Note that, for $\sigma < 0.1$ BER nearly vanishes

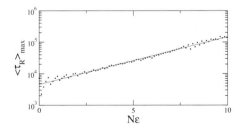

FIGURE 15.7
The upper limit for the relaxation time. The line stands for Equation (15.9).

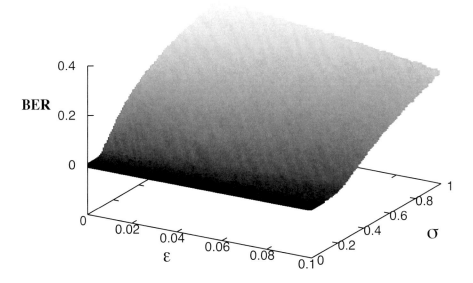

FIGURE 15.8
Bit error ratio (grayscale) as a function of the intra-network coupling strength
and the noise level. The values represent an average over 500 randomly chosen
messages of length 51 bits.

for all values of ε and, thus, there is no difference between the decoded and
emitted messages for a time $n =< \tau_R >$. Even for a stronger noise level the
BER was found to be small, showing that the communication process is robust
in the presence of noise.

15.5   Application to a specific example

We now exemplify the use of the communication system described in the
previous sections to transmit a specific binary message. Let us suppose that
the message contains $N = 21$ bits and is given by the binary string $m =$
101110010111111110011 [Figure 15.9(a)]. Hence both the emitter and receiver
networks should have $N = 21$ sites. Besides having the same number of sites
the networks should share a symmetric security key that is represented by
the intra-network coupling strength ε, and the inter-network, from which we
calculated the time to synchronize. The value of the inter-network coupling
strength has been fixed as $\gamma = 1/2$. Assuming a value $\varepsilon = 0.1$, Equations
(15.4) and (15.9) result in $\langle \tau_s \rangle = 323$ and $\langle \tau_R \rangle \approx 10^4$, respectively, for the

average synchronization and relaxation times. Since these are average values, we can consider here $\tau_s = 10^3$ e $\tau_R = 2 \times 10^4$.

We couple the E and R networks following Equation (15.10) and during $n = \tau_S$ iterates in the first stage of the process. At this point we expect to obtain a synchronization error $w_n < 10^{-14}$ and, if so, we truncate the state variables such that $\mathbf{e}_n = \mathbf{r}_n$ and switch off the inter-network coupling. In the beginning of the second state we use Equation (15.5) to obtain a signal s_n which is transformed by Equation (15.7) in a binary sequence $\tilde{s}_n$.

The receiver network keeps this binary signal during subsequent $n = \tau_R$ iterates and, through Equation (15.6), computes the transfer entropy for all network sites [Figure 15.9(b)]. Finally, using Equation (15.8) as a high-pass filter, the receiver network decodes the message sent [Figure 15.9(c)], which is clearly identical to the sent message. We remark that this process is extremely safe, since even if an eavesdropper would be able to snatch the signal that has been sent, the probability of this eavesdropper to strike all the variables $\tilde{r}_n^{(i)}$ during the time τ_R would be $P = 2^{-N\tau_R} = 2^{-63000}$. This is also the probability of striking the message sent, which is utterly insignificant.

15.6 Improving security and efficiency

Here we study another example in order to examine the security and the efficiency of our method. For the model presented in this section, we assume that in each time step both networks exchange information just from one of their variables. Besides, the variable whose information is received is different from that whose information is emitted. If, for instance, the network X sends information about the state variable k, the network Y sends information about the variable $k \neq k'$. We also assume that variables k e j change over time. The interaction between the networks is given by

$$\begin{cases} x_{n+1}^{(i)} = F(x_n^{(i)}) + \gamma_n^{(x,i)}(F(y_n^{(i)}) - F(x_n^{(i)})) \\ y_{n+1}^{(i)} = F(y_n^{(i)}) + \gamma_n^{(y,i)}(F(x_n^{(i)}) - F(y_n^{(i)})) \end{cases} \tag{15.10}$$

in which $x_n^{(i)}$ and $y_n^{(i)}$ are the state variable of X and Y, respectively, and $\gamma_n^{(z,i)}$ is coupling strength between each pair of sites. If $i \neq k$ ($i \neq k'$), then $\gamma_n^{(x,i)}(\gamma_n^{(y,i)})$ is zero; otherwise $\gamma_n^{(z,i)}$ corresponds to a random number $\xi \leq 1.0$ defined in each time step.

Although no restriction has been imposed on the temporal evolution, in order to favor the synchronization between the networks, we must ensure that all sites of X exchange information with all sites of Y in a short time interval. To solve this problem, one possibility is to assume that $k = n$ mod 8 and $k' = k - 1$ mod 8. In this case, after exactly 8 iterates both networks share information about all their states. Figure 15.10 presents the schematic of the

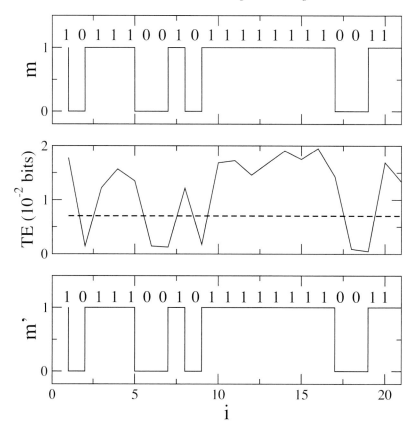

FIGURE 15.9
(a) 21-bit message sent; (b) transfer entropy between $\tilde{r}_n^{(i)}$ and $\tilde{s}_n$; (c) message received.

coupling prescription for the first and the second time step. Note that the elements of a network interact only with the elements with same index, and the interaction follows the temporal order. In fact, since the elements of both networks are identical and the initial conditions are random tagging of sites is accomplished during the process. More precisely, when a network receives the first signal, a variable is selected and associated with the label 1. At the same time, the network sends the information of the neighboring site which is associated with the index N. The second signal is associated with indices 2 and 1, and so on.

Networks interact with each other until there is complete synchronization between each pair of corresponding variables, i.e., $x_n^{(i)} = y_n^{(i)} \forall i = 1, 2, \cdots 8$. When this occurs, the networks stop sending information about their variables

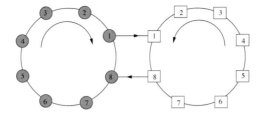

FIGURE 15.10
Snapshot of the system in the first moment of the first stage.

and start to send an arbitrary message. The synchronization occurs if the w_n given by the following equation is null:

$$w_n^{(z)} = \sum_{i=1}^{8} |F(y_{n-i}^{(i)}) - F(x_{n-i}^{(i)})| \qquad \text{for } n \geq 8. \qquad (15.11)$$

Practically, due to the invariance of the synchronization subspace, the condition $w_n = 0$ is never observed. We then define a quantity δ and establish synchronization when $w_n < \delta$. Although this procedure is standard, it requires that the coupling is maintained for all n; otherwise a small difference between the states of networks grows exponentially immediately after leaving the networks to share information about their state variables. However, since we wish that networks remain synchronized, we perform a truncation of order $\mathcal{O}(\delta)$ in the variables of each network immediately after the synchronization condition is checked.

Only if $\gamma > \gamma_c$ networks can synchronize, where γ_c depends on both the coupling and the local dynamic of the networks. In the case represented in Figure 15.10 we find

$$\gamma_c = 1 - e^{-4\Lambda}, \qquad (15.12)$$

in which Λ is the largest Lyapunov exponent of the uncoupled network. If we consider the local dynamics governed by the tent map, $f(z) = 1 - 2|z - 0.5|$ e $\varepsilon \in [0.0, 0.1]$, entãÅfo $\Lambda = \ln(2) - \alpha\varepsilon$, with $\alpha \approx 1.0$. Therefore, for the case considered, we have $\gamma_{c,max} = 0.9375$ e $\gamma_{c,min} \approx 0.90$. Thus, to ensure synchronization between networks we must assume $\xi \in [\gamma_{c,max}, 1.0]$.

Another important aspect to consider here is the synchronization time between networks. Using linear analysis we verified that the average synchronization time follows the equation

$$\langle \tau_s \rangle \geq -\frac{\ln(\delta)}{\alpha\varepsilon}. \qquad (15.13)$$

FIGURE 15.11
Temporal evolution of binary variables X network for the first 50 iterates. The vertical axis corresponds to the indices of each variable and the horizontal axis corresponds to time. The black refers to the value 1 and the white color to 0. The figure at the top represents the case $\tilde{x}_n^{(i)} = \tilde{y}_n^{(i)}$; figure at the bottom shows the same case for x with a 1% variation.

Therefore, after establishing the values of δ, from which the networks are considered synchronized, and ε, which represents the security key, it is possible to estimate the time of interaction necessary for their corresponding variables to synchronize with each other.

Synchronization between elements of the two systems is a necessary requirement for the transmission of messages. In this context, an intruder, trying to intercept the communication, could build a network E with the same number of elements and try to synchronize it with the networks X and Y, intercepting the values of the two variables sent every instant of time. However, we assume that each network sets a value of γ at random from each time step. Thus, the network E will not be able to synchronize with X and Y. The attacker would be able to determine all the variables only if the networks X and Y were already synchronized, in which case the parameter γ has no influence on the system. Since the interaction between networks ceases immediately after synchronization occurs, the attacker cannot determine all the variables of the network.

15.6.1   Establishing the reservoir

In this stage, each network evolved independently over a time interval known as the relaxation time τ_r. At each instant of time the network variables are checked and the following binary variable is defined:

$$\tilde{z}_n^{(i)} = \begin{cases} 0 & \text{se } z_n^{(i)} < 0.5 \\ 1 & \text{se } z_n^{(i)} \geq 0.5 \end{cases} \qquad (15.14)$$

in which $\tilde{z}_n^{(i)} = \{\tilde{x}_n^{(i)}, \tilde{y}_n^{(i)}\}$. Since networks are synchronized, we have $\tilde{x}_n^{(i)} = \tilde{y}_n^{(i)} \; \forall \; n$. Figure 15.11 represents the temporal evolution of the variables for the first 50 iterates. We also present the variables of a third system, but with a difference of 1% in ε. Note that during the evolution of networks, the patterns lose correlation quickly.

Each network stores the binary pattern in a kind of reservoir. This reservoir is identical for both networks and will be fundamental in the process of coding and decoding of messages, which we discuss below. We can regard that the reservoir contains the encryption key of all information exchanged between the networks during contact. The functionality of the reservoir resembles that of the Vernam cipher. If the networks always use the same value of ε in each contact, the pattern will be different due to randomly defined initial conditions on each network.

The binary pattern at each instant of time can also be stored in the form of characters, using some eight bits code scheme. For this case, the patterns shown in Figure 15.11 could be replaced by a string. Another option would be to save the binary pattern writing it in a decimal form; that is, calculating, and storing the number

$$\hat{a}_n = \sum_{i=0}^{8} \tilde{z}_n^{(-i)} 2^i. \tag{15.15}$$

The binary pattern shown in Figure 15.11 produces the time series presented in Figure 15.12. The information contained is the same in both figures. Only the representation is different. The use of one or the other depends on the convenience.

The relaxation time τ_r gives the size of the binary pattern, or, equivalently, the length of the time series $\hat{a}$. We will discuss this time interval next, but we can advance that is the minimum interval of observation for the message to be decoded.

15.6.2 Reading and decoding a message

A message is a sequence of characters $m = \{m^{(1)} m^{(2)} \cdots m^{(N')}\}$ taken from an alphabet $\mathcal{A}$, which is ordered according to a grammar rule. In this work, we will assume that the communication between the networks is by computers, and, therefore, the message corresponds to a sequence of characters generated by key presses over time. Each key pressed corresponds to an integer number or a binary sequence whose size depends on the coding scheme used. Consider that the proposed system utilizes an encoding scheme in which all characters are represented by a single string of 8 bits or 1 byte. An example of such a scheme is the extended ASCII; however, any other coding scheme of 8 bits may be employed.

Consider that the encoding scheme used is public. This means that we

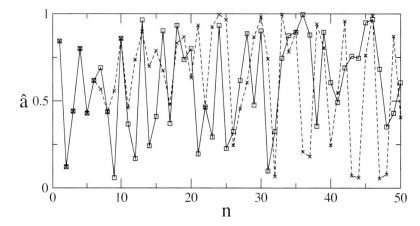

FIGURE 15.12
Time series of the binary pattern from Figure 15.11 that was converted to decimal number by Equation (15.15). The dashed (full) line corresponds to the binary pattern of the network X (E).

cannot send the message directly to the recipient safely. In this scenario, we need to implement a new encoding of the message, so that only the recipient can decrypt and retrieve the message. In other words, the sender and receiver must share a security key known only to both. At this point, we must remember that the two networks have identical reservoirs whose equality is due to the fact that the networks share the same coupling prescription. So this reservoir can be used to encode and decode the message safely.

One way of encrypting the message from the data of the reservoir is to use the XOR encryption algorithm. In this scheme, the message bits are XORed with the reservoir, and the result is the encrypted message. For example, if the emitter wants to transmit the letter "c," and the encoding scheme is ASCII, then pressing the corresponding key, we have the sequence 01100111. From Figure 15.11, we see that the first binary pattern of the reservoir is 00100111. Therefore, using the XOR algorithm we obtain

$$01100011 \oplus 00100111 = 01000100. \tag{15.16}$$

Upon receiving the encrypted message 01000100, the receiver reapplies the XOR operation using the reservoir and obtains

$$01000100 \oplus 00100111 = 01100011, \tag{15.17}$$

retrieving the message.

Although the XOR algorithm is quite efficient, we propose another encryption method that uses the entropy transfer and presents similarities with the computing process called reservoir computing (RC). Consider that the message is encrypted in a scalar s given by

$$s_n = \frac{1}{\eta} \sum_{i=1}^{8} m^{(i)} \tilde{x}_n^{(i)}, \tag{15.18}$$

in which $m^{(i)}$ is the i-th element of the sequence, $\eta_k = \sum_{i=1}^{8} m(i)$ is normalization factor, and $\tilde{x}_n^{(i)}$ is the i-th element of the binary sequence fo the number $\hat{a}_n$. We then define a binary variable

$$\tilde{s}_n = \begin{cases} 0 & \text{se } s_n < 0.5 \\ 1 & \text{se } s_n \geq 0.5 . \end{cases} \tag{15.19}$$

This variable represents the signal to be transmitted at time n. The operation is repeated for all variables in the reservoir. This means that 8 bits of a character, "c," for example, 10^3 bits, are encoded in the signal. This approach thus increases significantly the size of data and the operation RC. Moreover, as the standards contained in the reservoir depend on the initial conditions, which are random, the receiver, *a priori*, does not know which sequence is linked to each character. Then, as in the case of RC, the receiver determines the sequence corresponding to each character during the communication, performing a learning process.

15.6.3   Transmitting the signal and recovering the message

The signal in the binary format can be transmitted from any digital channel. To receive the signal the receiver starts the decoding procedure. First, the receiver calculates the transfer entropy of each variable $\tilde{y}_n^{(i)}$ for $\hat{s}_{n+1}$. The transfer entropy reads

$$T_{\tilde{y}_n^{(i)} \to \hat{s}_{n+1}} = \sum p(\hat{s}_{n+1}, \hat{s}_n, \tilde{y}_n^{(i)}) \ln \frac{p(\hat{s}_{n+1}|\hat{s}_n, \tilde{y}_n^{(i)})}{p(\hat{s}_{n+1}|\hat{s}_n)}. \tag{15.20}$$

For the sake of simplicity, we write $T_{\tilde{y}_n^{(i)} \to \hat{s}_{n+1}} \equiv T_i$. After calculating all T_i, the receiver retrieves the message by using the following formula:

$$\hat{m}^{(i)} = \Theta(T_i - \sigma) \tag{15.21}$$

in which Θ is the step one Heaviside function, and σ is the standard deviation of the transfer entropy.

15.7   Conclusions

We proposed in this work a robust and safe device for transmitting binary messages using replicas of coupled map networks. The communication system uses transfer entropy and as a consequence allows the construction of a communication system where information can only be decoded if the emitter and receiver are completely synchronous and they are exactly identical. Since nodes in each network are not synchronous, that enables the generation of a decorrelated encoded signal. This provides an extra security for the communication system, since it makes virtually impossible for an eavesdropper to discover the dynamics of the emitter.

In addition, imagine that the eavesdropper is very clever and it knows exactly the time the emitter and the receiver network take to synchronize. If it does not know exactly the connecting topology of the receiver network, it will not maintain synchronization. The eavesdropper will also not be able to verify the existence of synchronization, since all nodes in its network will be desynchronous. Suppose now that the eavesdropper knows all the secret keys (intra e inter couplings and connecting topology). Still all that is needed for the communication system to regain security is that the emitter and the receiver change at a given time their network connecting topology, after the message has started being transmitted. This will maintain synchronization between E and R, but will make the network of the eavesdropper to become desynchronous. Both networks must have the same size N as the message itself (in number of bits).

The proposed system works in two stages: in the first one we synchronize the emitter and receiver networks. Only emitter and receiver know how much time it takes to achieve synchronization. After the first stage the inter-network coupling is switched off, but the networks remain synchronized.

In the second stage we encode a given message into a signal which is read by the receiver network using the transfer entropy between the receiver network and the signal, with a high-pass filter based on the standard deviation of the transfer entropy. The message can be decoded after a relaxation time. We obtained an expression for an upper bound of the relaxation time as a function of the intra-network coupling strength and the message size. Hence, given a message of arbitrary size, the intra-network coupling strength can be chosen in order to minimize the transmission time.

The proposed device is similar to a method developed by Hung and Hu [11] but differs from it in this way: in our method we compact N bits of the message into a bit-stream of the signal, whereas in the Hung and Hu method every bit of the message is encoded in a different bit of the signal. Hence our method represents a considerable increase in the channel capacity attainable.

We considered a specific example to test this method and, comparing the message read by the receiver with the message emitted, we verified that the method is reliable. Besides this advantage, the method we propose is robust since, even in the presence of external noise, the bit error ratio can be kept in sufficiently low levels, varying the intra-network coupling and noise level.

Finally, we have shown that even if an eavesdropper could intercept the signal, it could not strike the message (more precisely, the probability of it is negligible). We suggest that other continuous maps as well as other intra- and inter-network couplings may also be used. Some tests with other intra-network couplings have suggested to us that, for instance, the synchronization time power-law behavior remains unaltered. This was also verified when considering the logistic map for the dynamics of each network site.

Bibliography

[1] V. Ahlers and A. Pikovsky. Critical properties of the synchronization transition in space-time chaos. *Phys. Rev. Lett.*, 88(25):254101(4), June 2002.

[2] E. Alvarez, A. Fernandez, J. Garcia, P. Jimenez, and A. Marcano. New approach to chaotic encryption. *Physics Letters A*, 263(4-6):373–375, 1999.

[3] G. Alvarez and S. J. Li. Some basic cryptographic requirements for chaos-based cryptosystems. *International Journal of Bifurcation and Chaos*, 16:2129–2151, 2006.

[4] L. Appeltant, M. C. Soriano, G. van der G. Sande, J. Danckaert, S. Massar, B. Schrauwen, C. R. Mirasso, and I. Fischer. Information processing using a single dynamical node as complex system. *Nature Communications*, 2:468, 2010.

[5] M. S. Baptista. Cryptography with chaos. *Physics Letters A*, 240:50–54, 1998.

[6] M. Cencini, C. J. Tessone, and A. Torcini. Chaotic synchronization of spatially extended systems as nonequilibrium phase transitions. *Chaos*, 18:037125, 2008.

[7] T. M. Cover and J. A. Thomas. *Elements of Information Theory.* John Wiley & Sons: New York, 2006.

[8] J. P. Crutchfield and K. Kaneko. *Phenomenology of Spatio-Temporal Chaos*, volume 1. World Scientific, Singapore, December 1987.

[9] K. M. Cuomo and A. V. Oppenheim. Circuit implementation of synchronized chaos with applications to communications. *Phys. Rev. Lett.*, 71:65–68, 1993.

[10] M. Ding and E. Ott. Enhancing synchronism of chaotic systems. *Phys. Rev. E*, 49(2):R945, February 1994.

[11] Y. C. Hung and C. K. Hu. Chaotic communication via temporal transfer entropy. *Phys. Rev. Lett.*, 101:244102(4), 2008.

[12] H. Jaeger. The "echo state" approach to analysing and training recurrent neural networks - with an erratum note. *Technical Report GMD, German National Research Center for Information Technology*, 148, 2001.

[13] B. Jovic. *Synchronization Techniques for Chaotic Communication Systems*. Springer, Auckland: New Zeland, 2011.

[14] W. Kinzel, A. Englert, and I. Kanter. On chaos synchronization and secure communication. *Phil. Trans. R. Soc. A*, 368:379–389, October 2010.

[15] L. Kocarev. Chaos-based cryptography: a brief overview. *IEEE Circuits and Systems Magazine*, 1:6–21, 2002.

[16] L. Kocarev, G. Jakimoski, T. Stojanovski, and U. Parlitz. From chaotic maps to encryption schemes. In *Proc. IEEE International Symposium Circuits and Systems*, California, 1998. Naval Postgraduate School. Paper presented at the Conference on the International Symposium on Circuits and System.

[17] S. J. Li, G. R. Chen, K. W. Wong, et al. Baptista-type chaotic cryptosystems: problems and countermeasures. *Physics Letters A*, 332:368–375, 2004.

[18] E. Ott. *Chaos in Dynamical Systems*. Cambridge University Press, Cambridge, 1993.

[19] L. M. Pecora and T. L. Carroll. Synchronization in chaotic systems. *Phys. Rev. Lett.*, 64:821–824, February 1990.

[20] R. F. Pereira, S. E. de S. Pinto, and S. R. Lopes. Synchronization time in a hyperbolic dynamical system with long-range interactions. *Physica A: Statistical Mechanics and Its Applications*, 389(22):5279–5286, November 15, 2010.

[21] G. Perez and H. A. Cerdeira. Extracting messages masked by chaos. *Phys. Rev. Lett.*, 74:1970–1973, March 1995.

[22] T. Schreiber. Measuring information transfer. *Phys. Rev. Lett.*, 85:4, 2000.

[23] C. Shannon. Communication theory of secrecy systems. *Bell Syst. Tech. Journal*, 28:656–715, 1949.

[24] C. E. Shannon and W. Weaver. *The Mathematical Theory of Communication*. The University of Illinois Press, Champaign, 1963.

[25] P. Stavroulakis, editor. *Chaos Applications in Telecommunications*. CRC Press: New York, 2005.

[26] I. G. Szendro, M. A. Rodrigues, and J. M. López. Spatial correlations of synchronization errors in extended chaotic systems. *Europhys. Lett.*, 86:2008, 2009.

[27] D. B. Vasconcelos, R. L. Viana, S. R. Lopes, A. M. Batista, and S. E. D. Pinto. Spatial correlations and synchronization in coupled map lattices with long-range interactions. *Physica A: Statistical Mechanics and Its Applications*, 343:201–218, November 15, 2004.

[28] G. S. Vernam. Cipher printing telegraph systems for secret wire and radio telegraphic communications. *Journal of the IEEE*, 55:109–115, 1926.

[29] G. Vidal, M. S. Baptista, and H. Mancini. Fundamentals of a classical chaos-based cryptosystem with some quantum cryptography features. *International Journal of Bifurcation and Chaos*, 22, 1250243, 2012.

16

Robustness of chaos to multipath propagation media

Hai-Peng Ren

Department of Information and Control Engineering
Xi'an University of Technology

Murilo S. Baptista and Celso Grebogi

Institute for Complex System and Mathematical Biology
King's College, University of Aberdeen

CONTENTS

Multipath propagation and amplitude damping are unavoidable physical phenomena in wireless communication systems. The first causes intersymbol interference in the information signal leading to problems in the decoding of information at the receiver end. The second signifies loss of communication at receiver end. In this work, we present numerical evidences that the multipath propagation and amplitude damping do not alter the largest positive Lyapunov exponent of some chaotic discrete maps. We also discuss results of simulations illustrating how one could use this property to create a chaos-based communication system that is robust to multi-path propagation and amplitude damping.

16.1   Introduction

The OGY control method [12] and the drive-response chaos synchronization method [14] have made us realize that chaos has potential applications to communication. In addition to that, due to special inherent properties that chaotic signals have, such as broadband, orthogonality, and sensitivity to the initial conditions, chaotic signals are being considered to form the basis of reliable and secure communication systems [1, 5, 6, 22, 23].

So far, there are four main schemes to use chaos for communication at the physical level. The schemes are the chaotic masking [5], in which a transmitter (the master) synchronizes with a receiver (the slave) and information is masked in this synchronous signal, the chaotic modulation [22, 24], the chaotic shift-key [6], and the symbolic message bearing [4, 8], where the signal transmitted encodes some symbolic stream. Most of the research in this topic focuses on the transmission of information in ideal channels of communication and on the weaknesses of the schemes for the intruders to break the security [13, 15, 16, 19, 25–27]. However, the channel is not ideal in many cases, such as in wireless communication. A signal traveling in this kind of channel might suffer the effect of filtering (limited bandwidth), amplitude damping effect, multipath propagation, and noise [21]. If chaos is used in such type of more realistic physical media, a major concern that arises is how the signal is modified by the channel constraints.

Among all the channel constraints, the effect of filtering was first noted. In [2], the authors have shown that the fractal dimension of a filtered chaotic signal can be modified depending on the relationship between the filter coefficient and the spectra of the Lyapunov exponents of the original system. The effect of filtering on the synchronization of two master-slave coupled chaotic oscillators was addressed in [17]. The authors showed that synchronization can be affected by the filtering: the slave system might settle down to a different attractor than the one of the master, depending on the bandwidth of the filter. This results in an unstable communication system. With the help of maps, the authors of [7] have shown that a filter with not very low bandwidth can still allow information to be transmitted. The effect of filtering on the symbolic message bearing scheme was investigated in [18, 28]. It was shown in [28] that if the filter does not change the fractal dimension of the transmitter, the information can easily be decoded from the filtered signal. If the bandwidth of the filter is low enough to change the fractal dimension, it is still possible to partially recover the information by using some strategies. In [18], the authors pointed out that the reason for not being able to fully recover the filtered signal is because of a phenomenon called "symbol missing," responsible for inducing the receiver to decode less symbols than the ones transmitted. They also proposed a methodology to overcome this problem, recover the missing symbols, and transmit information with low bit error rate.

FIGURE 16.1
General structure of symbolic message bearing communication method.

A channel with multipath propagation causes the same signal to arrive many times with different time-delays. This distortion results in a serious interference in the signal, thus leading to problems in its decoding. In the particular case of the underwater wireless acoustic communication, in order to avoid interference an on-off protocol is often used, where information is intermittently sent followed by a moment of silence. This in turn contributes negatively to the bit transmission rate, which is already low due to the low speed of sound in water.

Little attention has being paid to the effect of multipath propagation using chaotic signals. In [9, 10], the authors study this channel constrain in the chaotic spread frequency communication scheme. They show that the performance of the differential chaos shift keying communication is better than conventional communication schemes functioning in media that present multipath propagation.

There are many works showing how to filter noise from a chaotic signal, in particular the works of [3, 20]. In this work, we study the modification suffered by a chaotic signal when it is transmitted through a physical media presenting multi-path propagation and amplitude damping. We present numerical evidence that the multi-path propagation and amplitude damping do not alter the largest positive Lyapunov exponent of some discrete chaotic maps. We then discuss some simulations illustrating how one could use this property to create a chaos-based communication system that is robust to multi-path propagation and amplitude damping.

16.2 Symbolic message bearing communication in the physical media with multipath and damping

The chaotic symbolic message bearing communication scheme is schematically represented in Figure 16.1. In this scheme, the binary bitstream to be transmitted is encoded into the chaotic trajectory by small perturbations [8]. Then, this encoded chaotic signal is transmitted through the communication channel to the receiver. At the receiver end, the information is decoded using the same protocol used in the encoding process.

The physical media considered here present multi-path propagation and exponential amplitude damping. Denoting the transmitted signal as $x_t(t)$, the received signal after being transmitted through such type of media is assumed to be described by

$$x_R(t) = \sum_{j=1}^{N} e^{-\beta \tau_j} x_t(t - \tau_j), \qquad (16.1)$$

where β is the damping coefficient of the media, τ_j is the time-delay of the j-th path, and N is the number of paths considered. The first path is the one along which the signal travels the fastest. The time for the signal to travel from the transmitter to the receiver along the first path is denoted by τ_1. In general, it is appropriate to consider that the larger delay τ is for the signal to arrive at the receiver, the longer the path is and the larger is the damping caused by the media. Hence, we assume that the signal has its amplitude decreased by $e^{-\beta \tau_j}$.

16.2.1   Information transmission in an ideal media

Here, we show an example of the message bearing communication scheme using Chua's circuit [11] given by

$$\begin{aligned}
\dot{x} &= \alpha(y - h(x)), \\
\dot{y} &= x - y + z, \\
\dot{z} &= -\beta y,
\end{aligned} \qquad (16.2)$$

where $h(x) = m_1 x + 0.5(m_0 - m_1)(|x+1| - |x-1|), \alpha = 8.7, \beta = 14.2886, m_0 = -1/7, m_1 = 2/7$. The variable x is chosen as the transmitted signal x_t. Using the time-delay embedding method, we can derive the attractor shown in Figure 16.2.

In the symbolic message bearing communication, control is used to create a chaotic signal that produces the desired symbolic sequence the message. In this example, we consider that the message is generated by the non-controlled chaotic system. The message to be transmitted can be encoded by a code constructed from the local minima of the signal. From the time series of x, we obtain all its local minima, denoted by $x_L(i)$. These local minima are represented in Figure 16.3 by the crosses.

The return map of the chaotic attractor is built using these local minima $x_L(k)$. The horizontal axis shows $x_L(i)$ and the vertical axis represents $x_L(i+1)$, as shown in Figure 16.4. The trajectory encodes a symbol '0' at time i if $x_L(i) \le x_L^*$ (points marked in the figure by stars), and a symbol '1,' otherwise, i.e., if $x_L(i) > x_L^*$ (points marked in the figure by empty boxes). Since this map is unimodal, the critical point x_L^* that defines a natural phase space partition to encode the signal is the value of $x_L(i)$ for which $x_L(i+1)$ is the minimum.

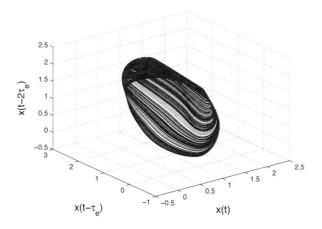

FIGURE 16.2
The attractor reconstructed using time-delay embedding coordinates. τ_e is the embedding time-delay.

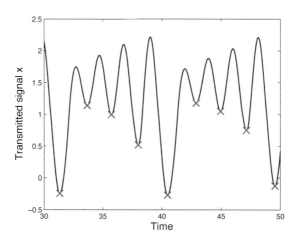

FIGURE 16.3
Nine local minima $(x_L(i))$ extracted from $x(t)$ are represented by crosses.

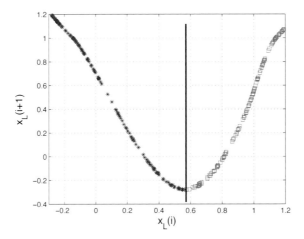

FIGURE 16.4
The return map built using local minima of the transmitted signal.

If the communication channel is ideal, the received signal is the same as the transmitted signal. The decoding protocol is the same one used to encode, i.e., the receiver uses the same value of the critical point x_L^* used by the transmitter. By this way the information can be easily decoded.

16.2.2 Difficulty of decoding caused by multipath propagation and amplitude damping

We now assume that the received signal is described by Equation (16.1). For illustration purposes, we assume the existence of only two paths and that the time-delay of the first path is $\tau_1 = 2$, the time-delay of the second path is $\tau_2 = 3$, and the damping coefficient is $\beta = 0.13$. In Figure 16.5 we show the reconstructed attractor obtained from the received signal, x_R. In Figure 16.6 we show the return map of the local minima of the received signal. There is not an apparent location where one can define well a phase partition to code the signal. Determining the code to decode the received signal, i.e., to discriminate between whether a '0' or a '1' was sent, seems to be unfeasible. The return map is not unimodal any longer, but rather composed by a set of disconnected branches.

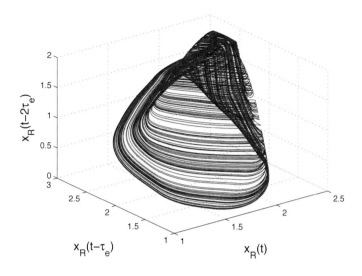

FIGURE 16.5
The attractor built from the received signal after being transmitted through
the media with multi-path propagation and amplitude damping.

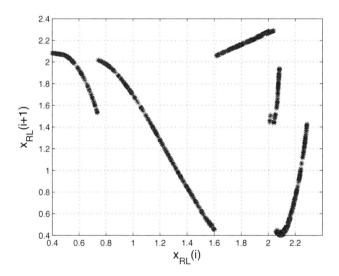

FIGURE 16.6
The return map obtained using the local minima of the received signal after
being transmitted through the media with multipath propagation and amplitude damping.

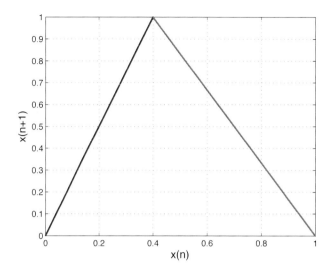

FIGURE 16.7
The return map of the tent map in Equation (16.3).

16.3   Invariance of the largest positive Lyapunov exponent

To better understand what is being preserved under the transformation in Equation (16.1), we consider the generalized tent map, given by

$$x(n+1) = \begin{cases} x(n)/p, & x(n) \le p, \\ (1-x(n))/(1-p), & x(n) > p, \end{cases} \qquad (16.3)$$

where we choose $p = 0.4$. The return map of this tent map is shown in Figure 16.7. In this figure, the points to the left of the critical point, the maximum point, form the "left branch," and the points on the right of the critical point form the "right branch."

The Lyapunov exponent λ of this map can be analytically calculated and is equal to $\lambda = p \ln\left(\frac{1}{p}\right) + (1-p) \ln\left(\frac{1}{1-p}\right)$, a result that can be understood in that the trajectory of this map spends a fraction p of the time in the left branch that has a slope of $1/p$ and a fraction $(1-p)$ of the time in the right branch that has a slope of $1/(1-p)$. For $p = 0.4$, we obtain $\lambda = 0.673$.

The received signal at a time t can be expressed in terms of the transmitted signal at a time t, by normalizing Equation (16.1), i.e., by dividing both sides of this equation by $e^{-\beta \tau_1}$ and shifting the time by the offset τ_1. We obtain the nominal form as

$$x_R(t) = x_t(t) + \sum_{j=1}^{N-1} e^{-\beta \Delta \tau_j} x_t(t - \Delta \tau_j), \qquad (16.4)$$

where $\Delta \tau_j$ is the excess delay with respect to the first path, i.e., $\Delta \tau_j = \tau_{j+1} - \tau_1$. The second term in the right-hand side of Equation (16.4) corresponds to the change in the transmitted signal caused by the propagation media. Its discrete form appropriate to describe the received signal of discrete signals can be written as

$$x_R(n) = x_t(n) + \sum_{j=1}^{N-1} e^{-\beta \Delta n_j} x_t(n - \Delta n_j), \qquad (16.5)$$

where Δn_j is discrete excess delay.

Considering the two path propagation problem and the excess time delay equal to 1, the received signal of the tent map is described by $x_R(n) = x(n) + e^{-0.13} x(n-1)$; its return map can be seen in Figure 16.8. It consists of four branches whose slopes from left to right are 2.5 (left down branch), 1/2.1 (left up branch), 35/4 (right down branch), and 5/3 (right up branch). By calculating the probabilities of having the trajectory visiting the domain defining each branch, the Lyapunov exponent of the received signal is given by $0.16 ln(2.5) + 0.24 ln(1/2.1) + 0.096 ln(35/4) + 0.218 ln(5/3) = 0.673$. It is surprisingly unaltered compared to the original signal $x(n)$.

Considering another example, changing the excess time to 5, the received signal can be described by $x_R(n) = x(n) + e^{-0.13 \times 5} x(n-5)$. The return map of this received signal is shown in Figure 16.9. Many branches can be observed. The number of branches is the result of a multiplicative Markov process that depends on the number of paths, the number of branches of the original map, and on the excess delay. This can be understood from Figure 16.10 with the received signal given by $x_R(n) = x(n) + e^{-0.13 \times 8} x(n-8) + e^{-0.13 \times 11} x(n-11)$. For received signals with many branches, the Lyapunov exponent is calculated numerically, and we obtain, surprisingly, that $\lambda = 0.673$.

16.4 Communication scheme based on the invariance of the Lyapunov exponent

The fact that the largest positive Lyapunov exponent is not changed because of the multi-path propagation and amplitude damping suggests that the information content carried by the chaotic signal might not be modified. However, decoding the received signal might not be trivial, specially if there is noise in the channel. From Figures 16.9 and 16.10, the branches that can be used to

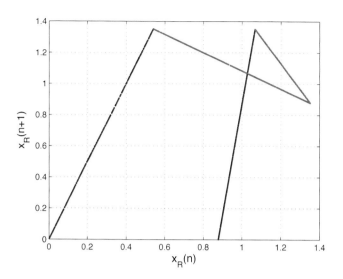

FIGURE 16.8
Return map of the received signal of the tent map after being transmitted in a media that has two-path propagation: $x_R(n) = x(n) + e^{-0.13}x(n-1)$.

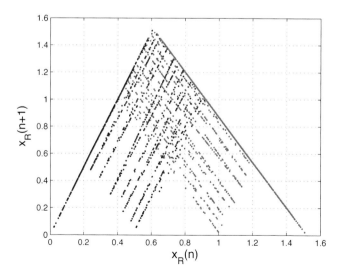

FIGURE 16.9
Return map of the received signal of the tent map after being transmitted having two-path propagation: $x_R(n) = x(n) + e^{-0.13 \times 5}x(n-5)$.

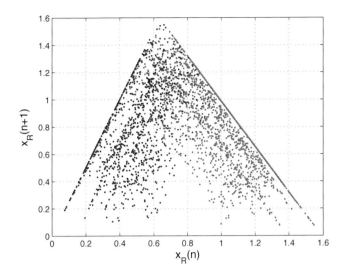

FIGURE 16.10
Return map of the received signal of the tent map after being transmitted in a media that has three-path propagation: $x_R(n) = x(n) + e^{-0.13 \times 8} x(n-8) + e^{-0.13 \times 11} x(n-11)$.

encode for '0' and '1' overlap. Consequently, discerning whether a '0' or a '1' was transmitted might be indeed impossible if any noise is present. The reason is because the slopes of the two branches are different. If the original chaotic system has a constant derivative and if the positive Lyapunov exponent of the received signal is the same as the transmitted signal, then the slope of the branches of the received signal is also constant. The result is that the return map of the received signal is composed of many parallel lines that do not overlap. In addition, it is advantageous to have a map whose branches are well separated, so that even when there is a high level of noise in the channel, decoding is still possible.

A map that satisfies this condition is the shift map, described by

$$x(n+1) = 2x(n)(mod1), \qquad (16.6)$$

where $(mod1)$ operation means modulus 1. The return map of Equation (16.6) is shown in Figure 16.11. It has two branches with equal slope, i.e., 2. As can be seen in Figure 16.12, the slope of the branches of the received signal after two-path propagation is constant and equal to 2.

A communication system can be constructed by identifying in the return map of the received signal the branches that correspond to a '0' or a '1.' For the shift map, the encoding of the transmitted trajectory is done by using the critical point $x(n)^* = 0.5$. Therefore, a '0' is transmitted at time n if

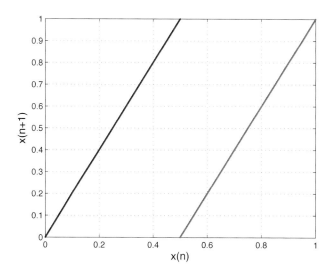

FIGURE 16.11
The return map of the shift map.

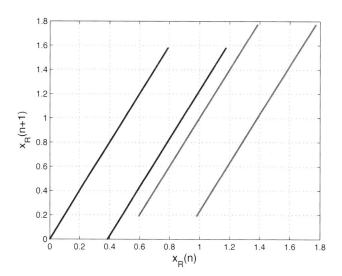

FIGURE 16.12
The return map of the received signal of the shift map after being transmitted
through a media with two-path propagation: $x_R(n) = x(n) + e^{-0.13 \times 2} x(n-2)$.

$0.5 < x(n) < 1$, and '1' otherwise. If the excess delay is not too large, the branches encoding a '0' (points on left two line in Figure 16.12) are located in the left side of the return map and the branches encoding a '1' (points on right two line in Figure 16.12) are located in the right side of the return map. To this end, we can use a line with slope 2 and horizontal axis cross point $x^* = 0.5$ to descriminate the different transmitted code; thus we can use such a chaotic discrete map to communicate the binary information through multi-path and damping media.

16.5 Conclusions

The understanding of how signals are modified by the action of non-ideal physical media is a fundamental issue to communication. In this chapter, we show that chaotic signals have some inherent properties to cope with physical media that have multi-path propagation and amplitude damping. These channel constraints do not affect the largest Lyapunov exponent of the chaotic signal. Consequently, if the transmitter uses a chaotic system that has a return map with constant derivative, the return map of the received signal also has constant derivative. Hence, the return map is formed by a set of parallel lines that do not intersect each other. Each branch can be associated with binary symbols being transmitted and leading to a full decoding of the information transmitted.

Compared to its air-based wireless communication counterpart, the underwater communication suffers from the large amplitude damping and from serious multi-path propagation. The damping effect can be diminished by the use of acoustic signals. However, due to the low sound speed, the multi-path propagation becomes a more serious issue. This work proposes a possible solution for this by the use of chaotic signals.

Bibliography

[1] A. Abel and W. Schwarz. Chaos communications — principles, schemes, and system analysis. *Proceedings of the IEEE*, 90:691–710, May 2002.

[2] R. Badii, G. Broggi, B. Derighetti, M. Ravani, S. Ciliberto, A. Politi, and M. A. Rubio. Dimension increase in filtered chaotic signals. *Phys. Rev. Lett.*, 60:979–982, March 1988.

[3] M. S. Baptista and L. López. Information transfer in chaos-based communication. *Phys. Rev. E*, 65:055201, May 2002.

[4] E. Bollt, Y.-C. Lai, and C. Grebogi. Coding, channel capacity, and noise resistance in communicating with chaos. *Phys. Rev. Lett.*, 79:3787–3790, November 1997.

[5] K. M. Cuomo and A. V. Oppenheim. Circuit implementation of synchronized chaos with applications to communications. *Phys. Rev. Lett.*, 71(1):65–68, July 1993.

[6] H. Dedieu, M. P. Kennedy, and M. Hasler. Chaos shift keying: modulation and demodulation of a chaotic carrier using self-synchronizing chua's circuits. *IEEE Transactions on Circuits and Systems II*, 40:634–642, October 1993.

[7] M. Eisencraft, R. D. Fanganiello, and L. A. Baccala. Synchronization of discrete-time chaotic systems in bandlimited channels. *Mathematical Problems in Engineering*, 2009:207971, May 2009.

[8] S. Hayes, C. Grebogi, and E. Ott. Communicating with chaos. *Phys. Rev. Lett.*, 70:3031–3034, May 1993.

[9] G. Kaddoum, D. Roviras, P. Charge, and D. Fournier-Prunarety. Performance of multi-user chaos-based ds-cdma system over multipath channel. In *Proc. IEEE International Symposium on Circuits and Systems*, pages 2637–2640. IEEE, May 2009.

[10] M. P. Kennedy, G. Kolumban, G. Kis, and Z. Jako. Performance evaluation of fm-dcsk modulation in multipath environments. *IEEE Transactions on Circuits and Systems I*, 47:1702–1711, May 2000.

[11] T. Matsumoto. Chaos in electronic circuits. *Proceedings of the IEEE*, 75:1033–1058, August 1987.

[12] E. Ott, C. Grebogi, and J. A. Yorke. Controlling chaos. *Phys. Rev. Lett.*, 64:1196–1199, March 1990.

[13] A. T. Parker and K. M. Short. Reconstructing the keystream from a chaotic encryption scheme. *IEEE Transactions on Circuits and Systems I*, 48:624–630, May 2001.

[14] L. M. Pecora and T. L. Carroll. Synchronization in chaotic systems. *Phys. Rev. Lett.*, 64:821–824, February 1990.

[15] G. Perez and H. A. Cerdeira. Extracting messages masked by chaos. *Phys. Rev. Lett.*, 74:1970–1973, March 1995.

[16] V. I. Ponomarenko and M. D. Prokhorov. Extracting information masked by the chaotic signal of a time-delay system. *Phys. Rev. E*, 66:026215, August 2002.

[17] A. A. Prokhorov and E. S. Mchedlova. Chaos synchronization with signal distortion in communication channel: experiment and numerical simulation. *Technical Physics*, 53:1463–1470, May 2008.

[18] H.-P. Ren, M. S. Baptista, and C. Grebogi. Uncovering missing symbols in communication with filtered chaotic signals. *International Journal of Bifurcation and Chaos*, 22:1250199, August 2012.

[19] H.-P. Ren, C. Han, and D. Liu. Breaking chaotic shift key communication via adaptive key identification. *Chin. Phys. B*, 17:1202–1208, April 2008.

[20] E. Rosa Jr., S. Hayes, and C. Grebogi. Noise filtering in communication with chaos. *Phys. Rev. Lett.*, 78:1247–1250, February 1997.

[21] M. Stojanovic and J. Preisig. Underwater acoustic communication channels: propagation models and statistical characterization. *IEEE Communications Magazine*, 47:84–89, May 2009.

[22] C. W. Wu and L. O. Chua. A simple way to synchronize chaotic systems with applications to secure communication systems. *International Journal of Bifurcation and Chaos*, 3(6):1619–1627, December 1993.

[23] T. Yang. A survey of chaotic secure communication systems. *International Journal of Computational Cognition*, 2:81–130, June 2004.

[24] T. Yang and L. O. Chua. Secure communication via chaotic parameter modulation. *IEEE Transactions on Circuits and Systems I*, 43:817–819, September 1996.

[25] T. Yang, L. B. Yang, and C. M. Yang. Breaking chaotic secure communication using a spectrogram. *Physics Letters A*, 247:105–111, October 1998.

[26] T. Yang, L. B. Yang, and C. M. Yang. Breaking chaotic switching using generalized synchronization: examples. *IEEE Transactions on Circuits and Systems I*, 45:1062–1067, October 1998.

[27] C. Zhou and C.-H. Lai. Extracting messages masked by chaotic signals of time-delay systems. *Phys. Rev. E*, 60:320–323, July 1999.

[28] L. Q. Zhu, Y. C. Lai, F. C. Hoppersteadt, and E. M. Bollt. Numerical and experimental investigation of the effect of filtering on chaotic symbolic dynamics. *Chaos*, 13:410–419, February 2003.

Index

Printed and bound by CPI Group (UK) Ltd, Croydon, CR0 4YY

18/10/2024

01776270-0013